ENCYCLOPAEDIA OF

ASTRONOMY AND SPACE SCIENCE

ENCYCLOPAEDIA OF
ASTRONOMY AND SPACE SCIENCE

Anandi Bhardwaj

ANMOL PUBLICATIONS PVT. LTD.
NEW DELHI - 110 002 (INDIA)

ANMOL PUBLICATIONS PVT. LTD.
H.O.: 4374/4B, Ansari Road, Daryaganj
New Delhi - 110 002
Ph.: 23278000, 23261597
B.O.: No. 1015, Ist Main Road, BSK III Stage
III Phase, III Block
Bangalore - 560 085 (India)
Visit us at: www.anmolpublications.com

Encyclopaedia of Astronomy and Space Science

First Published, 2006

ISBN 81-261-2885-2

PRINTED IN INDIA

Printed at Mehra Offset Press, Delhi.

Preface

The aim of this encyclopaedia is to give, in the light of present knowledge and opinion, definitions of terms that are of importance in Astronomy. It is an encyclopaedia of Astronomy terms, written by specialists, partly for other specialists but more particularly for the man-in-the-street, and for students and interested workers of all kinds and ages: indeed, for all who wish to understand what Astronomy is.

The terms defined in this book include laws, relationships, equations, basic principles and concepts, as well as the most widely used instruments and apparatus in Astronomy. In short, this treatment comprises the terms both of pure science and of its applications. The fields covered include planets, asteroids, satellites and satellite technology, technical advancements, low temperature Astronomy, the properties of heavenly bodies, nuclear Astronomy; mathematical Astronomy, radio Astronomy and representative topics in relativity and other more advanced and specified fields.

Further to serve the needs of the teacher, student and worker in the field of Astronomy, as well as in related fields, a considerable number of terms have been included from the subject-areas bordering on Astronomy, not only mathematics, but physical chemistry, applied electronics etc. Clarification of the meaning of terms, precision in their use, and uniformity in usage has been especially taken care of. It is worthwhile to make Astronomy literature intelligible to as wide a field of readers as possible.

The aim of this volume is to provide a concise repository of number of major concepts in this field. In preparation of this book, the author has freely consulted large number of books and journals so no authenticity is claimed. Author is especially thankful to M/S Anmol Publications for shaping this book in its final form. Suggestions for further improvement of this book are not only welcome but also greatly appreciated.

Author

A

A band A broad Fraunhofer line in the Sun's spectrum at around 760 nm, due to absorption by oxygen in the Earth's atmosphere. Because the oxygen is in molecular form, the A band is actually a group of close, regularly spaced lines over the range 759-768 nm, unresolved at low resolution.

A star A star of spectral type A, whose spectrum is dominated by absorption lines of hydrogen in fact, their hydrogen absorption lines are the strongest of any normal stars. They appear blue-white in colour. A-type stars on the main sequence have temperatures in the range 7200-9500 K, are 7-50 times more luminous than the Sun, and are of 1.5-3 solar masses. Sirius, the brightest star in the sky, is of spectral type Al, and Vega is type A0. A-type supergiants, such as Deneb, are more massive stars evolving off the main sequence, with temperatures up to 9700 K, and luminosities over 35 000 times the Sun's. There are a number of peculiar groups among stars of this spectral type, particularly the Ae stars, Am stars, and Ap stars. Also, two of the principal types of pulsating variable are evolved stars whose surface temperatures are in the A-star range.

Anthropic principle The proposition that the presence of human life is connected with the properties of the Universe. There are various forms of the anthropic principle. The least controversial is the weak anthropic principle, according to which human life occupies a special place in the Universe because it can evolve only where and when the conditions are suitable for it. This selection effect must be taken into account when interpreting the properties of the Universe. A more speculative version, the strong anthropic principle, asserts that the laws of physics must have those properties that allow life to evolve. The implication that the Universe was somehow designed to accommodate human life makes the strong anthropic principle highly controversial.

Abell cluster A cluster of galaxies listed in the Abell Catalogue. To appear in the catalogue, a cluster must satisfy selection criteria which include containing more than 50 galaxies and having a dense concentration. The clusters are classified as regular or irregular in appearance, ranked in increasing richness from 1 to 5, and increasing distance from 1 to 6. The approximate frequency of Abell clusters is one per 2.4×10^5 cubic megaparsecs.

Abell radius A radius of about 2 megaparsecs within which at least 50 galaxies of a particular range of brightness must be found if the cluster is to qualify as an Abell cluster. Aberration A defect in an optical instrument or system. There are several types of abberation, descriptions of a few follow: Chromatic abberation occurs in lenses and is caused by different wavelengths of light focusing at different points and results in coloured fringes around objects; Spherical abberation is caused when the innner and outer parts of a lens or mirror have different focal lengths and results in blurred images; Astigmatism is where an image is focused into a cross or ellipse and is caused by optics that are not a uniform, spherical shape; Coma causes elongated or fanned images toward the edge of the field of view.

Aberration The aberration of light is an astronomical phenomenon which produces an apparent motion of celestial objects. It is caused by the twin facts that the speed of light is finite, and that an observer on Earth is moving in an inertial frame. It does not require Earth to carry an observer to some other position after some period of time—only that Earth have some instantaneous velocity. A change in the position of an observer causes parallax, which must not be confused with the aberration of light.

Stellar aberration is independent of the distance of a celestial object from the observer. It should also be distinguished from light-time correction, which is due to the displacement of a solar system object, like a planet, through space during the time taken by its light to reach an observer on Earth. Light-time correction is independent of the motion of Earth. Planetary aberration (of solar system objects) is the combination of the aberration of light (due to Earth's velocity) and light-time correction (due to a planet's displacement). Both are determined at the instant when the object's light reaches Earth.

Types of aberration

There are a number of types of stellar aberration, caused by the differing components of the Earth's motion:

- Annual Aberration is due to the revolution of the Earth around the Sun.
- Diurnal Aberration is due to the rotation of the Earth about its own axis.
- Secular Aberration is due to the motion of the Sun and solar system relative to other stars in the galaxy.

Annual and diurnal aberration cause stars to appear to vary in position on a periodic basis, and their effect must be included when computing the apparent position of a star at any given time. Secular aberration can be regarded as constant for all practical purposes, and so is usually ignored.

Annual aberration

As the Earth revolves around the Sun, it is moving at a velocity of approximately 30 km/s. The speed of light is approximately 300,000 km/s. In the special case where the earth is moving perpendicularly to the direction of the star, the ratio of the two velocities, i.e. 1/10000 or about 20.5 arcseconds.

This quantity is known as the constant of aberration. Its precise accepted value is 20".49552 (at J2000).

The plane of the Earth's orbit is known as the ecliptic. Annual aberration causes stars exactly on the ecliptic to appear to move back and forth along a straight line, varying by ê either side of their true position. A star that is precisely at one of the ecliptic poles will appear to move in a circle of radius ê about its true position, and stars at intermediate ecliptic latitudes will appear to move along a small ellipse.

A special case of annual aberration is the nearly constant deflection of the Sun from its true position by ê towards the west (as viewed from Earth), opposite to the apparent motion of the Sun along the ecliptic. This constant deflection is often erroneously explained as due to the motion of the Earth during the 8.3 minutes that it takes light to travel from the

Sun to Earth. The latter is a type of parallax, and actually causes the apparent motion of the Sun along the ecliptic towards the east relative to the fixed stars. (8.316746 minutes divided by one sidereal year (365.25636 days) is 20".49265, very close, but of opposite sign, east vs. west.) Nor is this the Sun's light-time correction because the Sun is almost motionless, moving around the barycenter (center of mass) of the solar system by usually much less than 0".03 (as viewed from Earth) during 8.3 minutes.

Aberration can be resolved into an east-west and north-south component on the celestial sphere. The former is larger, but the latter, present because of the 23.4° tilt of the Earth's axis (obliquity of the ecliptic), was the first to be detected. This is because accurate clocks are needed to measure the East-West component, but only a good plumb line is needed for the north-south component.

Diurnal aberration

Diurnal aberration is caused by the velocity of the observer on the surface of the rotating Earth. It is therefore dependent not only on the time of the observation, but also the location of the observer. Its effect is much smaller than that of annual aberration, and is only 0".32 in the case of an observer at the equator, where the rotational velocity is greatest.

Ablation The wearing away of the outer layers of a body by melting, erosion, vaporization, or some other process due to aerodynamic effects as the body moves at high speed through a planetary atmosphere. Ablation can affect natural bodies such as meteoroids, or artificial objects such as spacecraft. Ablation of a spacecraft's protective heat shield prevents overheating of the spacecraft's interior during atmospheric entry.

Absorption line A dark feature in the spectrum of a star, formed by cooler gas in the star's outer layers that absorbs radiation emitted by hotter gas below. The Fraunhofer lines in the solar spectrum are the best-known examples. Each line is a unique signature of the element or molecule that forms it, which enables the chemical composition of the star to be determined.

Absorption spectrum An absorption spectrum (also called a dark-line spectrum) consists of dark absorption lines superimposed on a bright continuous spectrum. An absorption spectrum is created when light from an incandescent source passes through a cooler gas that absorbs photons. Each different element and molecule absorbs light at a unique set of frequencies. Astronomers can determine the composition of gases in stars by looking for their characteristic frequencies (this is called spectroscopy).

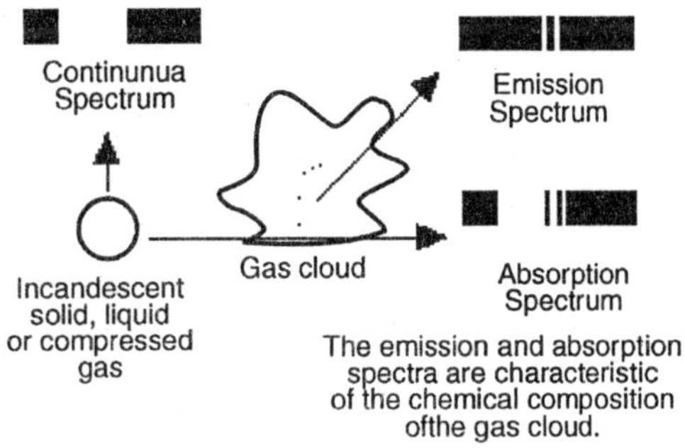

Fig. Absorption spectrum

Accretion The process by which the mass of a body increases by the accumulation of matter, in the form of either gas or small solid bodies which collide with and adhere to the body. The bodies in the Solar System are thought to have grown by accretion; some stars are surrounded by an accretion disk.

Accretion disk A structure that forms around a compact object when matter flows towards

Fig. Accretion disk

it. Accretion disks are found in interacting binary stars, and are assumed to exist in active galactic nuclei and quasars. In a binary, mass lost from the secondary star forms a disk of gas around the compact object. The disk may have a hot spot where the stream of material hits its outer edge. Material is fed from the inner edge of the disk through a boundary layer on to the compact object. When the compact object has an extremely strong magnetic field, as in an AM Herculis star, the material may form an accretion column over each magnetic pole, rather than a disk. The gravitational energy that is released can cause high ultraviolet or X-ray luminosities, and may accelerate jets of material from the disk to very high speeds.

Achondrite A class of stony meteorite usually lacking the tiny, rounded inclusions known as chondrules found in chondrites. Achondrites make up about 9% of all meteorite falls. They consist principally of one or more of the minerals plagioclase, pyroxene, and olivine. The main distinction between achondrites and chondrites is that the achondrites have different abundances of calcium and similar elements, and almost no metal or sulphide. Achondrites are thought to have crystallized from a magma in the same way as terrestrial rocks. The achondrites are divided into five main classes. The two main calcium-rich classes are the pigeonite-plagioclase achondrites and the plagioclase-hypersthene achondrites There are three main calcium-poor classes the hypersthene achondrites the olivine-pigeonite achondrites and the enstatite achondrites The eucrites, howardites, and diogenites are often collectively referred to as the basaltic achondrites.

Achondritic meteorite An achondritic meteorite is a stony meteorite, coarsely crystallized, with large bits of various minerals visible to the naked eye.

Acid rain Acid rain is polluted and harmful to the environment. Acid rain may have been a component of the K-T extinction.

Fig. Acid rain

A-class asteroid A rare class of asteroid that has both a moderately high albedo and an extremely reddish spectrum at wavelengths shorter than 0.7 μm. Strong absorption in the near-infrared is interpreted as indicating the presence of the mineral olivine. Members of this class include Asporina, diameter 70 km, and Aeternitas, diameter 52 km.

Acronical Referring to the rising or setting of a celestial object at or shortly after sunset. A planet's rising is acronical when it is at opposition.

Acrux Acrux (also known as alpha Crux) is a first magnitude star in the Southern Hemisphere constellation Crux (the Southern Cross).

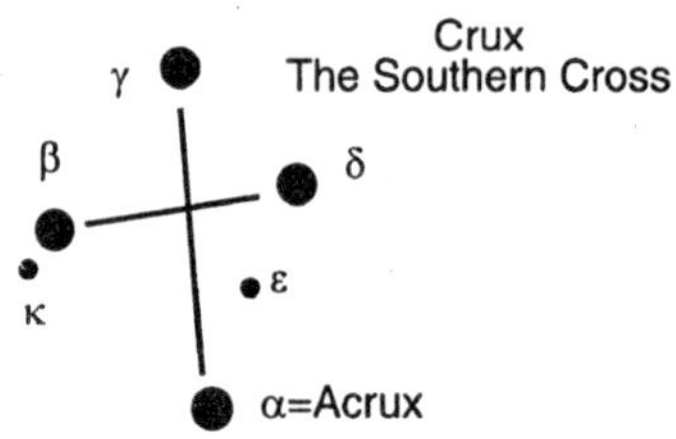

Fig. Acrux

Action This technical term is a historic relic of the 17th century, before energy and momentum were understood. In modern

Adams ring The outermost of Neptune's rings, named after J. C. Adams. It lies 62 950 km from Neptune's centre, and is less than 50 km wide. It has three denser arcs, 4°, 4°, and 10° long, separated by 14° and 12°. One of these arcs has six moonlets within it, each 10-20 km across.

Adams, John Couch English mathematical astronomer. In 1845 he calculated the orbit of a new planet whose gravitational effects would explain why Uranus did not follow its predicted path. A search from England was delayed, mainly because of the reluctance of the Astronomer Royal, G. B. Airy. In 1846 J. G. Galle sighted the new planet, subsequently named Neptune, from independent calculations by U.J.J. Le Verrier. Adams and Le Verrier were eventually both credited with predicting Neptune's existence. Adams's later work included calculation of the Moon's secular acceleration, and of the orbital elements of the Leonid meteor swarm.

Adams, Walter Sydney American spectroscopist, born in Syria. He was the first to detect systematic differences in the spectra of giant and dwarf stars. From 1914, with the German astronomer Arnold Kohlschütter Adams developed methods of establishing the surface temperature, luminosity, and distance of stars from their spectra. In 1918 he showed that the density of the white dwarf Sirius B is 50 000 g/cm^3. Adams also carried out spectrographic studies of the atmospheres of Mars and Venus.

Adaptive optics Adaptive optics (AO) is a process in which distortions (like those from the Earth's atmosphere) are removed from a telescope's image in real time. First, a wavefront sensor uses a reference star to measure the distortions that are occurring. The distortions are then removed with a phase corrector. AO was recently installed on the Keck II telescope in Hawaii.

Adhara The star Epsilon Canis Majoris, magnitude 1.5. It is a bright giant of type B2, 570 L.Y away, with a companion of magnitude 7.4.

Adonis Asteroid 2101 The second of the Apollo group to be discovered, by the Belgian astronomer Eugène Joseph Delporte in 1936, when it passed within 0.015 All of the Earth. It was not seen again until 1977. Adonis is about 1 km in diameter. Its orbit has a semimajor axis of 1.874 AU, period 2.57 years, perihelion 0.44 AU, aphelion 3.31 AU, and inclination 1°.4.

Adrastea The second-closest satellite of Jupiter, distance 129000 km, orbital period 0.298 days; also known as Jupiter XV. Adrastea is 25 × 20 × 15 km in size. It was discovered in 1979 by the Voyager spacecraft. It lies less than one Jovian radius above Jupiter's cloud tops, and very close to the outer edge of Jupiter's ring. Adrastea's gravity probably keeps the ring particles within its orbit—it acts as a shepherd moon.

Advance of perihelion A gradual turning of the major axis of an orbit in the same direction as the body moves along the orbit; also known as apsidal motion. An advance of perihelion means that the longitude of perihelion of the orbit increases. In the Solar System, the effect is caused mostly by the gravitational attractions of the planets. A similar effect, the advance of peticentre, is seen in the orbits of binary stars, caused by the oblateness of the stars themselves. In the 19th century, a small unexplained advance of Mercury's perihelion was attributed to an undiscovered planet within the orbit of Mercury This effect is now known to be caused by the curvature of space near the Sun, as predicted by the general theory of relativity.

Ae star A star of spectral type A which exhibits emission lines in its spectrum Usually these are sharp emission lines of hydrogen, and are superimposed on an otherwise normal spectrum. These lines arise in a surrounding expanding shell or disk of material. Ae stars are young stars still in the process of formation.

Aerobraking The technique of using the atmospheric drag of a planet to modify the orbit of a spacecraft, thereby saving propellant. It was first used at another planet by the US Venus probe Magellan in 1993. Aerobraking lowered the high point of Magellan's orbit above Venus from 8500 to 600 km and shortened its orbital period from 195 to 94 min. Aerobraking can also be used to help space probes enter orbit around a planet, as with the Mars Global Surveyor craft in 1997. aerolite An alternative name for a stony meteorite, now largely obsolete.

Aerolite An aerolite is a meteorite that is consists mostly of stony matter.

Aeronomy The study of processes and phenomena in the Earth's upper atmosphere, principally near its uppermost limit which, for most practical purposes, can be taken to lie at an altitude of around 300 km. Among the processes studied are airglow, reactions in the chemosphere, and the formation of noctilucent clouds and the ionosphere.

AGB stars Cool, luminous, and pulsating red giant stars. Most stars in the Universe that have left the main sequence will reach their final evolutionary stage as stars on the asymptotic giant branch (AGB).

Agglutinate A small object consisting of impact glass and fragments of minerals or rocks, all welded together into an aggregate. Agglutinates are produced by the impact of micrometeorites into the lunar regolith or other planetary surface.

AGK Abbr for Astronomischen Gesellschaft Katalog, a series of catalogues of star positions. AGK1 covered most of the sky, observed by meridian circles around the world; it was published between 1890 and 1954. AGK2 was a repetition of AGK1, with additional stars, from the north pole to -2° declination observed photographically from Hamburg and Bonn in about 1930. It contains 181581 stars, mostly to about magnitude 9 plus some fainter stars. AGK2A contains positions of 13 747 reference stars for AGK2, mostly of 8th and 9th magnitudes, derived from meridian observations made from Germany and Pulkovo, Russia. AGK3 was a reobservation of AGK2; the prefix AGK was retained even though the catalogue was not published under the auspices of the Astronomischen Gesellschaft. AGK3R contains the positions of 21499 reference stars for AGK3, observed in an international programme from about a dozen observatories.

Aitken double star catalogue Popular name for the New General Catalogue of Double Stars Within 120° of the North Pole by R. G. Aitken, published in 1932, containing measurements of 17180 double stars. It succeeded the General Catalogue of Double Stars published in 1906 by the American observer Sherburne Wesley Burnham. Double stars are often referred to by their ADS number as listed in Aitken's catalogue.

Al volorum star A pulsating variable star of short period that closely resembles a Delta Scuti star, although with a greater amplitude and somewhat higher luminosity. The Al Velorum stars may be the older of the two groups. They are sometimes referred to as dwarf Cepheids.

Albedo The fraction of the total light or other radiation falling on a nonluminous body, such as a planet, or on a planetary surface feature, that is reflected from it. In general, the albedo is equal to the amount of light reflected divided by the amount of light received. Albedo values range from 0.0 for a totally black surface that absorbs all incident light, to 1.0 for a perfect reflector. Planets or planetary satellites with dense atmospheres have much higher albedos than those with transparent or no atmospheres. The albedo may vary over the surface, so for practical purposes an average albedo is specified. Natural surfaces reflect different amounts of light in different directions, and albedo can be expressed in several ways according to whether the measurement is made in one direction or averaged over all directions.

Albireo Albireo (beta Cygni) is a third magnitude star in the constellation called Cygnus (Albireo is the head of the swan and the second-brightest star in Cygnus). Albireo is a double star; the primary (Albireo A, a K3II bright giant) is bright yellow and the companion (Albireo B) is blue. Albireo is roughly 386 ± 26 light years from us.

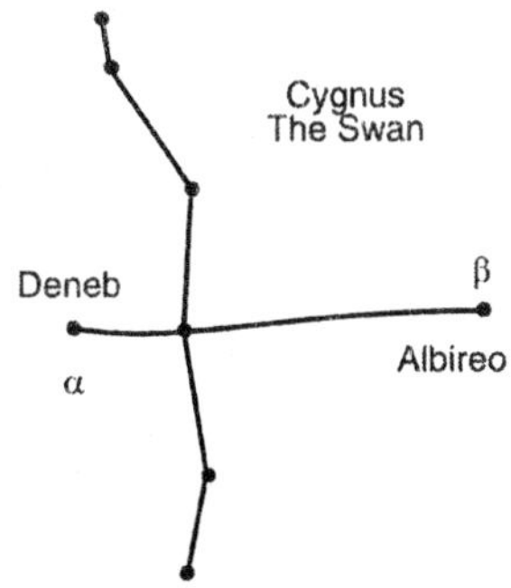

Fig. Albireo

Alcor The star 80 Ursae Majoris, an A5 dwarf of magnitude 4.0, distance 90 L.Y It forms a naked-eye double with Mizar.

Aldebaran Aldebaran is the brightest star in the constellation Taurus (it is one of the bull's eyes) and is the 13th brightest star in the sky. Aldebaran is seen along the ecliptic. Aldebaran means "the Follower" in Arabic (since it seems to follow the stars in the Pleiades. Aldebaran is an giant, old, orange star that is cooler than the Sun; it is under 4,000 Kelvin (the Sun is 5,800 Kelvin). Aldebaran is about 40 times as big as the Sun.

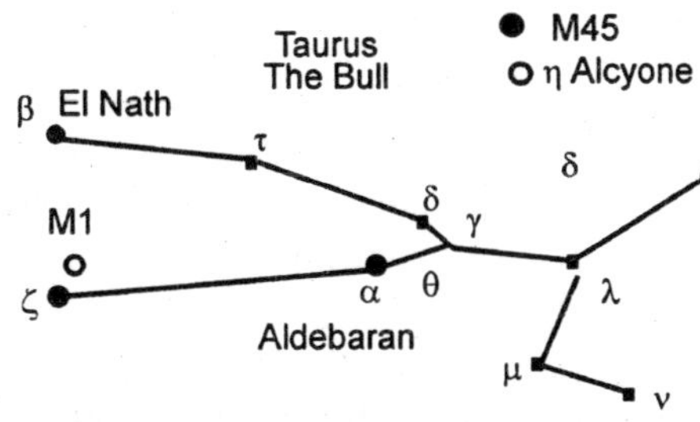

Fig. Aldebaran

Aldrin, Edwin E., jr Edwin Eugene Aldrin, Jr. (1930-) was the second man to walk on the moon. Also known as "Buzz" or "Dr. Rendezvous," Aldrin was the Lunar Module Pilot (July 16-24, 1969) on NASA's Apollo 11 mission. He and Neil Armstrong (the first man on the moon) were on the moon for about 2 hours, collecting rock samples and doing lunar surface experiments. Aldrin's first space mission was Gemini 12, which launched on November 11, 1966; he and Command Pilot James Lovell orbited Earth for 4 days. Aldrin has written many books about his life and his career as an astronaut.

Alexandra family A small family of asteroids at a mean distance of 2.6-2.7 AU from the Sun, with inclinations of 11-12°. The family is unusual in that its members are of varying composition The family is named after the C-class Alexandra, 176 km in diameter, discovered in 1858 by the German astronomer Hermann Mayer Salomon Alexandra's orbit has a semimajor axis of 2.712 AU, period 4.47 years, perihelion 2.18 AU, aphelion 3.25 AU, and inclination 11°.8.

Algol The star Beta Persei, the first eclipsing binary to be discovered. Algol was found to be variable in 1669 by the Italian astronomer

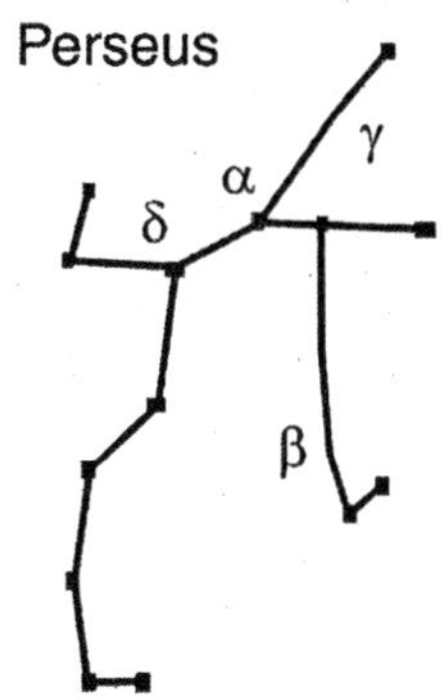

Fig. Algol

and mathematician Geminiano Montanari but the period was first determined by J. Goodricke in 1782-3. Algol varies from magnitude 2.1 to 3.4 in a period of 2.867 304 3 days, although the period has varied slightly. The variable period and the emission lines that are sometimes detectable in its spectrum are evidence for mass transfer, indicating that the system is a semidetached binary. The eclipsing pair is accompanied by a third component, magnitude 4.7, having an orbital period of 1.862 years. The system is a faint X-ray source and also emits radio bursts. Algol lies about 100 L.Y away.

Algol star A type of eclipsing binary with periods of constant brightness between well-defined eclipses; abbr. EA. This feature is an indication that the system is a detached binary or semidetached binary. The secondary minimum may be invisible. Periods range from 0.2 to 10 000 days, and the material is accreted directly on to the hot star, not via an accretion disk. Mass transfer via an accretion disk does occur in the W Serpentis stars, which have wider separations and may be pre-Algol stars. The terms W Serpentis star and Serpentid are sometimes used to include both groups of stars.

Algonquin space complex A radio observatory at Algonquin Park near Lake Traverse, Ontario, the site of a 45.7-m antenna originally opened in 1966. It was operated by the National Research Council of Canada under the name Algonquin Radio Observatory until 1986. Since 1989 the site has been operated under its present name jointly by the Institute for Space and Terrestrial Science at York University, Toronto, and the Geodetic Survey Division of the Canada Center for Surveying and Mapping, Ottawa.

Alhena The star Gamma Geminorum, magnitude 1.9. It is a subgiant of type Al lying 57 L.Y away.

Aliased response In radio interferometry, a spurious feature due to a source or a sidelobe that lies outside the field of view. Aliased responses can be suppressed by judicious choice of the gridding convolution function.

Allende meteorite A meteorite of carbonaceous chondrite type, which fell near Pueblito de Allende, northern Mexico, on 1969 February 8, scattering thousands of fragments over an area of 48 × 7 km. The parent body probably weighed over 30 tonnes. More than 2 tonnes of CV3-type material were collected, the largest piece weighing 110 kg. The formation age of this material is 4.6 billion years, making it some of the oldest primordial planetary material to be recovered.

All-sky camera A camera with a field of view that includes all, or nearly all, of the sky on one frame. The images from such cameras are circular, and have the zenith at the centre and the horizon around the edge. They are used in particular for meteor and fireball 'patrols'. A simple design of all-sky camera consists of a convex mirror which reflects the image of the sky to a camera above it, pointing downward. More advanced designs use ultra-wide-angle lenses.

Almach The star Gamma Andromedae; also spelled Almaak or Alamak. It is a double star consisting of a K3 bright giant and a B9 dwarf, magnitudes 2.3 and 4.8.

The fainter star is itself a close double with an orbital period of 61 years. Almach is 42 L.Y away.

Almagest A compendium of astronomical and mathematical knowledge written by Ptolemy in about AD 150. It incorporates the star catalogue compiled by Hipparchus, upon whose work Ptolemy may have drawn for other parts of the book. It is the most complete surviving treatise of ancient astronomy, and contains descriptions of the 48 Greek constellations on which our present-day constellation system is based. Its original Greek title was Syntaxis; Almagest, the name it acquired when translated into Arabic in about AD 820, means 'the greatest'.

Almanac A publication, usually issued yearly, listing predicted dates and times of forthcoming celestial phenomena and positions of celestial objects, along with other information of interest to astronomers, navigators, and surveyors. Examples are The Astronomical Almanac and The Nautical Almanac.

Almucantar An almucantar (also called a "parallel of altitude") is a circle on the celestial sphere that is parallel to the horizon. It consists of all points at a given altitude - if two points are on the same almucantar, they have the same altitude.

Alnilam The star Epsilon Orionis, magnitude 1.7, the middle star of the belt of Orion. It is a B0 supergiant, estimated to be 1400 L.Y away.

Alnitak The star Zeta Orionis, one of the three stars of the belt of Orion. It is a supergiant of type 09.5, magnitude 2.0, at an estimated distance of 1400 L.Y It has a close companion of magnitude 4.2.

A Symbol for right ascension.

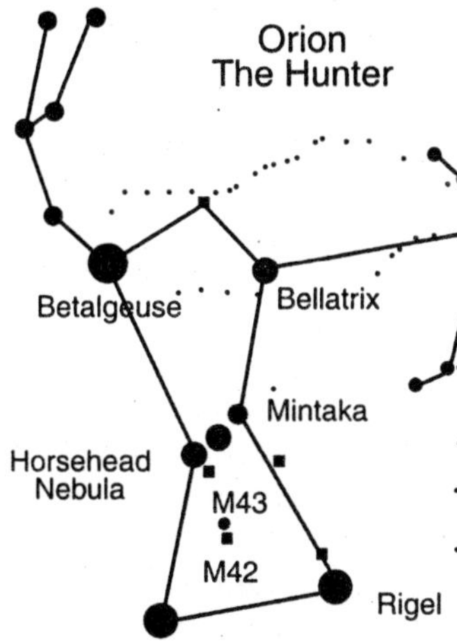

Fig. Alnitak

Alpha 2 Canum Venaticorum star A type of main-sequence, extrinsic variable star; abbr. ACV. Stellar rotation produces brightness variations of 0.01-0.1 mag., accompanied by changes in the strengths of the spectral lines and the magnetic fields. The rotational periods range from 0.5 days to more than 160 days. The spectraexhibit abnormally strong lines of silicon, strontium, chromium, and rare-earth elements. Stars in the ACVO subtype undergo non-radial pulsation (periods 0.004-0.01 days), with small (0.01-mag.) variations superimposed on the fluctuations caused by the rotation.

Alpha beta gamma theory An explanation of how primordial hydrogen was partly converted into helium during the Big Bang to provide the raw material from which stars were made. This theory predicts the existence of background radiation, which was later discovered to fill the universe.

Alpha capricomid meteors A meteor shower of generally low activity between July 15 and August 20. There appear to be several maxima, the principal one around August 2. Alpha Capricornid meteors are often long, slow, and bright. At maximum, the radiant lies at RA 20h 36m, dec. -10°.

Alpha centauri The closest star to the Sun, also known as Rigil Kentaurus. It is actually a triple system, consisting of a bright binary with a period of 80 years and a faint red dwarf 2° away called Proxima Centauri. The binary consists of a G2 dwarf of magnitude -0.01 and a K1 dwarf of magnitude 1.3. To the naked eye they appear as a single star of magnitude -0.27, the third-brightest in the sky. The binary pair lie 4.3 L.Y from the Sun, about 0.1 L.Y farther than Proxima Centauri.

Alpha crucis The brightest star in the constellation Crux, also called Acrux. It is a double star consisting of a B0.5 subgiant, magnitude 1.3, and a Bl dwarf, magnitude 1.7. To the naked eye they appear as a single star of magnitude 0.8. Alpha Crucis is 510 L.Y distant.

Alpha cygni star A type of supergiant pulsating variable star exhibiting non-radial pulsation; abbr. ACYG. The spectral types are Be-Ae Ia, and the optical amplitude is approximately 0.1 mag. Multiple pulsation frequencies are superimposed, giving rise to light-curves that often appear highly irregular. The periods range from a few days to several weeks.

Alpha cygnid meteors A meteor shower with ill-defined activity limits, emanating throughout July and August from an apparently stationary radiant at RA 21h 00m, dec. +48° near Deneb. Observed rates seldom exceed 1-3 meteors/h, and there are doubts as to the shower's authenticity.

Alt-azimuth telescope mount An Alt-Azimuth telescope mount is a "U"-shaped device that allows the telescope to move both up and down (Altitude) and left to right (Azimuth).

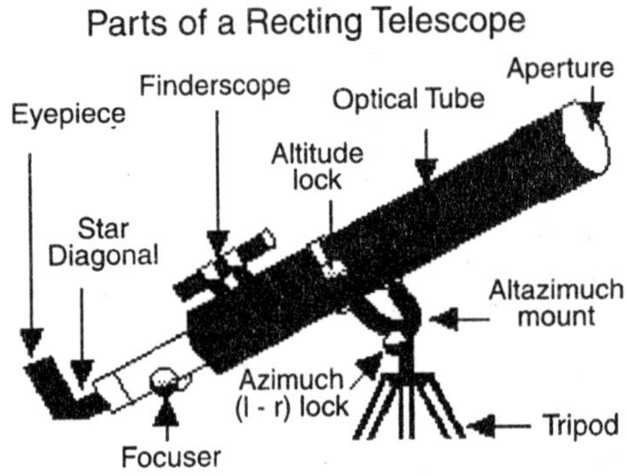

Fig. Alt-azimuth telescope mount

Altitude Altitude is a measurement in mapping astronomical objects on the celestial sphere (the sky as visible from Earth). Altitude is the angle of the object from the observer's horizon. If an object is on the horizon, its altitude is 0 degrees. If it is at the observer's zenith, its altitude os 90 degrees. To find an object in the sky, two coordinates are needed, its azimuth and its altitude.

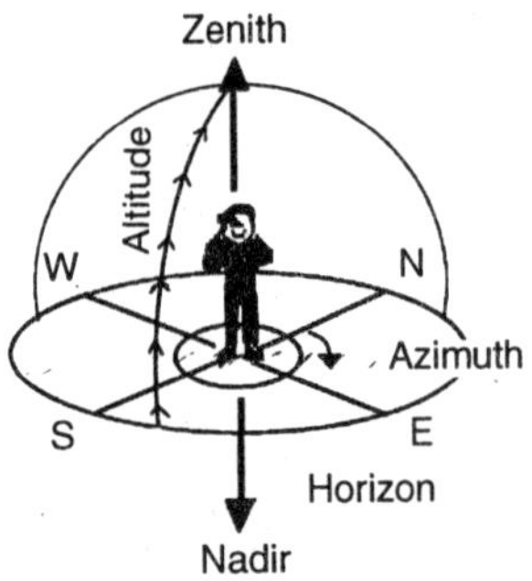

Fig. Altitude

Alvarez, Luis alvarez (1911-1988) He was a physicist who, with his son Walter Alvarez (a geologist), hypothesized that the a huge asteroid hit Earth the Earth 65 million years ago, causing a mass extinction. The Alvarez Theory of Extinction is widely accepted. Luis Alvarez received a Nobel Prize in physics (1968) for his work on subatomic particles.

AM canum venaticorum star A rare form of cataclysmic binary that appears to consist of a pair of white dwarfs. The few known are extremely hydrogendeficient and have practically pure helium spectra. There are no outbursts; the variations are like those of a W Ursae Majoris star or ellipsoidal variable, accompanied by rapid flickering. There may be mass transfer to an accretion disk around the primary, and the system may represent an extinct nova. The type star has the shortest period of any known eclipsing binary.

Am star A star of spectral type A whose spectrum additionally contains very strong metallic lines (hence the suffix 'm') more typical of type F. Am stars are usually members of close binary systems. They rotate more slowly than normal A stars, thereby allowing some elements to sink and some to rise in their atmospheres, and this produces the observed abundance anomalies. Sirius is an Am star.

Amazonian epoch The geologic history of Mars has been divided into three broad time periods, or Epochs. From oldest to youngest, these are Noachian, Hesperian, and Amazonian (named after places on Mars). These Epochs are defined by the number of impact craters on the ground surface; older surfaces have more craters. The Amazonian Epoch extends from about 1.8 billion years ago to the present day. The actual timing of the Epochs is not known because we have different models of the rates of meteorite falls on Mars.

Ambartsumian, Viktor amazaspovich Armenian astrophysicist. His major work was on the origin and evolution of stars, for which he was the first to take proper account of their physical properties. He discovered and named stellar associations, and did important early work on radio galaxies. He also studied mass ejection from novae and planetary nebulae. Ambartsumian was instrumental in the founding of Byurakan Astrophysical Observatory.

Ambipolar diffusion The process by which a gas cloud that is collapsing to form stars can detach itself from the interstellar magnetic field, so avoiding the build-up of magnetic pressure which would otherwise halt the collapse. During a slow collapse the neutral atoms and molecules can slip past the magnetic field lines in the gas. In this way the bulk of the cloud collapses but does not take the magnetic field all the way with it.

American association of variable star observers An amateur society founded in 1911. with headquarters in Cambridge, Massachusetts. Its membership is international and its activities include solar observations.

American astronomical society An organization founded in 1899 for the promotion of astronomy and related branches of science, with its headquarters in Washington, DC. It publishes the Astronomical Journal, Astrophysical Journal, and a quarterly Bulletin

Amor group A group of asteroids that cross the orbit of Mars, but not that of Earth; also known as Earth-approaching asteroids. Their orbits have perihelia between 1.017 AU and 1.3 AU perihelion of Mars). Close approaches to Mars and Earth can turn Amors into Earth-crossers temporarily, and vice versa The group is named after Amor, diameter 1 km, discovered in 1932 by the Belgian astronomer Eugène Joseph Delporte Amor's orbit has a semimajor axis of 1.921 AU, period 2.66 years, perihelion 1.09 AU, aphelion 2.76 AU, and inclination 11°.9. Amor group members show a broad variety of compositional types, evidently having originated from several sources. Asteroids may be perturbed into Amor-type orbits either by Jupiter, from near the 3: 1 and 5: 2 Kirkwood gaps in the main asteroid belt, or by Mars, from near the inner edge of the main belt. The two largest members of the Amor group are the S-class Ganymed, diameter 40 km, and Eros.

Amplitude 1. The overall range of brightness of a variable star: the difference between its maximum and minimum magnitudes. 2. In physics, the maximum difference of a cyclical quantity from its mean value, i.e. half the peak-to-peak value.

Anaemic spiral galaxy A type of spiral galaxy with characteristics intermediate between normal gas-rich spirals and gas-poor lenticular galaxies. They are seen most frequently in rich clusters of galaxies. Examples are NGC 4941 and 4866 in the Virgo Cluster, and NGC 4921 in the Coma Cluster.

Anaglyph A stereoscopic, composite image in which the right component (usually red in color) is superimposed on the left component (usually blue) to produce a three dimensional effect when viewed through correspondingly colored filters

Analemma A curve that depicts the changing altitude of the Sun during the year at noon, and also the difference between apparent solar time and mean solar time. The analemma is shaped like a thin figure of eight, its height corresponding to the Sun's altitude and its width to the time difference. It was originally a refinement to sundials that allowed corrections to be made for the equation of time.

Ananke Ananke is Jupiter's thirteenth moon. Ananke is 12.5 miles (20 km) in diameter and orbits 13,100,000 miles (21,200,000 km) from Jupiter. Ananke has a mass of 4 x 10^{16}kg. It orbits Jupiter in 631 (Earth) days and is in a retrograde orbit (orbiting opposite to the direction that Jupiter rotates). Very little is known about Ananke. Ananke was discovered by S. Nicholson in 1951.

Anastigmatic Describing a lens or optical system that has zero spherical aberration, coma, and astigmatism. Certain designs of Maksutov and Schmidt-Cassegrain telescope are anastigmats, as are all photographic lenses. Anastigmats usually have at least three optical elements, and often many more.

Anaximander Greek philosopher, born in modern Turkey. He developed a cosmogony in which universes come into existence from an ageless and eternal reservoir, into which they are eventually reabsorbed. In our universe rotation has put heavy material at the centre and fire at the periphery The significance of this world-view is its all-embracing principle governing everything, which can be regarded as the first exposition of a universal law. Anaximander has been wrongly credited with discovering the equinoxes and the obliquity of the ecliptic.

Andesite Dark-colored, fine-grained extrusive igneous rock with about 52 to 63 weight percent silica (SiO_2). Andesite consists mainly of plagioclase and one or more mafic minerals. The word andesite is derived from the Andes Mountains, located along the western edge of South America, where andesite rock is common.

Andromeda A constellation of the northern sky, representing Princess Andromeda of Greek

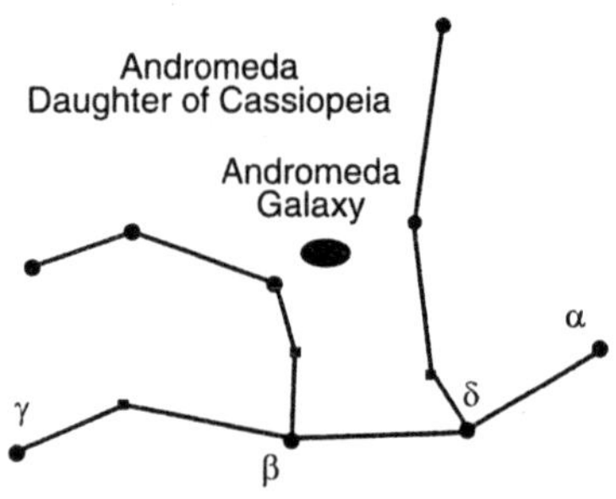

Fig. Andromeda

mythology. Its brightest stars are Alpha Andromedae and Beta Andromedae. It contains the notable double star Gamma Andromedae and the 9th-magnitude planetary nebula NGC 7662. The constellation's best-known feature is the Andromeda Galaxy.

Andromeda galaxy The spiral galaxy nearest to our own, and the largest member of the Local Group; also called M31 and NGC 224. It lies 2.3 million L.Y away and is visible to the naked eye as an elongated patch of light of total magnitude 3.4 in the constellation Andromeda. The Andromeda Galaxy has two arms and is classified as an Sb spiral. Its total mass is over 400 billion solar masses, slightly more massive than our own Galaxy. On long-exposure photographs it can be traced across over 4° of sky, corresponding to a diameter of about 150000 L.Y It has almost twice as many globular clusters as our own Galaxy. There are two small close companion galaxies, both of 8th magnitude: the elliptical M32 and the elliptical/lenticular NGC 205 plus at least three dwarf spheroidal systems, Andromeda I, II, and III.

Andromedid meteors A meteor shower seen in 1872, 1885, 1899, and 1904; also known as the Bielids. Activity resulted from the Earth's passage through debris from Comet Biela, which is presumed to have disintegrated in the mid-19th century. The Earth's approach to the node of the comet's orbit on 1872 November 27 was marked by a meteor storm of around 6000 meteors/h. The 1885 return was even more spectacular, with rates of perhaps 75 000 meteors/h. Gravitational perturbations have pulled the meteor stream orbit away from that of the Earth, so that few, if any, Andromedids are now seen.

Anglo-Australian observatory An observatory at an altitude of 1150 m on Siding Spring Mountain near Coonabarabran, New South Wales, jointly owned and operated by the UK and Australia. Its headquarters are in Epping, NSW. Its instruments are the 3.9-m Anglo-Australian Telescope and the 1.2-m United Kingdom Schmidt Telescope.

Anglo-Australian telescope A 3.9-m reflector opened in 1974 at the Anglo-Australian Observatory, New South Wales. In 1995 it was fitted with a corrector lens at the prime focus which gives it a two-degree field of view an array of fibre optics at the prime focus allows simultaneous spectroscopy of up to 400 stars or galaxies over this wide field of view.

Angrite A type of achondrite meteorite consisting of pyroxene, olivine, and plagioclase.

Angstrom A unit of length equal to 10^{-10} m. It was used mainly for specifying the wavelength of radiation, but has been replaced by the nanometre One angstrom is 0.1 nm. It is named after A. J. Ångström. Ångström, Anders Jonas Swedish physicist and astronomer. He noted that the dark Fraunhofer lines in the Sun's spectrum had the appearance of a reversed emission spectrum. In 1861 he began an intensive study of the solar spectrum, confirming the presence of hydrogen in the Sun. In 1868 he published an atlas of the solar spectrum and measurements of the wavelengths of over a thousand spectral lines. These measurements were expressed in units of 10^{-7} mm, subsequently named the angstrom unit in his honour.

Angular acceleration The rate of change in the angular velocity of a spinning body, or in the angular velocity of a celestial object in its orbit about another body.

Angular diameter The apparent size of an object, such as the diameter of a planet, expressed in degrees, minutes, or seconds of arc.

Angular distance Angular distance is the measure of an arc (a segment of the circumference of a circle). Angular distance measures the proportion of a circle that the arc in question consists of. Angular distance is expressed in degrees, radians, arc minutes (one-sixtieth of a degree), or arc seconds (one-sixtieth of a minute). For example, latitude is the angular distance north or south of the equator measured in degrees, and longitude is the angular distance east or west of the prime meridian (which goes through Greenwich, England), measured in degrees.

Angular momentum The momentum a body has by virtue of its rotation. A body such as a planet has two types of angular momentum: one type results from its motion in orbit around the Sun, and the other from its spin on its own axis. Angular momentum in orbit is given by the body's mass multiplied by its orbital angular velocity and by its distance from the Sun. Angular momentum of spin depends on the mass of the body's individual parts and their distances from its centre multiplied by its angular velocity of spin. As a consequence of conservation of angular momentum, a body spins faster as it gets smaller, such as when a gas cloud shrinks to become a star.

Angular resolution The distance, in angular measure, between two close objects that can just be separated by an optical system, such as a telescope.

Angular separation The apparent distance between two objects, such as two stars, expressed in degrees, minutes, or seconds of arc.

Angular size The apparent width of an object as seen by an observer, usually expressed in degrees, arcminutes, or arcseconds

Angular velocity The rate of rotation of a body, either about its own axis or in its orbit about another body. For example, the Earth rotates with a certain angular velocity about its polar axis, but it also has angular velocity in its revolution about the Sun.

Anisotropy Anisotropy is a state in which a physical characteristic (like the temperature of the cosmic background radiation) varies in value along axes in different directions - a physical measurement made in one direction differs from the measurement made in another direction. For example, the cosmic microwave background radiation (the radiant heat left over from the Big Bang) is anisotropic.

Ankaa The star Alpha Phoenicis, magnitude 2.4. It is a K0 giant 62 L.Y away.

Annealing In optical manufacture, the process of allowing glass or ceramic to cool slowly at a controlled rate, to avoid stress patterns within the material. A large mirror blank, for example, is cast from molten material. If cooling is too rapid, temperature variations will arise and some areas will solidify more rapidly than others, creating internal stresses that can cause the mirror blank to shatter when it is being ground to shape. Annealing of a large telescope mirror can take several months.

Annihilation Total destruction of matter in a burst of energy

Annual aberration The small displacement in position of a star's image during the year due to the motion of the Earth around the Sun. Annual aberration was discovered by J. Bradley in 1728 from observations of the changes in distance from the zenith of the star Gamma Draconis. The ratio of the Earth's mean velocity to the speed of light gives the constant of aberration. This is the maximum amount by which a star can appear to be displaced from its mean position. During the course of a year, the star appears to move around its mean position in a shape that ranges from a circle for a star at the ecliptic pole, via a progressively flattened ellipse, to a straight line for a star on the ecliptic..

Annual equation A periodic disturbance in the celestial longitude of the Moon, resulting from the changing gravitational pull of the Sun around the Earth's elliptical orbit; also known as annual inequality. It has an amplitude of 11' and a period of one anomalistic year.

Annual parallax The maximum apparent difference in position of a star during the course of a year due to the changing position of the Earth in its orbit around the Sun; also known as heliocentric parallax. The amount of displacement is equal to the angular separation between the Earth and Sun as seen from the star.

Annual variation The yearly changes in the right ascension and declination of a star due to annual precession and proper motion. These quantities are tabulated in star catalogues.

Annular eclipse An annular eclipse is a type of solar eclipse. During an annular eclipse, the Sun looks like an "annulus" or ring. The ring is visible when the Moon does not entirely cover the disc of the Sun during the eclipse. This type of eclipse happens when the Sun is at perihelion (closest to the Earth, making the solar disc appear larger) and the Moon is at apogee (farthest from the Earth, making it look smaller).

Anomalistic month The mean interval of time between successive passages of the Moon through its closest point to Earth, equal to 27.55464 days. The Moon's orbit is not fixed in space but can be regarded as a slowly rotating ellipse, the rotation due principally to perturbation by the Sun. Hence the perigee point advances slowly around the orbit, completing one circuit in 8.85 years. As a consequence, the anomalistic month is about 5.5 hours longer than the sidereal month.

Anomalistic year The interval of time between successive passages of the Earth through the perihelion of its orbit, equal to 365.2596 days. Due to perturbations by the other planets, the Earth's orbit is not a fixed ellipse but one that is slowly advancing. As a result the anomalistic year is about 5 min longer than the Earth's orbital period

Anomalous iron An iron meteorite with a chemistry and structure unlike those of any of the primary groups into which other iron meteorites are divided.

Anomaly An angle that describes a body's position at a given time in an elliptical orbit. There are three types of anomaly: eccentric anomaly, mean anomaly, and true anomaly. All are measured in the direction of the body's motion and take as their starting-point the body's periapsis

Anorthosite A medium- to coarse-grained igneous rock, grey to white in colour. It is composed almost entirely of calcium-rich plagioclase feldspar one of the most common rock-forming minerals, together with minor amounts of pyroxene olivine, and iron oxides. Anorthosite is thought to comprise a significant fraction of the lunar crust, and is present at all the Apollo and Luna landing sites. Samples of anorthosites 4.4-4.5 billion years old were among the oldest lunar rocks returned by spacecraft.

Ansae The extensions of Saturn's rings either side of the planet; sing. ansa. The word is Latin for 'handles', since the rings project from the planet like handles.

Antares The star Alpha Scorpii, an Ml.5 supergiant 10 000 times more luminous than

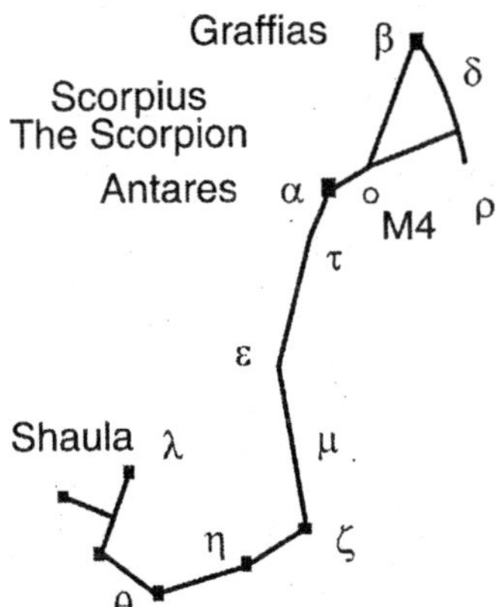

Fig. Antares

the Sun and with a probable diameter over 500 times the Sun's. It is a semiregular variable, ranging between magnitudes 0.9 and 1.8 with a period of about 5 years. Antares is estimated to lie 520 L.Y away. It has a B2 dwarf companion of magnitude 5.4 with an orbital period of about 900 years. The name Antares is Greek, and is usually translated as 'rival of Mars', because of its strong red colour, but sometimes as 'like Mars'.

Antenna In contrast to an "ordinary" telescope, which produces visible light images, a radio telescope "sees" radio waves emitted by radio sources, typically by means of a large parabolic ("dish") antenna, or arrays of them. The first of these was the 9m telescope constructed by Grote Reber in 1937. In the early 1950s the Cambridge Interferometer mapped the radio sky to produce the famous 2C and 3C surveys of radio sources. In the late 1950s and early 1960s the largest single-dish radio telescope in the world was the 76 metre telescope at Jodrell Bank, which became operational in 1957. This was just the latest of many radio telescopes constructed during the middle of the 20th Century, and has been surpassed by more modern telescopes and arrays of telescopes.

The largest individual radio telescope is the RATAN-600 (Russia) with 576 meter diameter of circular antenna (RATAN-600 description). The largest radio telescope in Europe is the 100 meter diameter antenna in Effelsberg, Germany, which also was the largest fully steerable telecope for 30 years until the Green Bank Telescope was opened in 2000. The largest radio telescope in the United States until 1998 was Ohio State University's The Big Ear. Other well known disk radio telescopes include the Arecibo radio telescope located in Arecibo, Puerto Rico, which is steerable within about 20° of the zenith, and the fully steerable Lovell telescope at Jodrell Bank in the United Kingdom. A typical size of the single antenna of a radio telescope is 25 metre, dozens of radio telescopes with comparable sizes are operated in radio observatories all over the world.

An example of the array-type radio telescope is the Very Large Array (VLA), in Socorro, New Mexico, which consists of an interferometric array formed from several receivers. The largest exisiting radio telescope array is the Giant Metrewave Radio Telescope, located in Pune, India. A larger array, LOFAR (the 'LOw Frequency ARray') is currently being constructed in western Europe, consisting of 25000 small antennas over an area of several 100s of kilometres in diameter.

The sub-field of astronomy related to observations made through radio telescopes is known as radio astronomy.

Many celestial objects, such as pulsars or active galaxies (like quasars), produce radio-frequency radiation and so are best "visible" or even only visible in the radio region of electromagnetic spectrum. By examining the frequency, power and timing of radio emissions from these objects, astronomers can improve our understanding of the Universe.

Radio telescopes are also the primary means to track space probes (see Deep Space Network), and are used in the SETI project.

Nikola Tesla in the Colorado Springs lab recorded cosmic waves emitting from interstellar clouds and red giant stars. He observed repeating signals conducted by his transceiver. He announced that he received extraterrestrial radio signals. Tesla stated that he received signals from planets in some of the scientific journals of the time. The scientific community did not believe him, primarily because research of cosmic signals did not exist (what is known today as radio astronomy), and the community of science rejected Tesla's data. Tesla spent the latter part of his life trying to signal Mars.

One of the earliest modern investigations into extraterrestrial sources of radio waves were by Karl Guthe Jansky, an engineer with Bell Telephone Laboratories, in the early 1930s. The first object actually detected was the center of the Milky Way, followed by the sun. Grote Reber (December 22, 1911 - December 20, 2002) was one of the pioneers of radio astronomy. He was instrumental in repeating Karl Guthe Jansky's pioneering but somewhat simple work, and conducted the first sky survey in the radio frequencies. After World War II, substantial improvements in radio astronomy technology were made by astronomers in Europe, Australia and the United States, and the field of radio astronomy began to blossom.

One of the most notable developments came in 1946 with the introduction of radio interferometry by Martin Ryle's group in Cambridge (who obtained a nobel prize for this and later aperture synthesis work), also the Lloyd's mirror interferometer developed independently in 1946 by Joseph Pawsey's group at the University of Sydney.

Antenna pattern A graph of the sensitivity or gain of a radio telescope in various directions. A typical antenna pattern shows several sensitivity maxima of which the strongest defines the direction in which the telescope is most sensitive. An antenna pattern with a narrow main lobe of circular cross-section and negligible side lobes is known as a pencil beam, while a pattern with a wide, flat main lobe is known as fan beam..

Antenna temperature A measure of signal strength in radio astronomy. It is defined as the temperature of a black-body enclosure which, if completely surrounding a radio telescope, would produce the same signal power as the source under observation. Antenna temperature is a property of the source, not of the antenna itself. Where a uniform extended source fills the telescope beam, the antenna temperature is equal to the brightness temperature of the source.

Antennae A pair of interacting galaxies, NGC 4038 and 4039, lying 60 million L.Y away in the constellation Corvus. They have two long, faint, curved, tails extending for 360 000 L.Y 20'on the sky), which give them their name. The tails are probably composed of stars pulled out by gravitational tidal interactions during the collision of a spiral and a lenticular galaxy.

Anthropic principle We see the universe the way it is because if it were different we would not be here to observe it

Anticentre..anti-dwarf nova.antimatter Matter composed of antiparticles: subatomic particles that have identical rest mass to corresponding particles of ordinary matter but opposite charge, and are opposites in other fundamental properties. For example, the antiparticle of the electron is the positron, which has a positive charge equal to the electron's negative charge; the antiproton has a negative charge equal to the proton's positive charge. When matter and antimatter meet, they annihilate each other, releasing energy. The Universe seems to be almost entirely in the form of matter rather than antimatter; why this should be so is presumably related to events shortly after the Big Bang.

Anti-coincidence detector shield A shield which is transparent to gamma rays but that detects when a cosmic ray passes through it; used to reject cosmic ray events in gamma-ray detectors

Antigravity A gravitational field that repels, rather than attracts, matter and light rays

Antimatter Antimatter is matter that is composed of the counterparts of ordinary matter. For example, positrons are the antimatter counterpart of electrons; positrons have a positive charge instead of a negative charge. The existence of an antimatter version of the electron was predicted by Paul Dirac in 1930. Carl Anderson, in 1932, found positron tracks that were produced by cosmic rays in a cloud chamber. Antimatter is also produced in some radioactive decays. A particle and its antimatter particle will annihilate when they meet, and energy is released (according to Einstein's $E = mc^2$). Antiparticles have the same mass as their counterpart, but the opposite charge. There seems to be very little antimatter in the universe.

Antiparticle An atomic particle that has exactly the opposite properties of its counter-part (e.g. a positron and an electron)

Antipodal point The point that is directly on the opposite side of the planet.

Roman Numeral Description.

Perfect seeing, without a quiver.

Slight undulations, with some moments of calm lasting several seconds.

Moderate seeing, with larger air tremors.

Poor seeing, with constant troublesome undulations.

Very poor seeing, scarcely allowing the making of a rough sketch.

Antitail A protruberance from a comet that appears to point towards the Sun, often looking like a spike. It is actually part of the comet's dust tail, and consists of larger particles lagging behind the comet in its orbit. The antitail does not actually point sunwards, but results from a perspective effect when the comet is seen from a certain angle. Antitails are uncommon. Where present, they are usually most prominent when the Earth passes through, or close to, the plane of the comet's orbit. The cometary dust, which lies in a thin sheet, is then seen edge-on.

Antlia An unremarkable constellation of the southern sky, representing an air pump. Its brightest star, Alpha Antliae, is magnitude 4.3.

Antoniadi scale A scale of seeing conditions, used by amateur astronomers, that describes the effect of atmospheric motions on an image, usually that of a planet. Its categories are:

I perfect, without a quiver

II slight undulations, but with calm periods lasting several seconds

III moderate, with some greater air movements

IV poor, with the image in constant movement

V very bad, making observations very difficult

It was devised by E. M. Antoniadi.

Antoniadi, Eugène Michael French astronomer, born in Turkey. He specialized in the planets, of which he was one of the most outstanding observers of his time. His highly detailed drawings show, for example, spots on Saturn's disk and 'spokes' in its rings, the South Tropical Disturbance on Jupiter, and surface details on Mercury. Antoniadi's chief interest, though, was Mars. He was sceptical about the Martian canals, and was able to demonstrate that they were illusory. He devised the Antoniadi scale for the quality of seeing.

Ap index An averaged planetary. A index based on data from a set of specific stations.

Ap star A peculiar star of spectral type A that has much stronger absorption lines than normal of elements such as manganese, mercury, silicon, chromium, strontium, and europium; also known as a peculiar A star. Such stars have unusually strong magnetic fields, thousands of times stronger than the Sun's typical surface field. This leads to certain elements being localized in spots, similar to sunspots. As the star rotates the spectral features therefore vary. Hotter and cooler versions of Ap stars are, respectively, Bp and Fp stars.

Apache point observatory An observatory at an altitude of 2780 m in the Sacramento Mountains near Sunspot, New Mexico. It houses a 3.5-m reflector opened in 1994; a 2.5-m reflector due to start operation in 1997 for the Sloan Digital Sky Survey; and a 1-m reflector owned by New Mexico State University. The observatory is owned and operated by the Astrophysical Research Consortium whose members are the University of Chicago, the Institute for Advanced Study at Princeton, Johns Hopkins University, New Mexico State University, Princeton University, the University of Washington, and Washington State University.

Apastron The point of greatest separation between two stars which are in orbit around each other. See binary stars. Opposite of periastron.

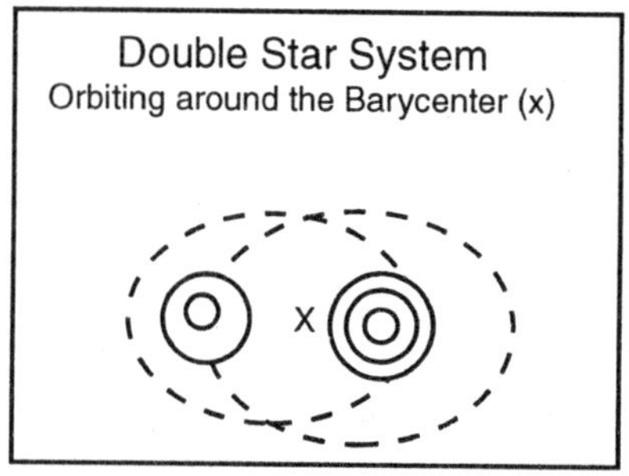

Fig. Apastron

Aperture The diameter of the main lens or mirror of a telescope, or, in radio astronomy, of the collecting dish. For Schmidt cameras, however, it is the diameter of the corrector plate.

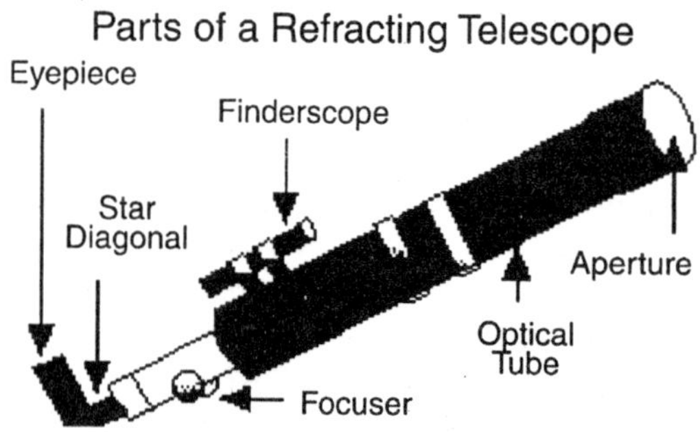

Fig. Aperture

Aperture efficiency A measure of the efficiency with which a telescope can collect radio waves. It is expressed as the effective aperture divided by the geometrical area. For a simple parabolic antenna, aperture efficiencies rarely exceed 65%, due to the difficulties of designing a primary feed to be equally sensitive to waves reflected from all points of the dish.

Aperture ratio The ratio of the aperture of a lens or mirror to its focal length, usually expressed as a true ratio, e.g. 1 : 8. It is more common, however, to refer to the focal ratio, which is the inverse of the aperture ratio and is expressed in the form f/8.

Aperture synthesis A technique used in radio astronomy to achieve high angular resolution. It uses an array of telescopes to simulate a single telescope of large aperture In principle, an aperture of any size can be synthesized by moving the two elements of an interferometer to occupy all possible positions in the desired aperture. In practice, all synthesis telescopes make use of the fact that over a period of 12 hours the Earth's rotation will move the elements to sweep out half a ring of the synthesized aperture the other half of the ring can be derived from the observations of the first half. The elements need then be moved only to sweep out successive rings. In practice, some aperture-synthesis telescopes employ several movable dishes to reduce observing time, while in others the dishes remain fixed. Aperture synthesis requires complex data-reduction techniques and powerful computers. Examples of aperture synthesis telescopes are MERLIN, the Ryle Telescope, the Very Large Array, and the Westerbork Synthe.

Aphelion The position of a planet (or other body) when it is at its furthest from the sun.

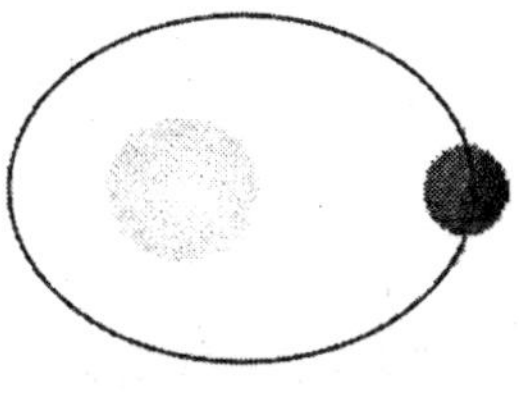

Fig. Aphelion

Aphrodite terra A large upland area on Venus, near the equator. It is over 10 000 km long, and has about the same area as Africa. It includes four smaller highland areas: Ovda Regione and Thetis Regione in its western half, and Atla Regione and Ulfrun Regione to the east. There are many troughs in Aphrodite, including Diana Chasma near the centre, and volcanoes such as Maat Mons, 8.4 km high.

Aplanatic Describing a lens or optical system that is free from both spherical aberration and coma Examples of aplanatic systems include some types of achromatic lens and the Ritchey-Chrétien telescope.

Apoapsis The point in an orbit when the two objects are farthest apart. Special names are given to this orbital point for commonly used systems: see apastron, aphelion, and apogee.

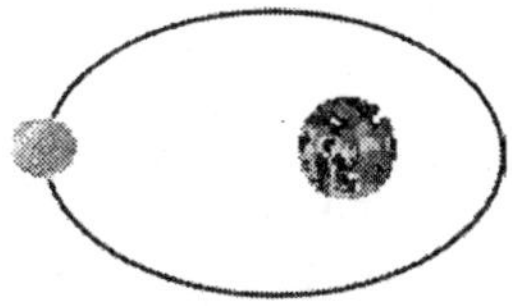

Fig. Apoapsis

Apocentre The point in an elliptical orbit that lies farthest from the centre of mass of the orbiting system, such as a binary star or a planet and satellite.

Apochromatic Describing a lens or optical system with very little chromatic aberration. In an apochromatic lens the chromatic aberration is completely cancelled out at three or more wavelengths, compared with two for a typical achromatic lens. Apochromats usually have at least three optical elements, made from different types of glass. Fluorite has become a popular material for apochromats, giving virtually perfect colour correction with only two elements, but at considerable cost.

Apogee The point in its orbit where an Earth

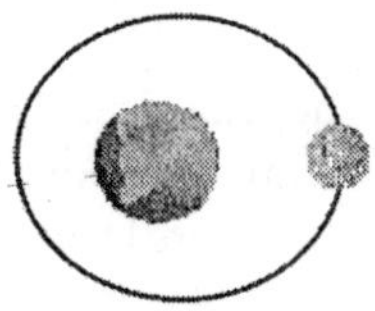

Fig. Apogee

satellite is farthest from the Earth. Opposite of perigee.

Apollo 1.The U.S. space program that sent astronauts to the moon in the 1960s and '70s2. an asteroid with a perihelion less than 1.017 AU.

Apollo group A group of asteroids that cross the Earth's orbit, but whose average distances from the Sun are greater than that of Earth; also known as Earth-crossing asteroids. Their perihelion distances are 1.017 AU or less. Most Apollo asteroids are small and highly irregular in shape. They are named after Apollo, the first of the group to be discovered, by the German astronomer Karl Reinmuth in 1932. Apollo is a Q-class asteroid of diameter 1.4 km. It came within 0.07 AU (10.5 million km) of Earth in 1932, but was then lost until 1973. Apollo's orbit has a semimajor axis of 1.471 AU, period 1.78 years, perihelion 0.65 AU, aphelion 2.30 AU, and inclination 6°.4. Apollo can approach to within 0.028 AU of Earth's orbit and can also make close approaches to Venus and Mars.

Apollo missions NASA's Apollo missions (1961 to 1975) were the only missions to send people to the moon. Apollo 11's LEM or LM (Lunar Excursion Module or Lunar Module) landed on the moon on July 20, 1969. Apollo 12 to 17 continued lunar exploration.

Apollo object or Apollo asteroid Apollo asteroids have an orbit that crosses the orbit of the Earth and have a period longer than 1 year. They have a semimajor axes greater than 1 astronomical unit (au), and perihelion distances less than 1.017 au. These asteroids were named for asteroid Apollo, #1862. Geographos is an Apollo.

Apollo project The American space programme to land humans on the Moon. The three-seater Apollo spacecraft were launched to the Moon by Saturn V rockets. In orbit around the Moon, two astronauts entered the Lunar Module and descended to the surface. After four manned test flights, the first manned landing was made on 1969 July 20 by Neil Alden Armstrong and Edwin Eugene 'Buzz' Aldrin during the Apollo 11 mission. On subsequent missions, astronauts spent longer on the lunar surface and were able to explore greater distances by means of an electrically powered lunar rover. A total of 381 kg of rock samples was brought back by the Apollo astronauts. Apollo spacecraft were also used to ferry crews to the Skylab space station and in the Apollo-Soyuz Test Project.

Apollo-Soyuz test project An international space mission between the USA and the former Soviet Union in which an Apollo spacecraft docked in orbit around the Earth with a Soyuz spacecraft in 1975 July. The purpose of the mission was to practice techniques for possible space rescue and for future joint flights, but its main significance was political.

Apparition The period of time during which an object in the Solar System is visible from Earth, such as an evening apparition of Venus or the apparition of a periodic comet. The term is not usually applied to objects that are more regularly visible, such as the Sun or stars.

Appulse The apparent close approach of two celestial bodies when one moves into the same line of sight as the other, such as when a planet appears to pass another planet or a star.

Aquarius Aquarius (the water bearer), abbreviated Aqr, is a constellation of the

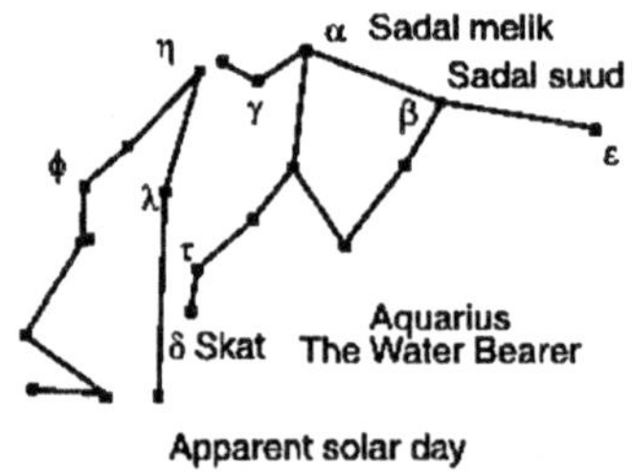

Fig. Aquarius

zodiac. This faint constellation is seen along the ecliptic. The sun passes through Aquarius from mid-February until mid-March. There are many interesting celestial objects in this large constellation, including globular clusters (M2 and M72), planetary nebulae (NGC 7009 and NGC 7293), the star group M73, a double star (zeta Aqr). Aquarius is the radiant for the meteor showers: March Aquarids, Delta Aquarids, Eta Aquarids, and Iota Aquarids. Alpha Aquarii (the brightest star in Aquarius) is called Sadalmelik (meaning the Lucky One of the King) and beta Aquarii (the second-brightest star in Aquarius) is named Sadalsuud (meaning the Luckiest of the Lucky); they are twin supergiants.

Arago, François French scientist and statesman. With the French physicist Augustin Jean Fresnel he established the wave theory of light and studied its polarization. In 1811 he invented the polariscope for measuring the degree of polarization of light. This instrument allows light from solid and fluid surfaces to be distinguished, and Arago used it in several astronomical studies. One was an extensive study of the total solar eclipse of 1842, in which he determined that the Sun's limb is gaseous, and examined polarized light from the chromosphere and corona.

Archaeoastronomy The study of the astronomical significance of archaeological remains and artefacts, particularly large stone structures Structures such as. Stonehenge in England and the Great Pyramid in Egypt have components which are oriented to align with the positions of celestial bodies at significant times of the year. This suggests that they were constructed at least partly for astronomical purposes, although further interpretation remains speculative.

Arcturus (Alpha Boötis) Arcturus is the brightest star in the constellation Boötes (the herdsman). It is a red giant (spectral type K1.5IIIp) that is the fourth brightest star in the sky. Arcturus is 34 light-years from Earth. It has an apparent magnitude of -0.04 and an absolute magnitude of 0.2.

Fig. Arcturus

Arend-Roland comet A long-period comet discovered in 1956 November by the Belgian astronomers Silvain Arend and Georges Roland formerly designated 1957 III. It reached perihelion, 0.32 AU, on 1957 April 8. Closest approach to Earth, 0.57 AU, was on April 21. At peak brightness, the comet reached magnitude -1, with a tail of 25-30°. It also had a prominent antitail. Comet Arend-Roland has a hyperbolic orbit of inclination 119°.9.

Areo Prefix referring to Mars; for example, areography is the study of the surface of Mars.

Argelander, Friedrich Wilhelm August German astronomer, born in modern Lithuania. From a study of the proper motions of several hundred stars he confirmed F. W. Herschel's calculated position of the Sun's apex. His main achievement, begun in 1852, was a survey of all stars down to 9th magnitude in the northern hemisphere. The resulting star atlas and catalogue were published in 1859-62 as the Bonner Durchmusterung. He subsequently initiated the first AGK catalogue. Argelander also introduced the subdivision of magnitudes into tenths, the step method for estimating stellar magnitudes, and the practice of assigning capital roman letters to variable stars. Argelander step method A visual method of estimating the magnitude of a variable star, described by F. W. A. Argelander.

Argo navis One of the 48 constellations known to the Greeks, representing the ship of the Argonauts, but broken up in the 18th century by N. L. de Lacaille into Carina, Puppis, and Vela.

Argument of perihelion An angle that defines the direction of the major axis of an orbit around the Sun. It is the angle between the ascending node and the perihelion of an orbit, measured in the plane of the orbit and in the direction of orbital motion.

Argyre planitia A circular feature in the southern hemisphere of Mars, visible from Earth as a white patch, centred at -49° lat., 43° long. Space probes have shown it to be the second-largest impact basin on Mars, about 870 km wide and 2 km deep.

Ariel The fourth-largest satellite of Uranus, 1158 km in diameter; also known as Uranus I. Ariel orbits Uranus in 2.52 days at a distance of 191020 km, keeping the same face towards the planet. It was discovered in 1851 by W. Lassell. Ariel has an icy surface scored by many giant steep-sided troughs, termed chasmata, between cratered regions. There are also smooth plains and ridged terrain. This fractured surface indicates that tectonic processes have been active in the past, perhaps a result of tidal forces.

Ariel satellites A series of UK scientific satellites launched by NASA between 1962 and 1979. Ariel 1, the first UK satellite, studied solar X-rays and the Earth's outer atmosphere. The next three studied the atmosphere and radio astronomy. The longest-lived mission was the X-ray astronomy satellite Ariel 5, which operated from 1974 to 1980. Ariel 6, the last UK satellite, was primarily concerned with cosmic rays but also carried X-ray instruments.

Aries A constellation of the zodiac, representing a ram. The Sun lies in Aries for the last 10 days of April until mid-May. Its brightest star is Alpha Arietis.Gamma Arietis is a pair of white stars of magnitudes 4.6 and 4.7.

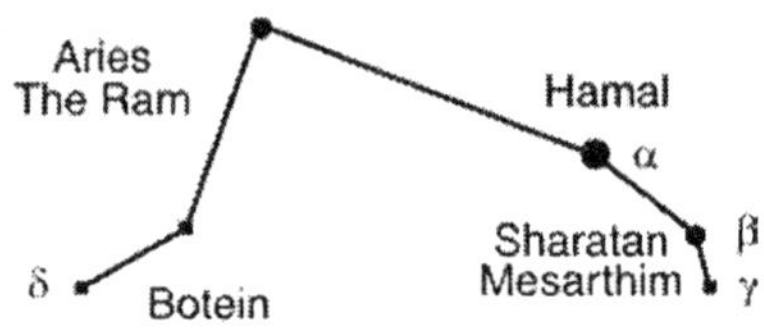

Fig. Aries

Aristotle Aristotle (384-322 BC) A Greek philosopher who theorized about astronomy. Using only philosophical speculation (he did no scientific observations), Aristotle believed that the universe is spherical, finite, and centered around the Earth. Aristotle, like many others of his time, believed that the circle was the "perfect" shape, so the universe must be spherical, and all the orbits in it must also be circular. He also believed that celestial bodies were composed of ether (in addition to the four other basic elements believed to exist at that time, earth, air, fire, water). Aristotle's ideas were adopted by the Church and were not tested for over a thousand years, until Galileo's experiments demonstrated errors in Aristotle's writings.

Armillary sphere An armillary sphere is an instrument from the 1500s that was used to

Fig. Armillary

determine the relative positions of the celestial equator, the ecliptic, the planetary orbits, etc. This device consisted of a series of concentric rings.

Armstrong, Neil Neil Alden Armstrong (1930-) The first person to walk on the moon. He piloted NASA's Apollo 11 mission, which took off on July 16, 1969. Armstrong and Edwin E. (Buzz) Aldrin, Jr., landed on the moon on July 20, 1969, in the lunar module (landing in the Mare Tranquilitatis), while Michael Collins orbited the moon in the command module. Upon his first step on the moon, Armstrong said, "One small step for man, one giant leap for mankind." Armstrong and Aldrin explored the Moon's surface for about 2 hours. Years earlier, in 1966, Armstrong and David R. Scott performed the first successful docking of two vehicles in space on the Gemini 8 mission.

Arrest, Heinrich Louis Danish astronomer who assisted Galle with the first observations of Neptune. After receiving its predicted position from Le Verrier, Galle and d'Arrest began searching. With Galle at the eyepiece and d'Arrest reading the chart, they scanned the sky and checked that each star seen was actually on the chart. Just a few minutes after their search began, d'Arrest cried out, "That star is not on the map!" and earned his place in the history books.

Artificial satellite A spacecraft sent into orbit around the Earth. The first artificial Earth satellite was Sputnik 1, launched by the former Soviet Union on 1957 October 4. A spacecraft can become an artificial satellite of the Moon or a planet if it goes into orbit around it.

ASM All Sky Monitor. An instrument designed to observe large areas of the sky for interesting astronomical phenomena. An ASM measures the intensity of many sources across the sky and looks for new sources. Many high-energy satellites have carried ASM detectors, including the ASM on Vela 5B, Ariel V, and the Rossi X-ray Timing Explorer.

Aspect The position of a body in the Solar System relative to the Sun, as seen from Earth. The main aspects are conjunction, greatest elongation, opposition, and quadrature.

Aspheric Describing a surface of a lens or mirror that has been given an optical figure which is not part of a sphere. The most common aspheric surface is the paraboloid, widely used for telescope mirrors. Another common aspheric surface is that used for the corrector plates of Schmidt-Cassegrain telescopes. In objective lenses most surfaces are spheroidal, one surface being aspheric to correct for aberrations.

Asteroid belt The region of the Solar System between the orbits of Mars and Jupiter in which most asteroids are found; also known as the main belt. It extends from 2.15 to 3.3 AU from the Sun, corresponding to orbital periods of 3.15-6.0 years. Asteroids are not uniformly distributed throughout the belt. There are sparsely populated zones, known as Kirkwood gaps, as well as concentrations of asteroids with similar orbital elements, forming groups and Hirayama families. The proportion of various asteroid classes changes markedly through the belt.

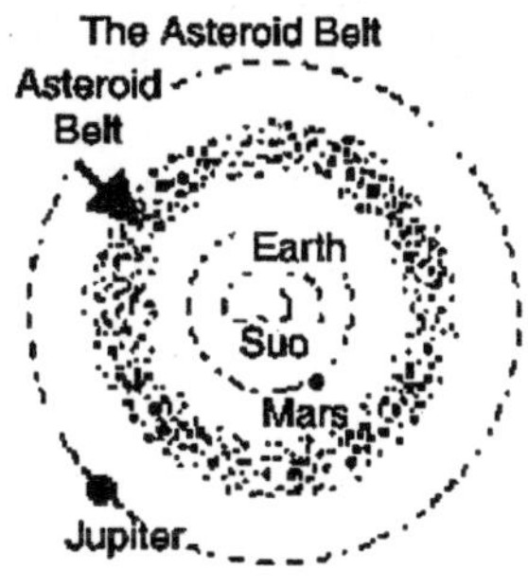

Fig. Asteroid belt

Asteroid family Group of asteroids with similar orbital elements (distance from the Sun, inclination of the orbital plane to the Earth's orbital plane, and eccentricity of the orbit) thought to be remnants of one larger asteroid parent.

Asteroid number Asteroids are assigned a serial number when they are discovered. It has no particular meaning except that asteroid N+1 was discovered after asteroid N.

Asteroseismology The study of global oscillations of stars by precise measurements of Doppler shifts in their light; it is the stellar equivalent of helioseismology. Observations of global oscillations have been claimed for some stars of spectral type Ap, with periods of 6-12 min.

Asthenosphere A weak layer within a planetary body, beneath the rigid outer layer known as the lithosphere. It is slightly weaker than the lithosphere and deformable over long periods, allowing continental drift to take place on the Earth. On other planets and satellites the asthenosphere allows the topography to smooth out slowly, highlands becoming lower and deep basins shallower. The depth of the asthenosphere varies according to the size, density, composition, and thermal structure of the planet or satellite. On Earth, the top of the asthenosphere is around 100 km below the surface, but on the Moon it lies 800 km below the surface, more than half-way to the centre.

Astrobiology Astrobiology is the science that addresses the question of whether or not there is life in space, and if so, then what it its origin, distribution, and evolution.

Astrobleme An eroded impact crater on Earth, identifiable by its geological structure and highly shocked rocks; the name means 'star wound'. About 150 terrestrial impact structures are known; about four new ones are identified each year. Evidence an impact origin includes signs of high-pressure shock waves, such as shatter cones, shock-induced minerals and microscopic linear patterns in quartz called shock lamellae. Among the largest are Sudbury in Ontario, Canada Popigai in Russia and Manicouagan in Quebec, Canada Virtually all are on land, but Montagnais is on the continental shelf off Nova Scotia, and Chicxulub extends into the Gulf of Mexico from the Yucatán Peninsula. The oldest are approaching 2 billion years old, but the majority are less than 200 million years old.

Astrology The supposed influence of the relative positions of the planets on people's personalities and events in their lives. In its modern form astrology is a pseudoscience, but in ancient times astrology and astronomy were intertwined. Often, the motive for keeping observational records was astrological. Ancient Chinese records of celestial events, from which the fortunes of entire dynasties were divined, are now of great value in the study of historical eclipses, novae, and comets.

Astrometry The measurement of positions, parallaxes, and proper motions on the sky. It can be divided broadly into two categories: global and small-field astrometry. Global astrometry is concerned with mapping and cataloguing positions and motions over large areas of sky. It was traditionally based on optical observations made with meridian instruments and astrolabes. Currently, optical interferometers are being developed to achieve greater accuracy. A stellar reference frame of bright stars is derived from such observations, and interpolation to fainter stars is achieved by photographic surveys. The accuracy of all ground-based optical astrometry is limited by thermal and mechanical instabilities in the telescopes, but also mainly by uncertainties in the amount of atmospheric refraction. The Hipparcos satellite was launched to overcome these limitations. Global astrometry at radio wavelengths is carried out by interferometers, with both short and very long baselines.

Astronaut Person engaged in or trained for spaceflight.

Astronautics Astronautics is the science of the design, construction, and operation of spacecraft.

Astronavigation The technique of position-finding by reference to celestial bodies. An observer uses a sextant to observe the altitude of a celestial body, such as the Sun, the Moon, a planet, or a bright star. From several observations of the same object at different times, or of different objects at the same time, the observer's position can be calculated. Traditional astronavigation on Earth has been largely superseded by methods using artificial satellites or radio beacons.

Astronomer An astronomer is someone who studies astronomy.

Astronomer royal An honorary title held by a prominent British astronomer, created in 1675 by King Charles II when the Royal Observatory was founded. Until 1971 the Astronomer Royal was also Director of the Royal Greenwich Observatory, but the post was subsequently made separate.

Astronomical almanac The A yearly publication containing tables of the Sun, Moon, and planets and their satellites, and other astronomical data. It is compiled and published jointly by the US Naval Observatory and the Royal Greenwich Observatory. It began in Britain in 1766 under the title The Nautical Almanac and Astronomical Ephemeris, but from 1914 two separate versions were issued, one for astronomers and one for navigators.In 1960 the name of the astronomical version was abridged to The Astronomical Ephemeris. The current title was adopted in 1980 when it merged with The American Ephemeris and Nautical Almanac, published in the US since 1852.

Astronomical unit A unit of length, formerly the mean distance of the Earth from the Sun, but now defined more technically, as below. In the original definition of the AU by C. F. Gauss, the Earth's mean distance from the Sun was given by the precise form of Kepler's third law as found by Newton:

$$n^2 a^3 = k^2$$

Where n is the Earth's mean motion a is the semimajor axis of the Earth's orbit m and m E are the masses of the Sun and Earth respectively and k is the Gaussian gravitational constant. The astronomical unit is now defined as the distance from the Sun of a massless particle moving in a circular orbit around the Sun with an orbital period of one Gaussian year of 2p/k ephemeris days. The Earth's mean distance from the Sun is 1.000 000 031 AU, where 1 AU = 149 597 870 km.

Astronomy The study of the space beyond the Earth and its contents; phenomena in the Earth's upper atmosphere that have their origin in space, such as aurorae and meteors, are also included. Before the application of the telescope to astronomy early in the 17th century, astronomy was concerned purely with measuring the positions and movements of celestial bodies, a branch now known as astrometry. The realization in the 17th century that gravitation governed the movements of celestial bodies led to mathematical methods for calculating orbits, now known as celestial mechanics. In the 19th century the development of spectroscopy, by which the composition of bodies could be determined by analysing their light, marked the birth of astrophysics. The recognition in the 20th century that galaxies exist outside our own and the discovery of the expansion of the Universe gave rise to modern cosmology, which deals with the origin and evolution of the Universe.

Astrophysics The study of the physical nature of the Universe and the objects in it, notably stars, galaxies, and the composition of the space between them. Astrophysics originated in the application of spectroscopy to the study of starlight in the 19th century. It complements the traditional branches of astronomy, astrometry and celestial mechanics, which are concerned with the positions and motions of objects. Observational astrophysics interprets the electromagnetic radiation emitted by celestial objects. Theoretical astrophysics attempts to explain the processes involved, which can lead to new understanding of the behaviour of matter under conditions not encountered on Earth. For example, nuclear physics had to develop before energy generation inside stars could be understood, and study of objects such as white dwarfs and neutron stars has helped confirm predictions about the behaviour of matter under extreme compression and intense gravitational fields.

Asymptotic giant branch The Asymptotic Giant Branch is the upper right hand corner of the Hertzsprung-Russell (H-R) diagram that represents stars in which Helium burning no longer occurs in their cores - red giants. As the carbon core of these old star collapses, the outer layers expand, and the star becomes huge red giants.

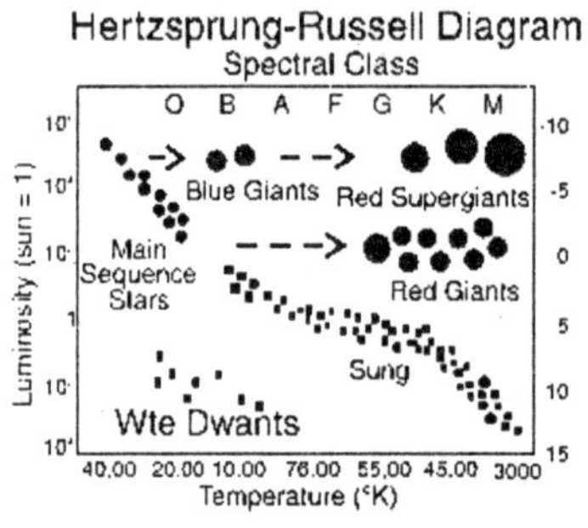

Fig. Asymptotic giant branch

Atlas Atlas is one of the smallest of the 18 moons of Saturn. Atlas is a shepherd satellite for Saturn's A-ring. It has a radius of about 18.5x17.2x13.5 km. Atlas orbits at about 137,670 km from Saturn. Its orbital period is 0.6019 earth days. It was discovered by R. Terrile/Voyager 1 in 1980.

Atmosphere The gaseous envelope surrounding an astronomical body. Several planets retain considerable atmospheres because of their strong gravity. Motions of the gas within planetary atmospheres in response to heating, coupled with rotational forces, generate weather systems. The planetary satellites Titan and Triton also possess atmospheres. Pluto has a 'seasonal' atmosphere, which forms when the planet is close to perihelion, condensing out at aphelion.

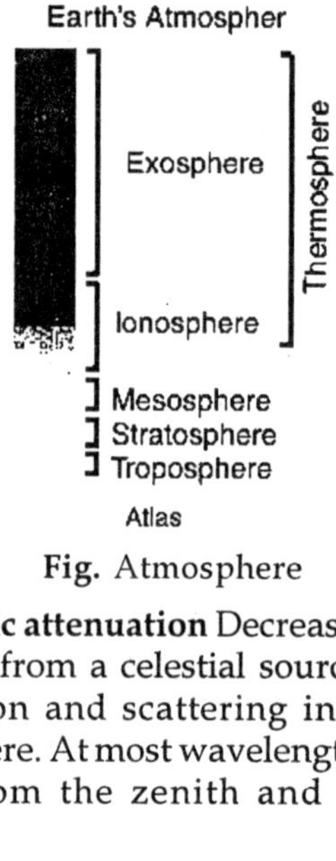

Fig. Atmosphere

Atmospheric attenuation Decrease in intensity of signal from a celestial source caused by absorption and scattering in the Earth's atmosphere. At most wavelengths, increases away from the zenith and toward the horizon.

Atmospheric extinction The loss of starlight in passing through the Earth's atmosphere. Most of the loss arises from Rayleigh scattering by molecules of nitrogen and oxygen. At certain wavelengths there is absorption from molecules of oxygen, ozone, and water vapour. Particles of dust and industrial pollutants can also contribute to extinction by Mie scattering. Atmospheric extinction is proportional to the airmass and the atmospheric pressure. At sea level and in a perfectly clear sky a star 80 from the zenith appears 1 mag. fainter than it would at the zenith. This in turn is 0.3 mag. fainter than if no atmosphere were present. These figures apply only to visual observations; in blue light, the extinction is greater, while in red light it is less. This is why the Sun and Moon in particular appear red when near the horizon.

Atom The basic structure of which all matter is made. It is made up of three particles; protons and neutrons (which make up the nucleus) and electrons (which orbit the nucleus). (The one exception to this is hydrogen which has only a proton at its nucleus. In its basic state an atom will form an element, the kind of element being dictated by the structure's atomic number. For example, a hydrogen atom will normally have 1 of each particle and its atomic number is 1. A helium atom, which has the atomic number 2, will have two of each particle, and so on. Atoms may be broken up (fission) or fused with another (fusion). Any combination of atoms will form a molecule.

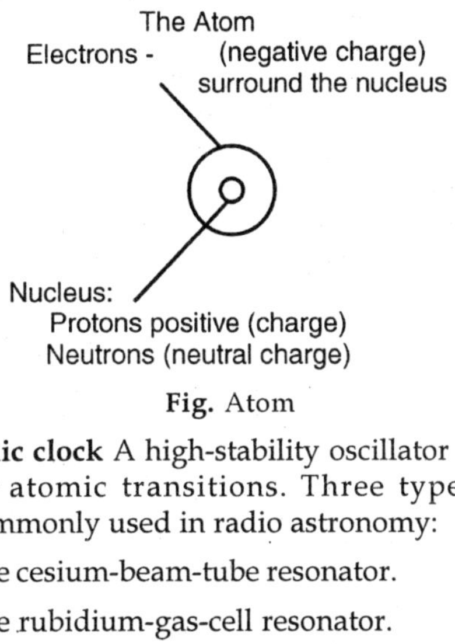

Fig. Atom

Atomic clock A high-stability oscillator based on atomic transitions. Three types are commonly used in radio astronomy:

The cesium-beam-tube resonator.

The rubidium-gas-cell resonator.

The atomic hydrogen maser oscillator.

Cesium-beam-tube resonators are used as the standard for atomic time because their absolute frequencies are relatively insensitive to environmental factors.

Rubidium-gas-cell resonators and hydrogen masers are more commonly used as local oscillators in VLBI experiments because they provide more stable signals when their environment is carefully controlled.

Au A measurement used by astronomers within the solar system; one astronomical unit (AU) is the average distance between Earth and the sun (about 93,000,000 miles or 150,000,000 kilometers)

Aurora An emission of light from the Earth's high atmosphere, caused principally by oxygen atoms or nitrogen molecules that are excited by electrons accelerated within the magnetosphere. The visible aurora is dominated by the green and red emissions of oxygen, occurring respectively at altitudes of 100 km and above about 400 km, and red nitrogen emissions at about 95 km. Violet purple nitrogen emissions are sometimes seen in the sunlit uppermost parts of aurorae at altitudes of 1000 km.

Popularly known as the northern lights the aurora takes a number of characteristic forms. These may range from a glow, low over the northern horizon through arcs and bands, which may be homogeneous, or may show vertical rays. Isolated rays and patches of auroral light may also be seen. Most spectacular of all is the corona, a perspective effect whereby rays appear to converge on a single region of the sky almost overhead during a particularly intense storm. During strong activity, the rays and other structures move, causing.

Aurora australis The southern lights.

Aurora borealis The "Northern Lights"; caused by the interaction between the solar wind, the Earth's magnetic field and the upper atmosphere. A similar effect happens in the southern hemisphere where it is known as the aurora australis.

Auroral oval A high-latitude ring of more or less permanent aurora, girdling the geomagnetic pole at a distance of 2000-2500 km under quiet geomagnetic conditions. There are two ovals, one in either hemisphere, each the mirror-image of the other. They are displaced such that the dayside edge is closer to the pole than that on the night side. Under disturbed geomagnetic conditions, the auroral ovals brighten, broaden, and expand towards the equator, particularly on the night side, so the aurora becomes visible at lower latitudes. The auroral ovals remain relatively fixed in space above the rotating Earth.

Autumnal equinox Equinoxes are days in which daytime and nighttime are of equal duration. The two yearly equinoxes occur when the Sun crosses the celestial equator. The autumnal equinox occurs around September 21; the vernal (spring) equinox occurs around March 21.

Azimuth Azimuth is a measurement in mapping astronomical objects on the celestial sphere (the sky as visible from Earth). Azimuth is the angle of the object from the observer's north point (projected onto the horizon). If an object is due north, its azimuth is 0 degrees. If it is due east, its azimuth is 90 degrees, etc. To find an object in the sky, two coordinates are needed, its altitude and its azimuth.

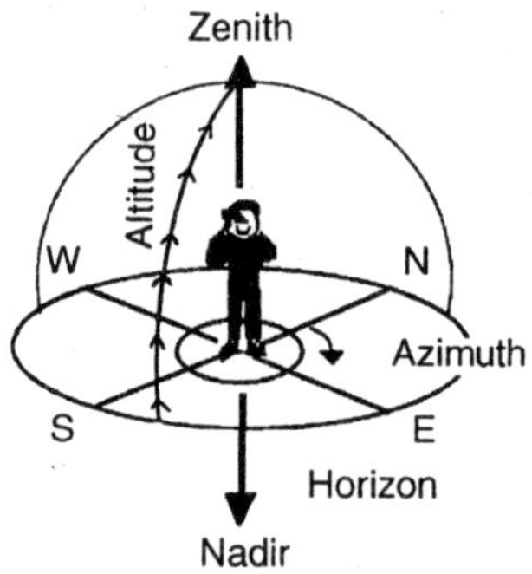

Fig. Azimuth

B

B band A broad Fraunhofer line in the Sun's spectrum at around 690 nm, due to absorption by molecular oxygen in the Earth's atmosphere.

B magnitude The magnitude of a star in blue light on the Johnson photometry system. The B filter is centred on a wavelength of 440 nm and has a bandwidth of 100 nm. It is the photoelectric equivalent of the older photographic magnitude, but modified to be free of ultraviolet light. There are also B filters in Geneva photometry, Walraven photometry, and the six-colour system, but the context should always make it clear when these are intended.

B star A star of spectral type B, whose spectrum contains predominantly absorption lines of hydrogen and neutral helium. B-type stars are hot and appear blue in colour, emitting strongly in the ultraviolet. Temperatures for B-type stars on the main sequence are 10500-30000 K. Main-sequence luminosities are 100-52000 times the Sun's, and masses are 3-18 solar masses. B-type supergiants are stars evolving from the top of the main sequence. As such, they have masses up to 25 solar masses and luminosities as high as 260000 times the Sun's. Regulus and Spica are B-type dwarfs, and Rigel is a B-type supergiant. Together with O stars they are the major constituents of OB associations, which delineate the spiral arms of galaxies. Being massive, they have lifetimes of only a few tens of millions of years.

Baade's window A small area of sky in the constellation Sagittarius which is relatively free from obscuring dust, and through which an optical telescope can see to the central region of our Galaxy and beyond. The window is at galactic longitude 0°.9 and latitude-3°.9, near the globular cluster NGC 6522. W. Baade used this window to observe RR Lyrae variable stars in the central bulge of our Galaxy, and determine their distance.

Baade-Wesselink method A method of determining the properties of certain types of pulsating variable star, including Cepheids. The changes in radial velocity magnitude, and spectrum are used to obtain the relative radii, effective temperature, and surface luminosity over the complete pulsation cycle. Absolute values can then be determined for the star's size and luminosity, independent of apparent luminosity, distance, or interstellar extinction. It is named after the American astronomer W. Baade and the Dutch astronomer Adriaan Jan Wesselink

Bailey type A variable star, a subtype of the RR-Lyrae stars, described by the American astronomer Solon Irving Bailey and distinguished by the shape and amplitude of their light-curves. In order of decreasing amplitude, the original types were designated a, b, and c, with mean periods of 0.48, 0.58, and 0.32 days, respectively. There is a continuous transition between types a and b, both of which have asymmetric light-curves, and so they are now regarded as one type, RRAB. Light-curves of type RRC are distinct, and frequently appear almost sinusoidal.

Balmer series A sequence of absorption or emission lines in the visible part of the spectrum, due to hydrogen; also known as Balmer lines. In order of decreasing wavelength they are Ha Ha H and so on, becoming closer together as they approach 364.6 nm the Balmer limit. Balmer absorption lines are caused by jumps of electrons from the second energy level to higher levels, and emission lines when the electrons drop back to the second energy level. They are named after the Swiss mathematician Johann Jakob Balmer.

Barium star A red-giant star of spectral type G or K in which heavier elements such as barium appear in the spectrum with unusually high abundance; also known as a Ba II star or a heavy-metal star. Helium burning in a shell around its core produces the heavier elements. Barium stars are similar to CH stars, but are more metal-rich and have insufficient carbon to be considered as carbon stars.

Barnard's star The second-closest star to the Sun, 6.0 L.Y away in Ophiuchus. It is an M5 dwarf of apparent magnitude 9.5, with a luminosity 2000 times less than the Sun's. It has the largest proper motion of any star known, 10".31 per year, as was discovered in 1916 by E. E. Barnard, after whom the star is named.

Barred spiral galaxy A type of disk galaxy with spiral arms extending from an almost rectangular or cigar-shaped bar of stars across its central regions, which can account for up to one-third of the galaxy's total light output; Hubble type SB. The central bars are typically between 2.5 and 5 times as long as they are wide. The masses of barred spirals range from about 109 to 5×10^{11} solar masses, and their diameters from about 10 000 to over 300 000 L.Y Almost half of all disk galaxies contain a noticeable bar; similar but much less prominent structures may exist undetected in most disk galaxies. The bars often show sharp, straight dust lanes on the edge leading the rotation. There is often a cluster of bright nebulae at the outer ends of the bar, where the spiral arms usually start. Many barred galaxies show a narrow ring of stars near the outer end of the bar.

Basin, impact A large impact crater within which one or more concentric inner rings of mountains are present. There are three morphological types: peak-ring basins, in which there is a single inner ring; central peak basins, in which both an inner ring and a central peak are present; and multi-tinged basins, in which there is more than one inner ring. The rings are thought to form when the central region of the crater floor rebounds upwards and then collapses again, producing the first inner ring. In larger craters the central region may continue to oscillate up and down, forming additional rings. As many as six concentric rings have been noted in some basins. The Orientale Basin on the Moon is a beautifully preserved multi-ringed basin 930 km in diameter. The largest basin on the Moon is the Imbrium Basin at nearly 1300 km, but much of it has been covered by younger lavas. The largest basin in the Solar System is Valhalla on Callisto, 2700 km across.

Battani, Muhammad Ibn Jabir Arab astronomer, born in modern Turkey; also known by the Latinized name Albategnius. He was one of the first Arab astronomers to grasp the importance of accurate observations. He produced a set of tables, including a catalogue of star positions more accurate than those in Ptolemy's Almagest, that was to influence medieval European astronomers. Al-Battani refined the values of the precession of the equinoxes, the obliquity of the ecliptic, and the length of the tropical year, and found that over the course of the year the Earth-Sun distance varies.

Bautz-Morgan class A category in a classification scheme for clusters of galaxies based on the contrast between the brightest member and the typical bright galaxies in the cluster. Class I clusters are dominated by an extremely large, luminous supergiant galaxy. In Class III, no member appears to stand out against the bright galaxy population. Class II is intermediate. The classification is named after the American astronomers Laura Patricia Bautz and W. W. Morgan, who published it in 1970.

Bayer system In each constellation, every star is classified using the Bayer system by assigning then Greek letters (alpha, beta, gamma, delta, epsilon, zeta, eta, theta, iota, kappa, lambda, mu, nu, xi, omicron, pi, rho, sigma, tau, upsilon, phi, chi, psi, and omega), in order of decreasing brightness. The Bayer system cannot go beyond 24 (the number of letters in the Greek alphabet). The brightest star in a constellation is alpha, the second-brightest is beta, the third is gamma, etc. Johannes Bayer (1564-1617) was a Bavarian (German) astronomer devised this system.

Beat Cepheid A Cepheid variable that is oscillating in both the fundamental and first overtone pulsation modes. The periods interact to produce a longer, beat period. Such stars, also known as double-mode Cepheids, account for nearly half of all Cepheids with periods of 2-4 days.

Becklin-Neugebauer object A powerful source of infrared radiation discovered by the American astronomers Eric Edward Becklin and Gerry Neugebauer in 1966 behind the Orion Nebula. It is thought to be a massive young star 10 000 times as luminous as the Sun. The star is embedded in a dense cloud of gas and dust which completely obscures it from view in visible light.

Belt of Orion A line of three bright stars, Alnilam, Alnitak, and Mintaka, in the constellation Orion, representing the hunter's belt.

Benetnasch An alternative name for the star Alkaid.

Bennett comet A long-period comet discovered on 1969 December 28 by the South African amateur John Caister Bennett formerly designated 1970 II. Comet Bennett reached perihelion, 0.54 AU, on 1970 March 20, and was prominent throughout 1970 April. Its maximum magnitude was 0, and its tail reached 20°. Spiral jets were seen emerging from the nucleus. Rapid changes in the gas tail were caused by turbulent conditions in the solar wind close to sunspot maximum. Its orbit has an eccentricity of 0.996, inclination 9°.0, and a period about 1700 years.

Besselian elements 1. Values used in computing an occultation of a star or planet by the Moon. They include the Universal Time of conjunction in right ascension between the Moon and the star; the Greenwich hour angle of the star at that time; and values of particular geometrical properties of the Moon, the star and the observer's location.

2. Eight values used in computing the circumstances of a solar eclipse. They describe the position and direction of the axis of the Moon's shadow, and the size of the shadow relative to the Earth. Besselian elements are named after F. W. Bessel, who devised the method of computation.

Besselian epoch An obsolete method of specifying the date of astronomical events. It was the standard before 1984, but has since been replaced by the simpler Julian epoch. The prefix 'B' is now added to Besselian epoch dates to avoid confusion with the Julian epoch. The Besselian epoch was expressed as the year plus a decimal part, e.g. 1975.2406. The time-scale used was Ephemeris Time, and the unit of time was the Besselian year. The standard epochs of 1900.0 and 1950.0 used earlier this century were Besselian epochs, differing by 50 tropical years. In fact,

B1950.0 = 1950 January 0.923d ET = 1949 December 31 22h 09m ET.

Besselian year The interval of time required for the right ascension of the fictitious mean sun to increase by 24h. It is defined as beginning when the mean sun's longitude is precisely 280°

8h 40m); this value was chosen as it corresponds roughly to January 1. The Besselian year is virtually identical to the tropical year and is 365.2422 days long. It was the basic unit of the nowsuperseded Besselian epoch.

Beta cephei star A type of pulsating variable, of early spectral type lying just above the main sequence; abbr. BCEP. They are also known as Beta Canis Majoris stars. Magnitude variations are 0.01-0.3, with periods of 0.1-0.6 days. There is no generally accepted explanation for the variations, but both radial pulsation and non-radial pulsation appear to be involved. They are sometimes called Beta Canis Majoris stars.

Beta taurid meteors A daytime meteor shower following a similar orbit to Comet Encke, occurring between June 5 and July 18. Peak activity, detected by radar, occurs around the end of June. The meteors cannot be seen visually because the radiant lies close to the Sun. The Tunguska event of 1908 has been attributed to a large meteoroid from the Beta Taurid stream.

Betelgeuse Betelgeuse (alpha Orionis) is the second-brightest star in the constellation called Orion and one of the brightest stars in the sky. It is a supergiant star, reddish in color, and over 600 million miles in diameter (almost 1,000 times bigger than the Sun but cooler than the Sun). Betelgeuse is about 14,000 times brighter than the Sun. If Betelgeuse were at the center of our Solar System, it would extend beyond the orbit of Jupiter. It is 520 light-years from Earth. It is a variable star, varying in magnitude from 0.3 to 1.2 over a period of about 7 years, averaging about 0.70. It is the only star (other than our sun) for which we have surface images.

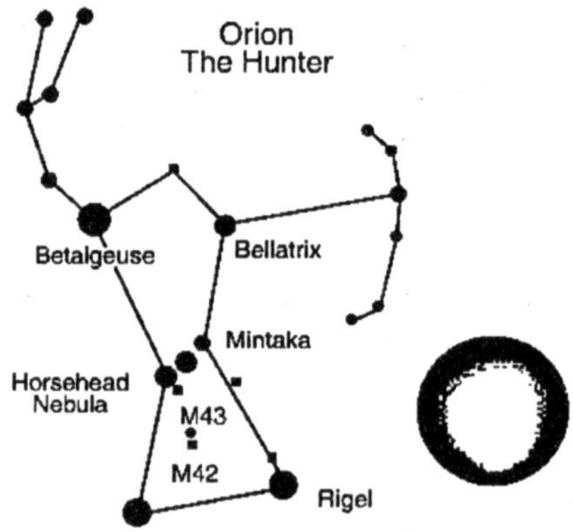

Bessel, friedrich

Fig. Betelgeuse

Bethe, Hans Albrecht German-American physicist. In 1938 he worked out the details of the proton-proton chain that had been proposed by the American physicist Charles Louis Critchfield and argued that it would account for energy production in stars less massive than the Sun. The same year, independently of C. F. von Weizsäcker, he suggested that the carbon-nitrogen cycle, in which hydrogen is converted into helium, is the means by which the Sun and more massive stars generate their energy. For this work he received the 1967 Nobel prize for physics. With the German-born theoretical physicist Walter Heinrich Heitler he developed the cascade theory of cosmic rays.

Bianca Bianca is the third moon of Uranus. It has a diameter of about 44 km and orbits at roughly 59,165 km from Uranus; this moon was discovered in 1986 by Voyager 2. Bianca was named for a character in Shakespeare's Taming of the Shrew.

Bianchi cosmology The study of universes which are homogeneous but not isotropic. Standard cosmological models assume that the Universe, on cosmological scales, is both homogeneous and isotropic that is, it conforms to the cosmological principle. If the assumption is relaxed, so that the former condition holds but not the latter, then the allowed solutions of the equations of general relativity are called Bianchi models, after the Italian mathematician Luigi Bianchi

Big bang theory In physical cosmology, the Big Bang is the scientific theory that the universe emerged from an enormously dense and hot state about 13.7 billion years ago. The Big Bang theory is based on the observed Hubble's law redshift of distant galaxies that when taken together with the cosmological principle indicate that space is expanding according to the Friedmann-Lemaître model of general relativity. Extrapolated into the past, these observations show that the universe has expanded from a state in which all the matter and energy in the universe was at an immense temperature and density. Physicists do not widely agree on what happened before this, although general relativity predicts a gravitational singularity.

The term *Big Bang* is used both in a narrow sense to refer to a point in time when the observed expansion of the universe (Hubble's law) began — calculated to be 13.7 billion (1.37×10^{10}) years ago (±2%) — and in a more general sense to refer to the prevailing cosmological paradigm explaining the origin and expansion of the universe, as well as the composition of primordial matter through nucleosynthesis as predicted by the Alpher-Bethe-Gamow theory.

One consequence of the Big Bang is that the conditions of today's universe are different from the conditions in the past or in the future. From this model, George Gamow in 1948 was able to predict, at least qualitatively, the existence of cosmic microwave background radiation (CMB). The CMB was discovered in the 1960s and further validated the Big Bang theory over its chief rival, the steady state theory.

History

The Big Bang theory developed from observations and theoretical considerations. Observationally, it was determined that most spiral nebulae were receding from Earth, but those who made the observation weren't aware of the cosmological implications, nor that the supposed nebulae were actually galaxies outside our own Milky Way. In 1927, the Belgian Catholic priest Georges Lemaître independently derived the Friedmann-Lemaître-Robertson-Walker equations from Albert Einstein's equations of general relativity and proposed, on the basis of the recession of spiral nebulae, that the universe began with the "explosion" of a "primeval atom" — what was later called the Big Bang.

In 1929, Edwin Hubble provided an observational basis for Lemaître's theory. He discovered that, relative to the Earth, the galaxies are receding in every direction at speeds directly proportional to their distance from the Earth. This fact is now known as Hubble's law. Given the cosmological principle whereby the universe, when viewed on sufficiently large distance scales, has no preferred directions or preferred places, Hubble's law suggested that the universe was expanding contradicting the infinite and unchanging static universe scenario developed by Einstein.

This idea allowed for two opposing possibilities. One was Lemaître's Big Bang theory, advocated and developed by George Gamow. The other possibility was Fred Hoyle's steady state model in which new matter would be created as the galaxies moved away from each other. In this model, the universe is roughly the same at any point in time. It was actually Hoyle who coined the name of Lemaître's theory, referring to it sarcastically as "this *big bang* idea" during a program broadcast on March 28, 1949 by the BBC Third Programme. Hoyle repeated the term in further broadcasts in early 1950, as part of a series of five lectures entitled *The Nature of Things*. The text of each lecture was published in *The Listener* a week after the broadcast, the first time that the term "big bang" appeared in print.

For a number of years the support for these theories was evenly divided. However, the observational evidence began to support the idea that the universe evolved from a hot dense state. Since the discovery of the cosmic microwave background radiation in 1965 it has been regarded as the best theory of the origin and evolution of the cosmos. Virtually all theoretical work in cosmology now involves extensions and refinements to the basic Big Bang theory. Much of the current work in cosmology includes understanding how galaxies form in the context of the Big Bang, understanding what happened at the Big Bang, and reconciling observations with the basic theory.

Huge advances in Big Bang cosmology were made in the late 1990s and the early 21st century as a result of major advances in telescope technology in combination with large amounts of satellite data such as that from COBE, the Hubble Space Telescope and WMAP. Such data has allowed cosmologists to calculate many of the parameters of the Big Bang to a new level of precision and led to the unexpected discovery that the expansion of the universe appears to be accelerating.

Based on measurements of the expansion of the universe using Type Ia supernovae, measurements of the lumpiness of the cosmic microwave background, and measurements of the correlation function of galaxies, the universe has a calculated age of 13.7 ± 0.2 billion years. The agreement of these three independent measurements is considered strong evidence for the so-called ËCDM model that describes the detailed nature of the contents of the universe.

The early universe was filled homogeneously and isotropically with an incredibly high energy density and concomitantly huge temperatures and pressures. It expanded and cooled, going through phase transitions analogous to the condensation of steam or freezing of water as it cools, but related to elementary particles.

Approximately 10^{-35} seconds after the Planck epoch a phase transition caused the universe to experience exponential growth during a period called cosmic inflation. After inflation stopped, the material components of the universe were in the form of a quark-gluon plasma (also including all other particles—and perhaps experimentally produced recently as a quark-gluon liquid) in which the constituent particles were all moving relativistically. As the universe continued growing in size, the temperature dropped. At a certain temperature, by an as-yet-unknown transition called baryogenesis, the quarks and gluons combined into baryons such as protons and neutrons, somehow producing the observed asymmetry between matter and antimatter. Still lower temperatures led to further symmetry breaking phase transitions that put the forces of physics and elementary particles into their present form. Later, some protons and neutrons combined to form the universe's deuterium and helium nuclei in a process called Big Bang nucleosynthesis. As the universe cooled, matter gradually stopped moving relativistically and its rest mass energy density came to gravitationally dominate that of radiation. After about 300,000 years the electrons and nuclei combined into atoms (mostly hydrogen); hence the radiation decoupled from matter and continued through space largely unimpeded. This relic radiation is the cosmic microwave background.

Over time, the slightly denser regions of the nearly uniformly distributed matter gravitationally attracted nearby matter and thus grew even denser, forming gas clouds, stars, galaxies, and the other astronomical structures observable today. The details of this process depend on the amount and type of matter in the universe. The three possible types are known as cold dark matter, hot dark matter, and baryonic matter. The best measurements available (from WMAP) show that the dominant form of matter in the universe is cold dark matter. The other two types of matter make up less than 20% of the matter in the universe.

The universe today appears to be dominated by a mysterious form of energy known as dark energy. Approximately 70% of the total energy density of today's universe is in this form. This component of the universe's composition is revealed by its property of causing the expansion of the universe to deviate from a linear velocity-distance relationship by causing spacetime to expand faster than expected at very large distances. Dark energy in its simplest formation takes the form of a cosmological constant term in Einstein's field equations of general relativity, but its composition is unknown and, more generally, the details of its equation of state and relationship with the standard model of particle physics continue to be investigated both observationally and theoretically.

All these observations are encapsulated in the ËCDM model of cosmology, which is a mathematical model of the Big Bang with six free parameters. Mysteries appear as one looks closer to the beginning, when particle energies were higher than can yet be studied by experiment. There is no compelling physical model for the first 10^{-33} seconds of the universe, before the phase transition called for by grand unification theory. At the "first instant", Einstein's theory of gravitation predicts a gravitational singularity where densities become infinite. To resolve this paradox, a theory of quantum gravitation is needed. Understanding this period of the history of the universe is one of the greatest unsolved problems in physics.

Theoretical underpinnings

As it stands today, the Big Bang is dependent on three assumptions:

1. The universality of physical laws
2. The cosmological principle
3. The Copernican principle

When first developed, these ideas were simply taken as postulates, but today there are efforts underway to test each of them. Tests of the universality of physical laws have found that the largest possible deviation of the fine structure constant over the age of the universe is of order 10^{-5}. The isotropy of the universe that defines the Cosmological Principle has been tested to a level of 10^{-5} and the universe has been measured to be homogeneous on the largest scales to the 10% level. There are efforts underway to test the Copernican Principle by means of looking at the interaction of galaxy groups and clusters with the CMB through the Sunyaev-Zel'dovich effect to a level of 1% accuracy.

The Big Bang theory uses Weyl's postulate to unambiguously measure time at any point as the "time since the Planck epoch". Measurements in this system rely on conformal coordinates in which so-called comoving distances and conformal times remove the expansion of the universe, parameterized by the cosmological scale factor, from consideration of spacetime measurements. The comoving distances and conformal times are defined so that objects moving with the cosmological flow are always the same comoving distance apart and the particle horizon or observational limit of the local universe is set by the conformal time.

As the universe can be described by such coordinates, the Big Bang is not an explosion of matter moving outward to fill an empty universe; what is expanding is spacetime itself. It is this expansion that causes the physical distance between any two fixed points in our universe to increase. Objects that are bound together (for example, by gravity) do not expand with spacetime's expansion because the physical laws that govern them are assumed to be uniform and independent of the metric expansion. Moreover, the expansion of the universe on today's local scales is so small that any dependence of physical laws on the expansion is unmeasurable by current techniques.

Observational evidence

It is generally stated that there are three observational pillars that support the Big Bang theory of cosmology. These are the Hubble-type expansion seen in the redshifts of galaxies, the detailed measurements of the cosmic microwave background, and the abundance of light elements. (See Big Bang nucleosynthesis.) Additionally, the observed correlation function of large-scale structure of the cosmos fits well with standard Big Bang theory.

The Big Bang theory predicted the existence of the cosmic microwave background radiation or CMB which is composed of photons emitted during baryogenesis. Because the early universe was in thermal equilibrium, the temperature of the radiation and the plasma were equal until the plasma recombined. Before atoms formed, radiation was constantly absorbed and reemitted in a process called Compton scattering: the early universe was opaque to light. However, cooling due to the expansion of the universe allowed the temperature to eventually fall below 3,000 K at which point electrons and nuclei combined to form atoms and the primordial plasma turned into a neutral gas. This is known as photon decoupling. A universe with only neutral atoms allows radiation to travel largely unimpeded.

Because the early universe was in thermal equilibrium, the radiation from this time had a blackbody spectrum and freely streamed through space until today, becoming redshifted because of the Hubble expansion. This reduces the high temperature of the blackbody spectrum. The radiation should be observable at every point in the universe to come from all directions of space.

In 1964, Arno Penzias and Robert Wilson, while conducting a series of diagnostic observations using a new microwave receiver owned by Bell Laboratories, discovered the cosmic background radiation. Their discovery provided substantial confirmation of the general CMB predictions—the radiation was found to be isotropic and consistent with a blackbody spectrum of about 3 K —and it pitched the balance of opinion in favor of the Big Bang hypothesis. Penzias and Wilson were awarded the Nobel Prize for their discovery.

In 1989, NASA launched the Cosmic Background Explorer satellite (COBE), and the initial findings, released in 1990, were consistent with the Big Bang's predictions regarding the CMB. COBE found a residual temperature of 2.726 K and determined that the CMB was isotropic to about one part in 10^5. During the 1990s, CMB anisotropies were further investigated by a large number of ground-based experiments and the universe was shown to be almost geometrically flat by measuring the typical angular size (the size on the sky) of the anisotropies.

In early 2003 the results of the Wilkinson Microwave Anisotropy satellite (WMAP) were released, yielding what were at the time the most accurate values for some of the cosmological parameters. This satellite also disproved several specific cosmic inflation models, but the results were consistent with the inflation theory in general.

Using the Big Bang model it is possible to calculate the concentration of helium-4, helium-3, deuterium and lithium-7 in the universe as ratios to the amount of ordinary hydrogen, H. All the abundances depend on a single parameter, the ratio of photons to baryons. The ratios predicted (by mass, not by number) are about 0.25 for $^4He/H$, about 10^{-3} for $^2H/H$, about 10^{-4} for $^3He/H$ and about 10^{-9} for $^7Li/H$.

The measured abundances all agree with those predicted from a single value of the baryon-to-photon ratio. The agreement is relatively poor for ^7Li and ^4He, the two elements for which the systematic uncertainties are least understood. This is considered strong evidence for the Big Bang, as the theory is the only known explanation for the relative abundances of light elements. Indeed there is no obvious reason outside of the Big Bang that, for example, the young universe (i.e. before star formation, as determined by studying matter essentially free of stellar nucleosynthesis products) should have more helium than deuterium or more deuterium than ^3He, and in constant ratios, too.

Detailed observations of the morphology and distribution of galaxies and quasars provide strong evidence for the Big Bang. A combination of observations and theory suggest that the first quasars and galaxies formed about a billion years after the Big Bang, and since then larger structures have been forming, such as galaxy clusters and superclusters. Populations of stars have been aging and evolving, so that distant galaxies (which are observed as they were in the early universe) appear very different from nearby galaxies (observed in a more recent state). Moreover, galaxies that formed relatively recently appear markedly different from galaxies formed at similar distances but shortly after the Big Bang. These observations are strong arguments against the steady-state model. Observations of star formation, galaxy and quasar distributions, and larger structures agree well with Big Bang simulations of the formation of structure in the universe and are helping to complete details of the theory.

Big dipper The Big Dipper is a group of 7 stars (an asterism) contained in the Northern Hemisphere constellation called Ursa Major (The Great Bear). The two brightest stars in the Big Dipper (Dubhe and Merak) "point" to the North Star, Polaris.

The Big Dipper, in Uraa Major

D'arrest, h.l.

Fig. Big dipper

Binary galaxy A pair of galaxies in orbit around each other. True binary galaxies are very difficult to distinguish from chance superpositions of two galaxies in the line of sight. The statistical study of the orbits of binary pairs is valuable in trying to estimate the total masses of particular types of galaxy, although the extremely long orbital periods prevent the accurate determination of masses of individual galaxies.

Binary star A pair of stars bound together by their mutual gravitation, and orbiting their common centre of mass. Such a system is distinct from an optical double, which is not gravitationally bound.

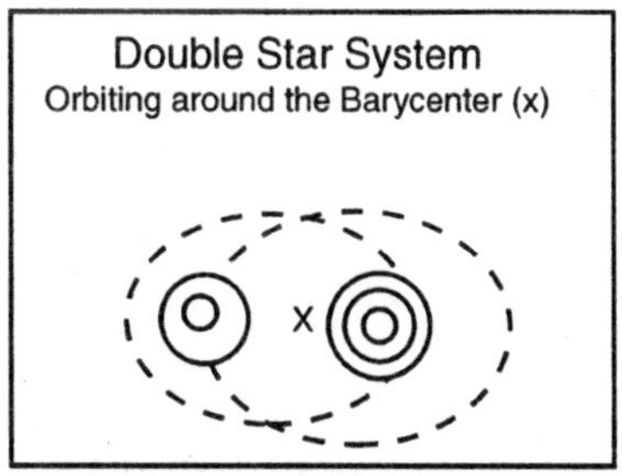

Fig. Binary star

Black body A black body is an idealized body that is a perfect radiator and perfect absorber of electromagnetic radiation. A black body not only absorbs all wavelengths of energy and radiates at all wavelengths, but it does this at the maximum possible intensity for any given temperature. It doesn't necessarily look black. A star is a good approximation to a black body since stellar gases are very good absorbers of energy.

Black-body radiation The thermal radiation that would be emitted by a black body. Black-body radiation has a characteristic spectrum, described by Planck's law, in which the peak of the emission moves to shorter wavelengths as the temperature of the black body rises The cosmic background radiation is black-body radiation, and stars often act as black-body radiators in the optical part of the spectrum.

Black drop A phenomenon seen during a transit of Venus across the Sun. The black drop is a dark area that appears to link the limb of the Sun to the following limb of Venus for a short time after second contact, and its preceding limb for a short time before third contact. It is caused by refraction through the very dense atmosphere of Venus, although it has also been attributed to optical effects and has been reported during transits of Mercury. The phenomenon means that accurate timing of Venus transits is impossible.

Black dwarf A black dwarf is a small, very dense, cold, dead star. It is made mostly of carbon. This dark star is what remains after a red giant star loses its outer layers, forming a planetary nebula and then a white dwarf. The nuclear core of a black dwarf is depleted. Black dwarfs are about the size of the Earth (but tremendously heavier). Our sun will someday turn into a black dwarf.

Black hole A black hole is a massive object (or region) in space that is so dense that within a certain radius (the Schwarzschild radius, which determines the event horizon), its gravitational field does not let anything escape from it, not even light. It is thought that giant stars (those with a mass over 3 times the mass of the Sun) will evolve into red supergiants, then supernova, and then black holes. It is thought that the typical black holes has a mass of roughly 10 times that of the Sun, but the range must be huge. For a typical black hole with a mass 10 times that of the Sun, the Schwarzschild radius would be roughly 18.6 miles (30 km). The phrase 'black hole' was coined by the physicist John Archibald Wheeler; before Wheeler, black holes were called 'frozen stars.' Astronomers think that there may be a black hole at the center of each galaxy.

Blinking In astronomical image processing, a method for comparing two similar but different images wherein a display is switched between the two images in rapid succession. Spatial differences appear to move back and forth, while intensity differences pulsate. When the switching rate, duty cycle and transfer functions of the displays are controllable by the user, blinking is a powerful tool for assessing subtle but systematic differences between images, e.g. those due to moving or variable objects in crowded fields, or to subtle artifacts in image formation.

Blue giant A massive star that has exhausted the hydrogen fuel in its core and left the main sequence. Blue giants have a surface temperature of around 30 000 K and a luminosity some 10 000 times that of the Sun. As they grow older they expand and cool, eventually becoming red giants.

Blue shift The blue shift is a decrease in the wavelength of the light that is emitted from an object that is moving toward us. This decrease in wavelength makes the object appear to be bluer than it actually is. For example, when a star is travelling towards Earth, its light appears bluer (the light waves are shortened, shortening the wavelength). Compare with red shift.

Bode's law Bode's "Law" (also known as the Titus-Bode Law) is an interesting mathematical coincidence, and not a physical law. It is a numerical series that matches planetary distances from the Sun. The Titus-Bode series predicts the positions of all the planets in our Solar System except Neptune (plus it predicts a planet where the asteroid belt is). This relationship is named for Johann Titus and the German astronomer Elert Bode, who did their work in the late 1700's. For the Titus-Bode series for our Solar System, start with 0.4 (A.U.), then form the series by adding 0, 0.3, 0.6, 1.2, 2.4, etc. (doubling each time after the first). The results it predicts are: Mercury: 0.4 A.U., Venus 0.7 A.U., Earth 1.0 A.U., Mars 1.6 A.U., Asteroid Belt 2.8 A.U., Jupiter 5.2 A.U., Saturn 10.0 A.U., Uranus 19.6 A.U.

Bolide A fireball accompanied by one or more audible explosions, often associated with meteorite falls. The initial explosion is the sonic boom of the leading mass; smaller detonations follow if there are several major fragments. Many shock waves may be set up if considerable fragmentation occurs, causing a rumbling noise which follows the first major fragmentation. bolometer A detector sensitive to radiation of all wavelengths. It has a blackened surface so that the incoming radiation raises its temperature, which is measured from a change in electrical resistance. While the bolometer can be used at any wavelength, it is best suited to measuring infrared wavelengths longer than 6 μm.

Bolometer A bolometer is an instrument that measures the total amount of radiant energy (not just visible light) received from a star or other celestial object.

Bonner durchmusterung A catalogue of 324198 stars down to 9th magnitude from the north celestial pole to declination 2° south. It was compiled at Bonn, Germany, by F. W. A. Argelander; most of the observations were made by Eduard Schönfeld and Adalbert Krüger It was published in three volumes in 1859-62, with an associated atlas in 1863. The catalogue was extended to 23° south in 1886 by Schönfeld, adding 133659 stars, with an atlas in 1887. Many stars are still known by their BD numbers as given in these catalogues.

Boötes A constellation of the northern sky, representing a herdsman. It contains Arcturus the brightest star north of the celestial equator. Epsilon Boötis is a double star with orange and blue components. The Quadrantid meteor shower radiates every January from northern Boötes.

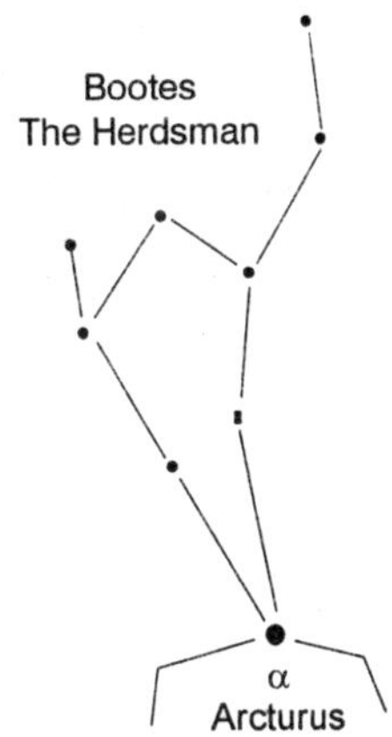

Fig. Boötes

Bose-Einstein Condensate Atoms crowded close together in ultra-low temperatures that behave as if they were one fluid-like superatom.

Boundary layer The level of the Earth's atmosphere in direct contact with its surface, in which friction between the surface and the air plays a significant role in determining atmospheric movements. The boundary layer lies at the base of the troposphere and varies in depth from a few hundred metres under stable conditions to 1-2 km when convection is strong.

Bp star A star of spectral type B which is peculiar in that it shows a deficiency relative to normal B stars of helium in its spectrum. Bp stars have strong magnetic fields, and are a hotter extension of Ap stars to about spectral type B8.

Brackett series A sequence of absorption or emission lines in the infrared part of the spectrum, due to hydrogen. The Brackett series of lines arises from electron jumps between the fourth energy level and higher levels. These lines have wavelengths of 1.46 μm to 4.05 μm They are named after the American physicist Frederick Sumner Brackett.

Bradley, James English astronomer. In collaboration with Samuel Molyneux he attempted to detect stellar parallax, but instead discovered the annual aberration of starlight. This discovery, announced in 1728, was the first direct observational verification that the Earth orbits the Sun. From the amount of aberration he calculated the speed of light to within 2% of the true value. In 1742 Bradley became the third Astronomer Royal. He subsequently discovered the nutation of the Earth's axis, announced in 1748.

Bright star catalogue A catalogue of stars brighter than about magnitude 6.5, published by Yale University Observatory and hence also known as the Yale Bright Star Catalogue. It contains positions, magnitudes, spectral types, and other data for the 9096 stars in the Harvard Revised Photometry. The first edition of the catalogue appeared in 1930 and the fourth in 1982; a fifth edition was issued in electronic format in 1991. A Supplement containing another 2603 stars down to magnitude 7.1 was published in 1983.

Brown dwarf An object that, because of its low mass never becomes hot enough to begin hydrogen fusion in its core; hence it is considered to be not a star, but a substellar object. Brown dwarfs thus have a very low luminosity and are difficult to detect. They are a possible component of the galactic dark matter. The first brown dwarf to be identified with certainty is a companion to the nearby red dwarf Gliese 229, photographed by the Hubble Space Telescope in 1995. An object below about 0.01 solar masses is regarded as a planet.

B-type star B-type stars are very hot and blue; they burn helium. Temperatures of B-type stars range from 11,000 - 25,000 K. Their mass averages 18 times the mass of the Sun and the average luminosity is 20,000 times that of the Sun. Examples include Rigel, Spica, and Regulus.

Butler matrix An optimum arrangement of cables used to connect the elements of a phased array of radio antennas to achieve the desired beam shape and direction. It is named after the American electronics engineer Jesse Lorenzo Butler

Butterfly cluster The 4th-magnitude open star cluster M6 in Scorpius, consisting of several dozen stars arranged in the shape of a butterfly. It is about ¼° wide and lies 2000 L.Y away.

BY draconis star A type of rotating variable star; abbr. BY. The variations are caused by non-uniform surface brightness and show a quasi-periodicity related to the rotation period. The latter ranges from a few hours to about 120 days, and the amplitude from a few thousandths to 0.5 mag. BY Draconis stars are G-M dwarfs. The non-uniform brightness is caused by star-spots and chromospheric activity, similar to that on the Sun. A BY Draconis star may simultaneously be a flare star.

Byurakan astrophysical observatory The observatory of the Armenian Academy of Sciences, founded in 1946 by V. A. Ambartsumian. It is located at Mount Aragatz near Yerevan, Armenia, at an altitude of 1400 m. Its main instruments are a 2.6-m reflector opened in 1976, and a 1-m Schmidt opened in 1961.

C

C line A Fraunhofer line in the Sun's spectrum at 656.3 nm, due to absorption by hydrogen. It is better known as the Ha line in the Balmer series.

C star The spectral type assigned to carbon stars in the Morgan-Keenan classification. These stars were previously known as R and N stars in the Harvard classification.

Caelum A small and faint constellation of the southern sky, representing a sculptor's chisel. Its brightest star, Alpha Caeli, is magnitude 4.5.

Caldera A large volcanic depression, usually several kilometres in diameter, formed primarily by collapse. A caldera often contains smaller craters, cones, and other types of volcanic vent. Small collapse craters less than 1 km in diameter are known as collapse pits. Calderas are often non-circular in shape. Many of the small number of volcanic craters on the Moon appear to be calderas, some of them at the heads of sinuous rilles. There are also huge calderas at the summits of many of the giant volcanoes of Mars. Olympus Mons has a complex of summit calderas more than 60 km in diameter, apparently the result of several collapse events. Similar features are found on the volcanoes of Venus.

Callipus Callipus of Cyzicus (370 - 300 BC) was an ancient Greek who accurately measured the length of the seasons. Callipus improved the Greek calendar, reconciling the lunar month with the solar year, by introducing a unit of time called the Callippic cycle (it was an improvement on the Metonic cycle of 6939.6 days or 19 solar years or 235 lunar months; the Callippic cycle was 4 Metonic cycles). He also added to the (incorrect) theory of the motions of the plants, as spheres within spheres, adding 7 more spheres to Eudoxus' system.

Callisto The second-largest satellite of Jupiter, and the outermost of the four Galilean satellites; also known as Jupiter IV. It orbits 1883 000 km from Jupiter in 16.689 days, keeping the same face turned towards the planet. Callisto is 4800 km in diameter but, at magnitude 5.7, is the faintest of the Galilean satellites, because of its low albedo of 0.2. Callisto's density is 1.86 g/cm^3. It has an icy surface uniformly covered with impact craters, with little sign of the tectonic activity seen on the other Galilean satellites. It has several multi-ringed basins, the two largest being Valhalla (4000 km across, the largest in the Solar System) and Asgard (about 1700 km across), which are surrounded by up to fifteen concentric rings.

Canals, Martian Supposed dark, straight lines on the surface of Mars seen through telescopes from Earth. They were first seen by G. V. Schiaparelli in 1877, and were reported by others until the 1960s when close-up pictures of Mars from the Mariner probes demonstrated them to be illusions. Early in the 20th century, P. Lowell and others considered the canals to be evidence for an advanced Martian civilization.

Cancer The faintest constellation of the zodiac, in which the Sun lies for three weeks from late July until mid-August. Cancer represents a crab. Its brightest star, Beta Cancri, is magnitude 3.5; Zeta Cancri is a double star of magnitudes 5.1 and 6.2. The main feature of the constellation is the open star cluster M44, Praesepe. There is also a smaller 7th-magnitude star cluster, M67, ½° wide.

Cape Canaveral The location of the Kennedy Space Center, on the eastern coast of Florida, USA, the main NASA launch site. From 1963 to 1973 it was known as Cape Kennedy.

Capella Capella is a multiple star system containing at least 9 stars. This bright system is in the Northern Hemisphere, 45 degrees from Polaris (the northern pole star); it is in the constellation Auriga. The two brightest stars in Capella are a binary star system. They are both yellow (like our Sun) with masses 2.6 times and 2.7 times that of the Sun. One is 9 times as large as the Sun, the other is 12 times as large. Each gives off roughy 78 times the light as the Sun. These two stars are about 43 light years from Earth.

Caph The star Beta Cassiopeiae, an F2 giant of magnitude 2.3 lying 45 L.Y away. It is a pulsating variable of the Delta Scuti type, but its variations are less than 0.1 mag.

Capricornid meteors A meteor shower active between July 5 and August 20. The shower is generally quite weak with possible maxima on July 8, July 15, and July 26. The radiant on July 15 lies at RA 21h 00m, dec. -15° Capricornid meteors are slow, yellowish, and sometimes bright.

Capricornus The smallest constellation of the zodiac, through which the Sun passes from late January to mid-February. Capricornus represents a goat with a fish's tail. Its brightest star is Delta Capricorni magnitude 2.9. Alpha Capricorni is a wide but unrelated double of magnitudes 3.6 and 4.2.

Captured rotation Rotation of an object that spins at the same rate as that object takes to orbit another object. Sometimes referred to as synchronous rotation. The Moon is a good example of an object that has captured rotation.

Carbon star A cool red-giant star in an advanced stage of evolution, displaying strong carbon features in the form of CN, CH, and C 2 bands in its spectrum; also known as spectral type C. In carbon stars, the abundance of carbon is greater than that of oxygen. The additional presence of lithium indicates that these elements have been produced by nuclear reactions in the star's core and are now being transported by convection to its surface. Since carbon can be produced only by the triple-alpha process at a very high temperature, these stars must be highly evolved. These rare but luminous objects include the former types R and N M-type giants which are cooler, about 3000 K) which were introduced in the Harvard classification. The N-type carbon stars can be up to 10 times as luminous as the R type.

Cassegrain telescope A reflecting telescope in which a convex secondary mirror reflects light to a focus through a hole in the primary mirror; it is an adaptation of the Gregorian telescope. The convex secondary has a hyperboloidal cross-section, while the primary mirror is paraboloidal. The primary mirror has a central hole through which the light is brought to the Cassegrain focus. A Cassegrain reflector is more compact than a Newtonian of the same focal length. It is also easier to mount heavy or bulky equipment at the Cassegrain focus than at the prime focus or Newtonian focus. The design was devised in the 17th century by an obscure Frenchman named Cassegrain, about whom nothing is reliably known.

Cassini division The widest and most prominent gap in the rings of Saturn, about 4700 km wide, separating the A and B Rings, discovered in 1675 by G. D. Cassini. Its inner edge is 117 500 km from Saturn's centre. Images from the Voyager probes show that it is not empty, but that several thin rings lie within it. The division could be accounted for by satellites 10-30 km in diameter clearing away ring particles, but no such satellites were seen in Voyager images.

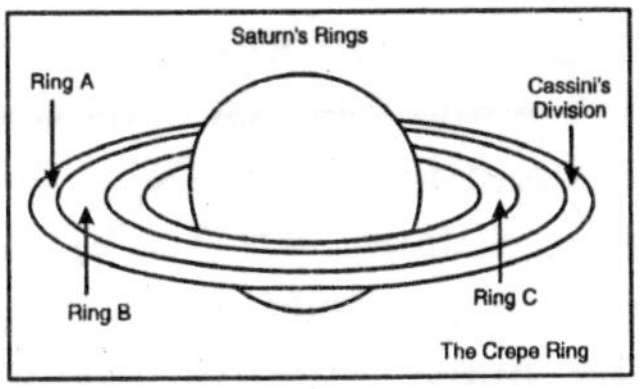

Fig. Cassini division

Cassini, Giovanni domenico Italian-born French astronomer, who moved to Paris in 1669; also known as Jean Dominique Cassini. He measured the rotation periods of Mars and Jupiter, discovered four of Saturn's satellites, and in 1675 observed the division in Saturn's rings now named after him. In 1672, with the help of observations by his countryman Jean Richer measured the distance to Mars by triangulation, and was able to refine the dimensions of the Solar System, obtaining a value for the astronomical unit that was short by only 7%. Cassini drew up improved tables for Jupiter's satellites, which played a part in O. C. Römer's determination of the velocity of light.

Cassini, Jacques French astronomer and geodesist, son of G. D. Cassini; also known as Giacomo Cassini. His accurate determination of the length of an arc of a meridian passing through France, later extended by his son César François, showed that the Earth is an oblate spheroid, flattened at the poles. He compiled the first tables of the motion of Saturn's satellites and in 1738 made one of the first definite determinations of a star's proper motion, that of Arcturus.

Cassiopeia Cassiopeia is an easily-seen constellation that is in the far northern sky. It circles the polestar (Polaris) throughout the year and also straddles the Milky Way. The five major stars of Cassiopeia (also known as "The Lady of the Chair") are shaped like a "W" (or an "M," depending on your orientation). All of the stars in Cassiopeia are all less than second magnitude brightness. The brightest star in Cassiopeia is Schedar (alpha CAS), which is a multiple star that is pale rose in color and varies in magnitude from 2.2 to 2.8 magnitudes. The second-brightest, called Caph (beta CAS), is a white star of magnitude 2.4. Cassiopeia contains two open clusters, M52 (magnitude 7.3) and M103 (magnitude 7.4). The strongest radio source, Cassiopeia A, emanates from Cassiopeia; it is the remnant of a supernova which ocurred about A.D. 1660, and is 10,000 light years from us. The constellation Cassiopeia was named for Cassiopeia, the mother of Andromeda (and the wife of Cephus) in Greek mythology.

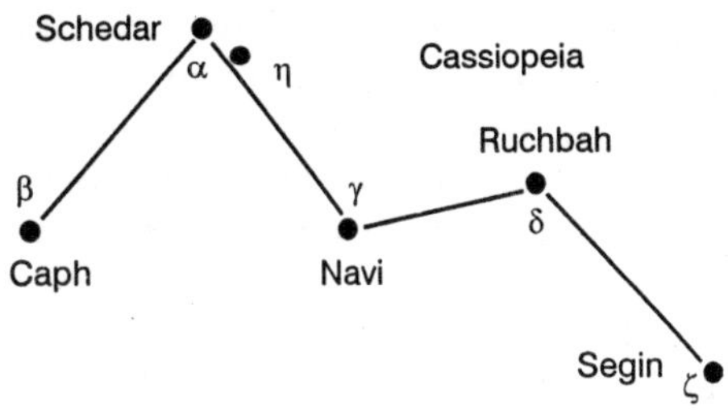

Fig. Cassiopeia

Cavendish, Henry Henry Cavendish (1731-1810) was an English chemist and physicist. Cavendish discovered that hydrogen gas was a substance different from ordinary air (whose components he analyzed), he described the composition of water (hydrogen and oxygen) and made other important discoveries. Cavendish was the first person to determine Isaac Newton's gravitational constant and accurately measured the Earth's mass and density.

C-class asteroid The most common class of asteroid in the outer main asteroid belt. C-class asteroids have low albedos and flat reflectance spectra at wavelengths longer than 0.4 μm. The 'C' is for carbonaceous, since they are believed to have surfaces that are similar in composition to carbonaceous chondrite meteorites. Their spectra show an ultraviolet absorption feature at wavelengths shorter than 0.4 μm, thought to be due to water of hydration. Examples include Hygiea and Davida. Classes B, F, and G are subclasses of C class.

Celestial coordinates Any system for locating the positions of objects on the celestial sphere. Various systems are defined, according to the chosen point of observation and the plane of reference they use. Geocentric coordinates give the position of an object as it would be seen from the centre of the Earth, the most commonly used point of origin. Heliocentric coordinates, sometimes used for objects in the Solar System, give the position as would be seen from the centre of the Sun; a variant system, barycentric coordinates, gives the position as would be seen from the centre of mass of the Solar System. Topocentric coordinates give positions as seen from a specific point on Earth. There are several different reference planes for celestial coordinates. Equatorial coordinates, the most common system, use the celestial equator as the reference plane; their origin can be geocentric or topocentric. Ecliptic coordinates take the ecliptic as their reference plane; they can be geocentric or heliocentric in origin.

Celestial mechanics The branch of astronomy that deals with the motions of bodies in orbit, such as planets, natural and artificial satellites, space probes, comets, and binary or multiple stars. The forces involved are gravitation and, for some artificial satellites, atmospheric drag and radiation pressure from sunlight. Although in most cases the orbit of a body is nominally elliptical and subject to Kepler's laws, the attraction of other bodies and the presence of other forces will cause small departures, known as perturbations, from the elliptical form.

Centaur group A group of unusual bodies in the outer Solar System, possibly giant icy planetesimals or cometary nuclei. Group members have orbits which lie roughly between those of Neptune and Saturn, with periods in the range 50-150 years. The first to be discovered were Chiron, in 1977, and Pholus, in 1992. Of the known Centaurs, 1993 HA2 has the greatest aphelion distance, 37.6 AU, while 1995 GO has the greatest orbital eccentricity, 0.622, and the smallest perihelion distance, 6.9 AU. Centaurs have diameters in the range 50-300 km, but there may be many smaller members still undiscovered. Centaur objects could be captured into the Jupiter comet family or Halley-type cometary orbits.

Central meridian An imaginary straight line joining the north and south poles of a planet's disk as seen from Earth. It is used as a reference for observers making longitude measurements of features on a planet. As the planet rotates, the time when a feature crosses, or transits, the central meridian is noted; its longitude can then be derived from published tables. Such observations are known as central meridian transit timings.

Cepheid A pulsating variable star. This type of star undergoes a rhythmic pulsation as indicated by its regular pattern of changing brightness as a function of time. The period of pulsation has been demonstrated to be directly related to a Cepheid's intrinsic brightness making observations of these stars one of the most powerful tools for determining distance known to modern day astronomy. The existence of this period-luminosity relationship was a point of contention during the 1920 Curtis-Shapley debate.

Cepheus A constellation of the north polar region of the sky, representing King Cepheus of Greek mythology. Its brightest star is Alpha Cephei but its best-known star is the double Delta Cephei, the brighter component of which is the prototype Cepheid variable, ranging from magnitude 3.5 to 4.4 in a period of 5.37 days. Beta Cephei is a double of unrelated stars, magnitudes 3.2 and 7.9, the brighter of which is slightly variable Mu Cephei is a variable red supergiant known as the Garnet Star.

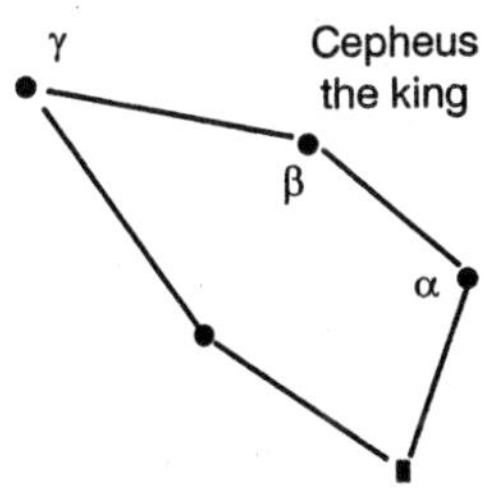

Fig. Cepheus

Cerenkov radiation The light emitted by charged particles such as protons or electrons when they pass through a transparent medium at a speed greater than the speed of light in that medium. The effect is the electromagnetic equivalent of a sonic boom. The radiation can occur at any wavelength, but increases in intensity with frequency and so is strongest in the blue and ultraviolet. The high-energy particles involved may be the secondary products of gamma rays or cosmic rays. For example, when gamma rays with energies of 10^{12} eV pass through the Earth's atmosphere they generate secondary electrons, and ground-based optical telescopes can detect the flash of blue light they emit. Cerenkov radiation is named after the Russian physicist Pavel Alexeyevich Cerenkov.

Ceres The first asteroid to be discovered, by G. Piazzi on 1801 January 1, and hence given the number 1. Ceres is by far the largest asteroid, diameter 940 km, mass 1.17 x 10^{21} kg, and mean density 2.7 gm/cm^3. Its rotation period is 9.075 hours. Ceres contains roughly one-third the mass of the entire asteroid belt. Its mean magnitude at opposition is 7.4; only Vesta can ever become brighter. Its orbit has a semimajor axis of 2.767 AU, period 4.60 years, perihelion 2.55 AU, aphelion 2.98 AU, and inclination 10°.6. Ceres is of G class, with a reflectance spectrum that implies a composition resembling the carbonaceous chondrite meteorites. Ceres is the largest member of a small family of asteroids with a mean distance of 2.76-2.80 AU from the Sun and orbital inclinations of 9-11°. Other Ceres family members include Laetitia, Libussa, Burgundia, and Aeternitas.

Cerro Pachón The site in Chile of the 8-m Gemini South telescope, one of the twin Gemini Telescopes. Cerro Pachón mountain lies 10 km southeast of Cerro Tololo.

Challenger space shuttle The Space Shuttle Challenger was the second NASA Space Shuttle to be used (Columbia was the first). Challenger's first launch was the sixth Space Shuttle launch, called STS-6 Challenger; it took off on April 4, 1983 (from Florida), and landed on April 16, 1983 (at Edwards Air Force Base, CA). The first crew consisted of Paul J. Weitz, Karol J. Bobko, Donald H. Peterson, and F. Story Musgrave. It orbited the Earth 81 times, flying for 5 days, 0 hours, 23 min, 42 seconds; the first Shuttle space walk occured on this mission (it lasted four hours, 17 minutes). After nine flights into space, STS-51L (the mission number) Challenger, exploded 73 seconds into its launch on January 28, 1986, killing the crew (Gregory Jarvis, S. Christa McAuliffe, Ronald McNair, Ellison Onizuka, Judy Resnik, Francis Scobee and Michael Smith).

Chandrasekhar limit The maximum possible mass of a degenerate star, above which it will be unable to support itself against the inward pull of its own gravity. For a star with no hydrogen content the limit is 1.44 solar masses, which is thus the maximum possible mass for a white dwarf. A degenerate star with a mass greater than this limit would collapse under gravity to become either a neutron star or a black hole. It is named after S. Chandrasekhar.

Chandrasekhar.S Subrahmanyan Chandrasekhar (born Lahore, India, in 1910 -died Chicago, USA, in 1995) was an Indian-American astrophysicist who studied stellar physics, evolution, and black holes. He realized that the fate of dying stars depended upon their mass, and above a certain point (1.4 times the mass of the Sun, now known as the "Chandrasekhar limit"), a star will undergo extreme collapse and not simply becomes a white dwarf. He won the Nobel prize in physics in 1983. The orbiting X-ray Observatory Chandra was named to honor S. Chandrasekhar.

Chandrasekhar-Schönberg limit The maximum mass of a star's helium core that can support the outer parts of the star against gravitational collapse, once the hydrogen at its centre has been exhausted. The limit is about 10-15% of the total mass of the star. If the mass of helium in the core exceeds this limit, the central parts collapse while the outer part expands rapidly to become a red giant. Calculations suggest that this happens only in massive stars. The limit is named after S. Chandrasekhar and the Brazilian astrophysicist Mario Schönberg.

Charles Messier A French astronomer and comet hunter who discovered 13 comets independently and codiscovered a half-dozen others. While hunting for comets, Messier compiled a list of fuzzy objects that were not comets in order to avoid them. These catalog entries were later identified as star clusters, nebulae, and galaxies and became the Messier Catalog. Published in various versions beginning in 1771, the catalog grew to 103 objects by 1781. Charles Messier lived from June 26, 1730, to April 12, 1817.

Charlier, Carl V. L. Carl Vilhelm Ludvig Charlier (1862-1932) was a Swedish astronomer who studied celestial mechanics, the calibration of photographic photometry, and the theory of lenses. Charlier also worked in statistics, including the theory of errors; he studied the distribution and motions of stars. Charlier showed that hotter stars and galactic clusters formed flattened systems. In Charlie's hierarchical model of the Universe, he argued that the Universe has infinite mass, and that the density of matter approaches zero as one goes farther into space (resolving Olber's paradox). A 100 km wide crater on Mars (at 68.6 south, 168.4 west, shown above) was named for Charlier in 1973.

Charon The only satellite of Pluto, distance 19600 km, orbital period 6.387 days. Its orbital period and its axial rotation are the same as the rotation period of Pluto itself, so Charon not only keeps one face permanently turned towards Pluto, but also hangs stationary over one point on the planet. Charon is 1186 km in diameter, by far the largest satellite compared to its primary. As its mass is approximately one-tenth that of Pluto, the pair can be regarded as a double planet. Charon's density is about 1.3 g/cm^3, suggesting a composition of rock and ice. Charon was discovered in 1978 by the American astronomer James Walter Christy.

Chondrite A type of stony meteorite, so named because most contain small spherules termed chondrules. Chondrites are the most abundant class of meteorite, accounting for around 86% of all falls. They are largely composed of iron- and magnesium-bearing silicate minerals. Except for the most volatile elements, their chemical composition closely resembles that of the Sun, and presumably the solar nebula. Chondrites are divided into three main classes primarily on the basis of their chemistry. Enstatite or E-chondrites contain the most refractory elements and are highly reduced. Carbonaceous or C-chondrites contain about 3% carbon; these have the highest proportions of volatile substances, and are the most oxidized. Ordinary chondrites, the most common type, are intermediate in their abundance of volatile elements and oxidation state. Each of the three main classes of chondrite is divided into several subgroups on the basis of composition.

Chromosphere A region of a star's atmosphere above its photosphere. The Sun's chromosphere extends from the temperature minimum, some 500 km above the base of the photosphere, outwards by 9000 km where it merges with the corona. For the first 1500 km the chromosphere is more or less continuous, but above this it breaks up into jagged spicules. The temperature of the chromosphere rises from 4400 K at 500 km to about 6000 K at 1000-2000 km. There is a rapid rise to coronal temperatures at heights of about 2500 km the exact height depending on the strength of the local magnetic field. At the top of the chromosphere, the density is a millionth of its value at the bottom. Immediately before and after a total solar eclipse, the chromosphere is visible either as a crescent or a diamond ring, with red coloration due to Ha emission from which it takes its name.Outside eclipses it is visible in Ha and calcium K-line filtergrams, and from space in ultraviolet emission lines.

Circumpolar constellation A circumpolar constellation is one whose apparent path seems to circle a celestial pole. A circumpolar constellation never sets; it is always above the observer's horizon.

Circumpolar star A circumpolar star is one whose apparent path seems to circle a celestial pole. A circumpolar star never sets; it is always above the observer's horizon.

Circumstellar matter A general term for material, usually gas and dust, surrounding a star. After a star's birth, the surrounding cloud absorbs the visible light, re-emitting it as infrared radiation. Most young stars go through a violent phase, ejecting material at speeds of hundreds of kilometres per second. When this ejected material collides with the surrounding gas, bipolar molecular outflows and Herbig-Haro objects can result. If the star is hot its ultraviolet radiation ionizes the gas, producing an emission nebula. These combined effects destroy the cloud, so that the star becomes visible. For most of the star's life a tenuous stellar wind blows. At the end of a star's life it ejects mass. Depending on the mass of the star and the stage in its evolution, this may result in a cool circumstellar envelope, a planetary nebula, or even a supernova remnant.

CLEAN A technique widely used in radio astronomy to enhance radio maps. It consists of a series of steps to identify and remove the effect of distortion caused by unwanted side lobes. The distortion can usually be calculated or measured for a particular telescope or interferometer. The technique is particularly useful for producing good maps from aperture synthesis observations in which only a fraction of the total aperture has been synthesized. The raw, or dirty, maps contain numerous false features caused by the side lobes. The CLEAN technique recognizes true features and separates them from the side-lobe pattern.

Clock star A bright star in the equatorial region of the sky with precisely known right ascension, for determining the error of clocks used to time meridian transits.

Closed universe A closed universe is a model of the universe in which the universe first expands (after the big bang), and then contracts, due to the gravitational forces of the matter in the universe. In this model, the universe is finite in space and time. Contrast with an open universe.

Cluster of galaxies An aggregation of galaxies, which may or may not be bound together by gravity. For example, our Galaxy, the Milky Way, is a member of the Local Group, a rather small cluster of which the only other large member is the Andromeda Galaxy. At the other extreme are the Abell clusters which contain many hundreds or even thousands of galaxies in a region just a few million light years across; prominent nearby examples are the Virgo and Coma Clusters. Between these two extremes, galaxies appear to be clustered in systems of varying density. The densest Abell clusters are held together by their own self-gravity. Such rich clusters are filled with X-ray emitting gas, at temperatures of up to 108 K, and tend to have giant elliptical galaxies at their centre. The less rich and more extended systems may not be bound by gravity variable.

Coalsack A conspicuous dark nebula in the constellation Crux, easily visible to the naked eye as a dark patch against the brighter background of the Milky Way. It lies about 600 L.Y away and covers nearly 7° × 5° of sky. Its true diameter is about 60 ly., and its mass is about 3500 solar masses. An open cluster, the Jewel Box, is visible next to the Coalsack, but is actually far more distant. The Northern Coalsack is another name for part of the Great Rift.

Cocoon star A star surrounded by a dense cloud of gas and dust which absorbs some of the star's radiant energy and re-emits it at infrared wavelengths. In extreme cases the star may be completely obscured optically, leaving only an infrared source. OH-IR sources are examples of cocoon stars.

Cold dark matter A particular type of non-baryonic matter that, according to some theories, is created in the early stages of the Big Bang and survives to the present time in sufficient numbers to contribute significantly to the present density of the Universe. The term 'cold' signifies that these particles move at speeds much less than that of light, usually because they are heavy. There are many possible candidates for cold dark matter, such as axions, photinos, and primordial black holes. Cold dark matter has, until recently, appeared to help solve the problem of galaxy formation and large-scale structure in the Universe. More recent observations, however, suggest that the simplest versions of this picture are not correct.

Columba Columba (the dove, abbreviated Col) is a small constellation in the Southern Hemisphere. This constellation pictures a dove holding an olive branch in its mouth. Columba is a member of the "Heavenly Waters," a constellation family. Columba was named by the astronomer Bayer, but was populatized by the French astronomer Augustin Royer in 1679 (before that, Colmba's stars have been included in the constellation Canis Major). Alpha Col (the brightest star in Columba) is called Phaet; beta Col (the second-brightest star in Columba) is called Wezn. The globular cluster NGC 1851 and the blue magnitude 5 star (mu Col, a "runaway star") are in Columba.

Columbia, space shuttle The first Space Shuttle, STS-1 Columbia (OV-102), launched on April 12, 1981 (from Florida), and landed April 14, 1981 (at Edwards Air Force Base, CA). The first crew consisted of Commander John W. Young and Pilot Robert L. Crippen. It orbited the Earth 37 times, flying 1,074,567 miles in 2 days, 6 hours, 20 min, 53 seconds. After dozens of flights into space, STS-107 (the mission number) Columbia, disintegrated during atmospheric re-entry on February 1, 2003, killing the seven crew members (Rick D. Husband, Kalpana Chawla, William C. McCool, David M. Brown, Laurel B. Clark, Michael P. Anderson, Ilan Ramon).

Coma berenices A faint but distinctive constellation of the northern sky, representing the hair of the Egyptian Queen Berenice. It is notable for a scattering of faint stars that make up the Coma Star Cluster. The constellation's brightest stars are Alpha and Beta Comae Berenices, both magnitude 4.3. Coma Berenices contains a number of galaxies which are members of the Virgo Cluster, as well as a considerably more distant cluster of galaxies, the Coma Cluster. Its two most famous galaxies, M64 and NGC 4565, a 10th-magnitude spiral seen edge-on, are both about 20 million L.Y away, closer to us than either cluster.

Coma, cometary The envelope of gas and dust that surrounds the solid nucleus of an active comet.The coma often appears as a teardrop, being largely shaped by the solar wind flowing around the comet. Near perihelion, the coma may be 100 000 km wide. A coma does not usually form until the comet is within 3-4 AU of the Sun. Coma production has, however, been recorded from Chiron at over 11 AU.

Coma cluster A large, regular cluster of more than 3000 galaxies, about 280 million L.Y away in Coma Berenices; also known as Abell 1656. It covers more than 4° of sky and has a true diameter of at least 20 million L.Y Most of the bright members are ellipticals or lenticulars. Two very bright giant ellipticals, NGC 4874 and NGC 4889, lie either side of the centre; these galaxies have about 10^{11} solar luminosities. The total mass of the cluster is estimated at 2×10^{15} solar masses.

Comet A chunk of frozen gasses, ice, and rocky debris that orbits the Sun. A comet nucleus is about the size of a mountain on earth. When a comet nears the Sun, heat vaporizes the icy material producing a cloud of gaseous material surrounding the nucleus, called a coma. As the nucleus begins to disintegrate, it also produces a trail of dust or dust tail in its orbital path and a gas or ion tail pointing away from the Sun. Comet comas can extend up to a million miles from the nucleus and comet tails can be millions of miles long. There are thought to be literally trillions of comets in our solar system out past Neptune and Pluto, but only once per decade or so does one become near and bright enough to see easily without binoculars or a telescope.

Commensurability The circumstance in which the ratio of the orbital periods of two bodies, such as satellites or planets, is an exact fraction such as one-half or two-thirds. In the Solar System there are many near-commensurabilities in orbital periods. For example, the orbital periods of Jupiter and Saturn have a ratio of nearly two-fifths, while the Saturnian satellite pairs Dione and Enceladus, and Mimas and Tethys, have ratios of almost one-half. Such situations give rise to resonances. In some cases the commensurabilities provide stability, while in others they lead to instability, as with the Kirkwood gaps in the asteroid belt where asteroids tend to avoid orbits in which they would have periods commensurable with that of Jupiter.

Common envelope binary A form of giant star that consists of two stellar cores surrounded by an extremely large, common envelope of gas. Studies of stellar evolution predict that a common envelope binary should result when one component of a W Ursae Majoris star expands in its red-giant phase; however, no such star has yet been positively identified.

Compact galaxy A type of galaxy which can only just be distinguished from stars on Schmidt camera sky survey plates. They have apparent diameters of 2-5 and a region of high surface brightness that can be due either to a bright nucleus or a region of active star formation. Some 2000 such objects were catalogued by F. Zwicky from Palomar Observatory Sky Survey plates.

Concordia family A small but well-defined Hirayama family of asteroids in the main asteroid belt, mean distance 2.70-2.75 AU from the Sun. Its members have orbital inclinations of 4-6°. There are only small differences between the C-class spectra of the three largest family members, Concordia, Nemesis, and Isabella, but another allocated member, Eduarda, has an S-class spectrum and may not be a true family member. Concordia itself, 104 km in diameter, was discovered in 1860 by the German astronomer Robert Luther.Concordia's orbit has a semimajor axis of 2.703 AU, period 4.44 years, perihelion 2.59 AU, aphelion 2.82 AU, and inclination 5°.1.

Conjunction Conjunction is the apparent close approach of a planet to the Sun (or another planet), from the point of view of an observer on the Earth. A planet is in conjunction when the Sun is exactly between that planet and the Earth or, for Mercury and Venus (the two inferior planets), when that planet, the Sun, and the Earth are lined up. Mercury and Venus have two positions of conjunction: when either planet is directly between the earth and the Sun, it is in inferior conjunction; when either planet is on the far side of the Sun from the earth, it is in superior conjunction. During conjunction, a planet cannot be seen from Earth (unless it is in transit); it is either behind the Sun or is lost in the glare of the Sun.

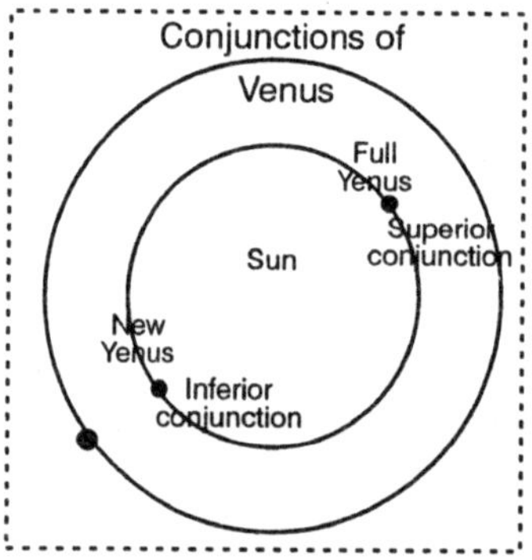

Fig. Conjunction

Constellation Any of the 88 areas into which the celestial sphere is divided for the purposes of identifying objects, as adopted by the International Astronomical Union in 1922.In 1930 the IAU went on to adopt official constellation boundaries, defined by lines of right ascension and declination for the epoch 1875. This epoch was chosen because it had already been used by B. A. Gould to draw boundaries for the southern constellations. The IAU boundaries, defined by the Belgian astronomer Eugène Joseph Delporte were published in Délimitation Scientifique des Constellations and the related Atlas Céleste. The brightest stars in a constellation are identified by a Greek letter or by a number referring to stars identified in this way, the genitive case of the constellation's name is always used, as in Alpha Orionis, 61 Cygni, or Zeta Ursae Majoris. Three-letter abbreviations of constellation names, as laid down by the IAU, are also frequently encountered (e.g. Ori, Cyg, or UMa).

Continuous spectrum A continuous spectrum is a spectrum of emitted light that contains all wavelengths of the colors that compose white light (red, orange, yellow, green, blue, indigo, violet, from long to short wavelength). Continuous spectra are emitted by incandescent solids, liquids, or compressed gases. If some discrete lines are missing from a spectrum, it is an absorption spectrum (indicating the presence of elements that absorb particular wavelengths).

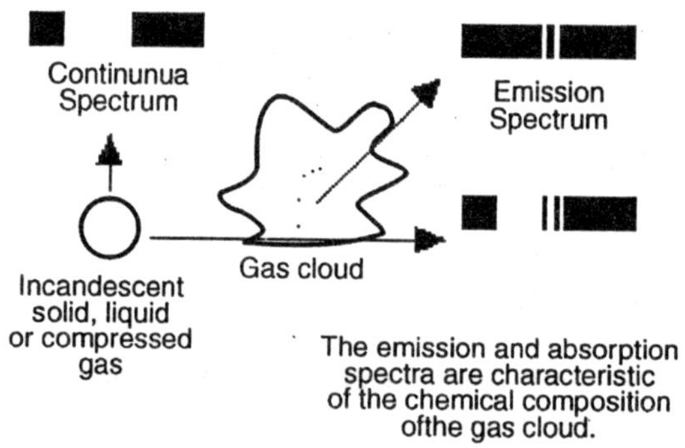

Fig. Continuous spectrum

Copernican system A model of the Solar System proposed by N. Copernicus in which the Sun lay at the centre with the planets orbiting around it; the stars lay at a vast distance beyond the planets. The model retained the circular orbits and epicycles of thePtolemaic system, but incorporated Copernicus's own observations. It also contained elements from variants of the Ptolemaic system proposed by the Arab astronomers al-Tausi and Ibn al Shauir which Copernicus apparently knew about. In Copernicus's model the motion of the sky results from the Earth's axial rotation.

The Copernican system reproduced planetary motion no better than the Ptolemaic system did, because the concept of elliptical orbits had not yet been introduced. However, it was significant in removing the Earth from the centre of the Universe.

Copernicus, Nicolaus Polish churchman and astronomer of German parentage.While a student he had realized that the Earth revolves around the Sun and not vice versa. Such a view was held to be heretical, the Church regarding the geocentric world-view of Ptolemy as consistent with its doctrines. Copernicus set down his basic ideas around 1510 in the Commentariolus, which was circulated anonymously. In 1512-29 he made the observations that he needed to support his theory, while carrying out ecclesiastic and local administrative duties. In 1539, the Austrian astronomer and mathematician Georg Joachim von Lauchen known as Rheticus, became a pupil of Copernicus and began to spread his ideas. The Copernican system was published openly in 1543 in the book De revolutionibus orbium coelestium. However, the reality of a heliocentric Solar System came to be accepted only after the work of Galileo and J. Kepler.

Coprates A dark, elongated albedo feature on Mars, visible in Earth-based telescopes, centred at -15° lat., 65° long. It corresponds to the Valles Marineris canyon system. The feature was previously known as Agathodaemon.

Core collapse 1. The collapse of a star's core, as in the formation of a neutron star. The core collapses when it has evolved to the point where it consists entirely of iron, which cannot be burnt in nuclear reactions and hence produces no energy to support itself against gravity. The core then contracts, releasing gravitational potential energy which tears the nuclei of iron atoms into their constituent protons and neutrons. As the density rises, protons combine with electrons to form neutrons. Collapse is halted when the pressure from the degeneracy of the neutron gas balances the inward pull of gravity. The entire process takes less than a second. 2. The collapse of the central part of a star cluster after energy is transferred to its outer parts through encounters between its constituent stars. The age of the cluster at which this should theoretically occur is, however, less than the true age of many globular clusters.

Core, planetary The densest, central part of a planetary body, differing markedly in composition from the layer above it. The core of the Earth is that part of the planet's interior over 2900 km or so below the surface. It consists of mainly iron-nickel alloy, and is further divided into a liquid outer core and a solid inner core.

Core, stellar The central region of a star, in which hydrogen burning occurs while the star is on the main sequence. After the star has left the main sequence, the core is the site of further nuclear burning of other fuels such as helium and oxygen. About 30% of the mass of a star of 10 solar masses is in its core. In such massive stars the cores are entirely convective, ensuring that the region is well mixed and of uniform composition. Stars of less than 0.4 solar masses are convective in both their cores and their envelopes. Stars of 0.4-1.2 solar masses have radiative cores.

Coriolis force An apparent force deflecting the motion of an object or a fluid moving over the surface of a rotating body such as a planet or star. For example, in the Earth's northern hemisphere objects are deflected to the right of their direction of travel, and in the southern hemisphere to the left. The effect applies to winds, water, and rocket launches. It is named after the French physicist Gustave Gaspard de Coriolis.

Corona 1. An extremely hot (about 2 million K), highly ionized gas surrounding the Sun. Certain other stars also have coronae. The

Sun's corona is visible at total eclipses as a white region extending out to several solar radii, displaying streamers, plumes, and bubbles or loops. The radiation of the white-light corona has components due to emission lines and scattering from electrons and dust particles.The corona's outer extension is the solar wind.

X-ray images of the Sun's corona show complex loop structures near sunspot groups, and near smaller X-ray bright points. The X-ray emission, plus emission lines due to highly ionized atoms shows that the temperature is about 2 million K; even higher temperatures of 4 million K or more are found in coronal condensations. Magnetic fields, with a strength of about 10^{-3} tesla, govern the corona's shape.

Cosmic background explorer A NASA satellite launched in 1989 November to map the cosmic background radiation.It carried three sets of instruments: a group of three Differential Microwave Radiometers to look for irregularities in the CBR; the Far-Infrared Absolute Spectrometer to measure the spectrum of the CBR; and the Diffuse Infrared Background Experiment to search for infrared radiation from forming galaxies. COBE's most important results were the discovery of irregularities in the CBR on the level of 1 part in 105, and the confirmation that the spectrum of the CBR is a black body with a temperature of 2.73 K.

Cosmic background radiation A faint, diffuse glow coming from all directions in the sky, most intense at a wavelength of around 1 mm; also known as the microwave background radiation or cosmic microwave background. Its existence was predicted from the Big Bang theory, and it was discovered in 1965 by A. A. Penzias and R. W. Wilson. The CBR originated when the Universe was 300000 years old and consisted of a plasma at a temperature of approximately 3000 K. The expansion of the Universe since then has redshifted the CBR photons to an apparent temperature of only about 3 K. Measurements by the Cosmic Background Explorer satellite showed that the CBR has a black-body spectrum with a temperature of 2.73 K. COBE detected random irregularities in temperature of the order of 1 part in 105. The CBR is regarded as one of the three most important pieces of evidence for the Big Bang, the others being the recession of the galaxies and the cosmic abundance of helium.

Cosmic rays Atomic and subatomic particles travelling through space at close to the speed of light. Protons form about 90% of cosmic rays, and alpha particles most of the rest. Heavier nuclei are very rare. Small numbers of electrons, positrons, antiprotons, neutrinos, and gamma rays are also present. Cosmic rays were first detected during a balloon flight in 1912 by V. F. Hess, and the term was coined in 1925 by the American physicist Robert Andrews Millikan.

Cosmic rays have extremely high energies, from 10^{8} to 10^{20} eV, which are a product of their velocity and mass. They continually bombard the Earth, but these primary cosmic rays rarely reach the ground because they collide with atoms in the Earth's atmosphere. The collisions produce secondary cosmic rays, which decay to form a cosmic-ray shower.

The origin of cosmic rays is still uncertain. Most of the lightest primary nuclei are thought to result from collisions between heavier nuclei as they travel from their source.

Cosmic-ray shower The cascade of secondary particles and photons produced when a primary cosmic ray enters the Earth's atmosphere and collides with air molecules; also known as an air shower or an Auger shower. The secondary cosmic rays are initially pions. Neutral pions decay into gamma rays, which in turn generate electrons and positrons by pair production. Deceleration of the electrons and positrons by the atmosphere produces more photons by bremsstrahlung radiation. The charged pions decay into muons which themselves decay into electrons and neutrinos. The primary cosmic ray and many of the secondaries may undergo further collisions, yielding more particles. The resulting shower can cover an area several kilometres wide at ground level, depending on the energy of the particle that caused it.

Cosmology The study of the structure and evolution of the Universe. Observational cosmology is concerned with the physical properties of the Universe, such as its chemical composition, density, and rate of expansion, as well as the distribution of galaxies and clusters of galaxies. Physical cosmology tries to understand these properties by applying known laws of physics and astrophysics. Theoretical cosmology involves making models that give a mathematical description of the observed properties of the Universe based on this physical understanding. Cosmology also has philosophical, or even theological, aspects in that it seeks to understand why the Universe has its observed properties. Active areas of research in cosmology include: the use of large-scale galaxy surveys to map the distribution of matter on cosmological scales, according to the redshift-distance relation; the investigation of fluctuations in the temperature of the cosmic background radiation and their implications for theories of galaxy formation.

Cosmonaut Cosmonaut is the Russian word for astronaut, a person who ventures into space.

Crab nebula A supernova remnant approximately 6500. L.Y away in the constellation Taurus; also known as Ml or NGC 1952, and as the radio source Taurus A. The Crab Nebula is the remains of a star that was seen to explode as a supernova in AD 1054, reaching a maximum apparent magnitude of -6. In telescopes it appears as an elliptical nebulosity of 8th magnitude. Its true dimensions are 6 × 4 L.Y Optically, the nebula has two components: an outer region of reddish, twisted filaments of hydrogen gas; and an inner, whitish core that shows no spectral features. The light from the core is synchrotron radiation, caused by high-speed electrons from the Crab Pulsar. It is highly polarized and is continuous at all wavelengths, from gamma-rays to radio waves. The expansion velocity of the outer filaments is about 1000 km/s. There is evidence that the expansion is accelerating, driven by the radiation from the pulsar.

Crab pulsar The pulsar at the heart of the Crab Nebula; also known as PSR 0531+21.The Crab Pulsar has a period of 33.3 ms and is slowing by 36.4 ns per day. The rotational energy it loses is turned into synchrotron emission in the surrounding nebula. Its spin-down age is consistent with its birth in the supernova of AD 1054 that gave rise to the Crab Nebula, making it one of the youngest pulsars known. The Crab Pulsar was discovered at radio wavelengths in 1968. In 1969 it became the first pulsar seen to be flashing optically. Pulses at the same frequency have since been detected at X-ray and gamma-ray wavelengths. Visually it appears as a star of 16th magnitude.

Crater A bowl-shaped depression on a planetary surface. Craters are usually of either impact or volcanic origin; however, some craters on the satellites of the outer planets may be the result of ice volcanism in which internal heating melts and vaporizes the ice to produce gas explosions and water flows. Volcanic craters tend to be less circular than impact ones, and may be formed either by explosion or collapse. Volcanic craters of purely explosive origin are normally under 1 km in diameter and are usually found at the top of a cone of ejected debris; collapse craters can be more than 100 km across, and may have no outer slope at all. Impact craters have been found on all bodies in the Solar

System photographed by space probes except Jupiter's volcanic satellite Io, but volcanic craters do not seem to exist on the smaller Solar System bodies

Critical density The mean density of matter that is required for gravity to halt the expansion of the Universe. A universe with a very low density will expand for ever, while one with a very high density will eventually collapse. A universe with exactly the critical density, around 10^{-29} g/cm^3, is described by the Einsteinde Sitter model which lies on the dividing line between these two extremes. The mean density of material that can be directly observed in our Universe is only about 20% of the critical value. There may, however, be a great deal of unseen dark matter that would raise the density to the critical value. Inflationary universe theories predict that the present density should be very close to the critical density; these theories require the existence of dark matter.

Crux Crux (or Crux Australis) is the scientific name of the Southern Cross constellation. This well-known, cross-shaped Southern Hemisphere constellation is on the Australian flag. The brightest star in Crux is Acrux (alpha Cru), a double-star system at the base of the cross. The second-brightest star is Becrux or Mimosa (Beta Cru); the third-brightest is Gacrux (Gamma Cru). The Jewel Box (also known as Kappa Crucis) is an open cluster of about 100 stars in the Southern Cross. Crux lies on the Milky Way and is surrounded by the constellation Centaurus on three sides.

Curvature of space-time A property of space-time in which the familiar laws of geometry no longer apply in regions where gravitational fields are strong. In general relativity the geometry of space-time is intimately connected with the distribution of matter. In a space of only two dimensions, such as a flat rubber sheet, Euclidean geometry applies so that the sum of the internal angles of a triangle on the sheet is 180°. If a massive object is placed on the rubber sheet, the sheet will distort and the paths of objects moving on the sheet will become curved. This is, in essence, what happens in general relativity.

In the simplest cosmological models, based on the Friedmann universe, the space-time curvature is simply related to the mean density of matter, and is described by a mathematical function called the Robertson-Walker metric.

Cybele group A group of asteroids, often called the Cybeles, in the outer part of the main asteroid belt at a mean distance of 3.4 AU from the Sun, between the 2: 1 and 5: 3 Jovian resonances. The Cybeles, in common with the Trojan asteroids, the Hilda group, and the asteroid Thule, are thought to be remnants of the primordial population at their respective distances. The group is named after the P-class asteroid Cybele, diameter 308 km, discovered in 1861 by the German astronomer Ernst Wilhelm Lebrecht Tempel.Cybele's orbit has a semimajor axis of 3.434 AU, period 6.36 years, perihelion 3.07 AU, aphelion 3.79 AU, and inclination 3°.5.

Cyclotron maser A maser source based on cyclotron radiation, observed in the magnetospheres of certain planets and in active areas of the Sun's atmosphere.

Cyclotron radiation Electromagnetic radiation emitted by a charged particle circling in a magnetic field substantially below the speed of light. It is named after the type of particle accelerator in which it was first observed. The radiation is circularly polarized and appears at a single frequency, the gyrofrequency, which is independent of the velocity of the particle. Cyclotron radiation has been detected from the magnetospheres of the outer planets.

Cygnus A A strong radio source approximately 109. L.Y away in the constellation Cygnus, identified with a giant elliptical galaxy of 15th magnitude. Cygnus A is the brightest extragalactic radio source. It is a classical radio galaxy with prominent radio-emitting lobes on either side of the central source, fed by faint, narrow jets. The optical parent galaxy, first identified in 1952, has an unusual structure which was originally interpreted as two spiral galaxies in collision. Although this model was subsequently rejected, it remains possible that giant elliptical galaxies such as Cygnus A may have arisen from the merger of smaller galaxies.

Cygnus loop A large supernova remnant in the constellation Cygnus at an estimated distance of 2000 L.Y The loop is 3° across and expanding at over 100 km/s. It is thought to be 30 000 years old. Radio, infrared, and X-ray maps show a complete loop structure. However, optical telescopes show only fragments of the loop, notably the Veil Nebula.

Cynthian Referring to the Moon. The term comes from an identification of the Moon goddess with Artemis, twin sister of Apollo, supposedly born on Mount Cynthus on the island of Delos.

D

D galaxy An extremely bright giant elliptical galaxy near the centre of a cluster of galaxies, in which it will be the first-ranked (i.e. brightest) member. They are assigned the letter D on Morgan's classification, which signifies that they are dustless. The more extreme supergiant examples at the centres of rich clusters are classified as cD galaxies. D galaxies are often radio sources, and are surrounded by extended envelopes of faint stars.

D star The spectral classification of white dwarf stars. The original sequence used was DB, DA, DF, DG, and DK, in order of declining effective temperature from 100 000 to 4000 K (i.e. the second letter follows the normal sequence of spectral types, B to K). White dwarfs with a continuous spectrum were classified as type DC. Subsequently, types DO (extremely hot, showing only ionized helium lines) and DQ(cointaining atomic or molecular carbon in addition to their mainly helium atmospheres) were added, and type DZ was introduced to combine the former types DF, DG, and DK. Also, a temperature index was added from 0 to 9. In the fully developed scheme, a DC9 white dwarf, for example, is very cool and contains no detectable absorption lines. The majority of white dwarfs are type DA showing only broad absorption lines of hydrogen, their breadth resulting from the star's enormous surface gravity. Most of the rest are type DB, which show only neutral helium lines. The metal lines in type DZ are thought to be caused

Dall-kirkham telescope A variant of the Cassegrain telescope, with a concave ellipsoidal primary mirror and a convex spheroidal secondary mirror. This combination produces an image free from astigmatism, although with greater coma off-axis. It is particularly suited to observations where good resolving power is more important than wide field of view, such as planetary observation. It was invented independently by the English amateur telescope-maker Horace Edward Stafford Dall and the American Alan R. Kirkham.

Damocles Asteroid 5335 A body with an unusual orbit that takes it from near the orbit of Mars out beyond the orbit of Uranus, discovered in 1991 by the British astronomer Robert Houston McNaught On the basis of its orbit, which has a period of 41 years and a high inclination, it would probably have been classed as a comet had it shown a trace of a coma or hint of a tail. However, Damocles appears to be a bare rocky body, although it could be an extinct Halley-family comet. Its diameter is estimated at 15-20 km. Although Damocles' present orbit does not cross that of the Earth, in a few tens of thousands of years from now it may well do so due to planetary perturbations. Currently its semimajor axis is 11.88 AU, perihelion 1.58 AU, aphelion 22.18 AU, and inclination 61.9.

Dark matter Material whose presence can be inferred from its effects on the motions of stars and galaxies, but which cannot be seen directly because it emits little or no radiation; also known as missing mass. It is thought that at least 90% of the mass in the Universe resides in some form of dark matter. Evidence for dark matter in spiral galaxies is provided by their rotation curves. The existence of dark matter in rich clusters of galaxies can be deduced from the motions of the member galaxies. A significant part of this dark matter may be in the form of low-mass stars or Jupiter-mass objects; such normal matter is described as baryonic. Dark matter may also exist in the space between galaxies, and could raise the average density of the Universe to the critical density required to reverse the current expansion. If the Big Bang theory is correct, then a large proportion of dark matter must be in nonbaryonic form, perhaps axions, photinos, or massive neutrinos, left over from the early stages of the Big Bang

Dark-line spectrum A dark-line spectrum (also called a absorption spectrum) consists of

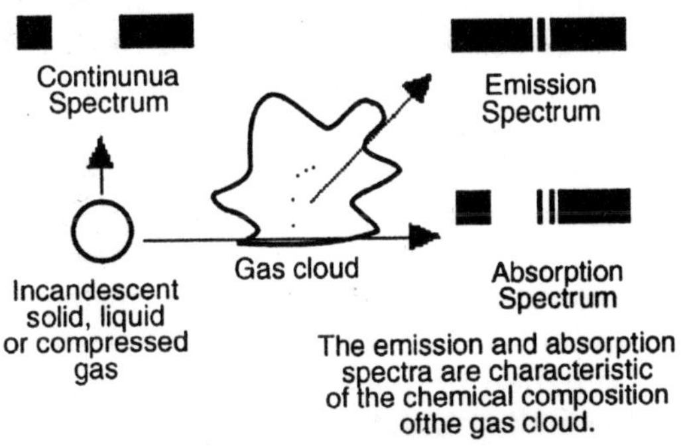

Fig. Dark-line spectrum

dark absorption lines superimposed on a bright continuous spectrum. An absorption spectrum is created when light from an incandescent source passes through a cooler gas that absorbs photons. Each different element and molecule absorbs light at a unique set of frequencies. Astronomers can determine the composition of gases in stars by looking for characteristic frequencies (this is called spectroscopy).

Day The rotation period of the Earth relative to some external point. Several types of day are defined according to the reference point chosen. For most purposes the day is defined as 86 400 s which corresponds to the mean solar day, but the mean solar day depends on the rotation of the Earth, which is neither completely uniform nor completely predictable. The apparent solar day, the interval from one meridian transit of the Sun to the next, has a considerable annual variation of about 30 s due to the changing value of the equation of time. This is averaged out to give the mean solar day. The mean sidereal day, on the other hand, is the Earth's period of axial rotation relative to the stars, with a minor correction of 0.0084 s to allow for precession. It is the interval between successive transits of the mean equinox, and is nearly 4 min shorter than the mean solar day. Due to the Earth's orbital motion around the Sun there is precisely one more sidereal day in the year than the number of solar days.

De Sitter universe A model of an expanding universe in which there is no matter or radiation but the expansion is driven by a cosmological constant. It was proposed in 1917 by W. de Sitter. Although this model is physically unrealistic, it introduced the idea that the real Universe might be expanding. An expansion phase which is very similar to that in the de Sitter model also plays an important role in modern theories of the inflationary universe.

Death star theory The Death Star Theory refers to the fact that mass extinctions are periodic, and may be caused by the Earth's passing through a cloud of comets (the Oort cloud) every 26 million years. Some people have hypothesized that there is a yet-to-be-discovered dark star or perhaps a planet (called, appropriately enough, Nemesis) orbiting in the outer reaches of our solar system. This body disrupts the Oort cloud (once every 26 million years), sending comets into the inner parts of the solar system, some of which hit Earth and cause mass extinctions.

Decaying orbit A decaying orbit is one in which the orbiting object is slowly spiraling towards the primary. The radius of the orbit is decreasing over time. For example, when one of the Earth's satellites is in a decaying orbit, it enters the Earth's atmosphere and burns up as it falls towards Earth.

Deceleration parameterA figure that describes the slowing-down of the expansion of the Universe. In a Friedmann universe, the deceleration parameter is simply half the density parameter, O A value of q greater than 0.5 indicates that the expansion is decelerating quickly enough for the Universe eventually to collapse. A value less than 0.5 indicates that the expansion will continue for ever. In models with a cosmological constant, q can even be negative, indicating an accelerated expansion, as in the inflationary universe.

Declination One of two coordinates for the celestial sphere, which are analogous to latitude and longitude for the Earth's surface. The declination of an object is how many degrees it is north or south of the celestial equator. The other coordinate is called right ascension, and it is measured eastward from a somewhat arbitrary "prime meridian" on the sky. The "prime meridian" passes through the position of the Sun at the time of the vernal equinox. Thus its position changes slowly over the years, due to the precession of the equinoxes. The position of the celestial poles also changes with precession. Thus, to locate an object from its right ascension and declination, you must also know the date for which those coordinates are valid; that date is called the epoch of the coordinates.

Declination axis The axis of an equatorial mounting around which a telescope can be moved in declination. The declination axis is at right angles to the polar axis.

Deconvolution The numerical inversion of a convolution equation, either continuous or discrete, in one or several variables. Except in trivial cases, deconvolution is an ill-posed problem: In the absence of constraints or extra side-conditions, and in the case of noiseless data---assuming that some solution exists— there usually will exist many solutions. In the case of noisy data, there usually will exist no exact solution, but a multitude of approximate solutions. In the latter case, if one is not careful in the choice of a numerical method, the computed approximate solution is likely not to have a continuous dependence on the given data. The so-called regularization method (of which the Maximum Entropy Method is a special case) is an effective tool for the deconvolution problem.

Discrete two-dimensional deconvolution is an everyday problem in radio interferometry,

Decoupling The stage early in the history of the Universe when, according to the Big Bang theory, particles of matter ceased to interact with radiation. Decoupling happened at different times, and therefore at different temperatures, for differen particles. Neutrinos, for example, decoupled from the background radiation at a temperature of about 10^{10} K (about 1s after the Big Bang) while ordinary matter decoupled at a temperature of a few thousand degrees K (after about 300 000 years). After matter and radiation decoupled, the background radiation propagated freely through the expanding Universe.

Dedispersion A technique used in radio astronomy to compensate for the smearing effect of interstellar dispersion. The radio band being observed is divided into several narrow channels, each of which is measured separately. Provided the dispersion measure of the source is known, appropriate time delays can be introduced between the channels to remove the dispersive delay across the band. The signals from the channels are then recombined to form a dedispersed signal. Dedispersion is widely used for the observation of pulsars, whose narrow pulses are readily smeared out by interstellar dispersion.

Deep space 1 Deep Space 1 (DS1) was a NASA mission that was launched on October 24, 1998. DS1 tested 12 new advanced technologies in deep space, including ion propulsion, solar arrays, a deep space transponder, autonomous navigation systems, a miniature integrated camera spectrometer, etc. DS1's primary mission ended in September 1999.

Deep space network A world-wide network of radio telescopes for tracking and communicating with space probes, owned by NASA and operated by the Jet Propulsion Laboratory. The network consists of three deep-space communications complexes, located on three continents to allow continuous coverage as the Earth rotates. The complexes are at Goldstone in the Mojave Desert of California, 72 km north of Barstow; Robledo de Chavella, 60 km west of Madrid, Spain; and the Tidbinbilla Nature Reserve, 40 km southwest of Canberra, Australia. Each complex consists of four main antennas: one of 70 m, two of 34 m, and one of 26 m. The 70-m dishes were enlarged from 64 m in 1989, before Voyager 2's encounter with Neptune. Additionally, Goldstone has a second 26-m dish and a 9-m antenna. The smaller dishes can be used for tracking some Earthorbiting scientific satellites. The dishes are also used for radio and radar astronomy observations.

Deflection of light The bending of light by the gravitational field of the Sun. At the Sun's limb, the deflection amounts to 1".75 radially away from the Sun. This effect is corrected for when reducing star positions from mean place to apparent place.

Degeneracy A state of matter attained when atomic particles are packed together as tightly as is physically possible, at densities of several thousand tonnes per cubic metre. Particles which are very close together are forbidden by the Pauli exclusion principle to have the same energy, and, as a result, the particles repel each other. This causes degeneracy pressure which, unlike thermal pressure, depends only on density and not on temperature. It provides the main support against gravity in white dwarfs and neutron.Degenerate matter is found also in the cores of low-mass stars that have exhausted their central hydrogen, in brown dwarfs, and in the central regions of the giant planets.

Deimos The outer satellite of Mars, distance 23 459 km, orbital period 1.262 days. Deimos is irregular in shape, 15 × 12 × 11 km across its three axes. It spins on its shortest axis in the same time that it takes for one orbit, and so keeps the same face towards Mars. Deimos was discovered in 1877 by A. Hall. Its surface is covered with impact craters, but it is smoother than Phobos as many of its craters appear to have been infilled. Deimos may be a captured asteroid.

Delta aquarid meteors A meteor shower producing moderate activity between July 15 and August 20. The meteor stream's orbit lies close to the ecliptic, and has been split into two major components. At maximum, on August 6, the Northern Delta Aquarids appear from a radiant at RA 23h 04m, dec. +02°. The Southern Delta Aquarids peak on July 29 from a radiant at RA 22h 36m, dec. -17°. The southern component is the more active, with peak ZHR around 25, compared with 10 for the northern component.

Delta aquarid meteor shower The Delta Aquarids are a meteor shower that occur each year from July 14-Aug.18, with a maximum on July 28-29. This meteor shower occurs each year as the Earth passes through the orbit of an unknown comet, and icy debris from the comet burns up as it enters the Earth's atmosphere. The meteors in this shower seem to emanate from the constellation Aquarius.

Delta delphini star A giant star of spectral type late A to early F, with weak Ca II lines, that may also be a Delta Scuti star. The type star is a spectroscopic binary with an orbital period of 40.58 days in which both components are Delta Scuti stars; the period of the primary is 0.158 day, and that of the secondary 0.134 day.

Delta scuti star A type of pulsating variable with a short period (0.01-0.2 day) and small amplitude (0.003-0.9 mag.): abbr. DSCT. The pulsation is driven primarily by instabilities in the hydrogen convection zone, and multiple pulsation modes occur simultaneously. These Population I stars have spectral types from AO to F5 and lie at the lower end of the Cepheid instability strip, either on the main sequence or among the subgiants and giants. In older literature, they were often described by the misleading terms dwarf Cepheid, RRs variable, or ultra-short-period variable.]

Dew cap An extension to the tube of a refracting telescope to prevent dew from forming on the objective lens. It works by reducing cooling of the objective by shielding it from cold air currents. The dew cap should be between two and three times longer than the telescope's aperture, with a diameter which does not restrict the field of view at low powers. If it is too long, however, it may impair the seeing because of air currents circulating within it.

Diamond ring The "Diamond Ring" is a large burst of light that appears a few seconds before and after totality during a solar eclipse. The "Diamond Ring" is caused by the last bit of sunlight shining through valleys on the edge of the moon.

Diana chasma A deep trough in the middle of Aphrodite Terra on Venus at -15° lat., 155° long. It is one of several deep troughs in the region, and contains some of the lowest elevations on Venus, over 1 km deep in places. It is 100 km wide and over 900 km long, and was discovered by radar from Arecibo Observatory.

Dichotomy The moment when a planetary disk appears exactly half-illuminated by the Sun, as seen from Earth. The term is used exclusively for the planets Venus and Mercury; the same phase for the Moon is known as first quarter when it occurs between new Moon and full Moon, and last quarter when it occurs between full Moon and new Moon.

Dichroic mirror A partially reflecting mirror that also transmits some light. Such mirrors are produced by the deposition of thin films of materials which reflect particular colours and transmit others by interference. A typical dichroic mirror may transmit blue light and reflect yellow light, for example. Nebula filters are produced by the same process.

Dicke radiometer A radio receiver designed to measure weak signals in the presence of noise; also known as a Dicke receiver. The input to the receiver is rapidly switched between the antenna and a reference noise source. It is useful where accurate measurements of absolute flux are required, and has been used to measure the very weak signal from the cosmic background radiation. It is named after R. H. Dicke.

Dicke switch A solid-state device designed to switch between the two inputs to a Dicke radiometer. The switching rate is normally around 10-1000 times per second.

Dicke, Robert Henry American physicist and astronomer. In 1961 he suggested that the gravitational constant varies with time.In 1964, with the Canadian-born American physicist Phillip James Edwin Peebles and others, he began to develop a hot Big Bang theory, independently of G. Gamow. The theory predicted the existence of the cosmic background radiation, discovered shortly after by A. A. Penzias and R. W. Wilson. He also invented the Dicke radiometer and Dicke switch, and in 1957 set out what has become known as the weak anthropic principle.

Dielectronic recombination A process by which an ion recombines with an electron in nebulae and hot interstellar gas, at temperatures above 10 000 K. The ion captures the electron into a particular energy level, but the electron's energy is transferred to another electron in a lower orbit, which is then excited to a higher level. As the electron returns to its lower level, it radiates away this energy, and the ion is said to have recombined. It is termed dielectronic recombination because two electrons take part in the process.

Differential rotation The rotation of a non-solid body in which different parts turn at different speeds. Stars and planetary atmospheres rotate differentially. For example, the equatorial regions of the Sun rotate about 25% faster than the polar regions, and the rotation of Jupiter is divided into System I (the equatorial region) and System II (the rest of the planet). Disk-like systems such as the rings of Saturn, accretion disks around stars, and disk galaxies also rotate differentially because they are composed of individual parts, each with a separate orbit.

Differentiation A geological process by which a magma separates into layers of igneous rock of differing density and composition within a body such as a planet or asteroid. Differentiation typically produces a core, a mantle, and a crust.

Diffraction The slight bending of light around the edge of an obstacle in its path. It is a consequence of the wave nature of light. Because of diffraction, a star image consists of an Airy disk surrounded by several diffraction rings, produced by diffraction at the edge of the telescope lens or mirror. In a reflecting telescope which has a secondary mirror supported by one or more arms, diffraction around the arms causes the images of stars to have spikes. Diffraction also occurs at other wavelengths, such as radio. A point radio source occulted by the Moon, for example, disappears or reappears with brightness oscillations, rather than abruptly, as a diffraction pattern sweeps across the observatory. The size of the diffraction pattern increases with the wavelength of the radiation involved.

Diffraction fringe Blurred fringe surrounding an image caused by the wave properties of light. No detail smaller than the fringe can be seen.

Diffraction grating A surface on which are ruled very fine and evenly spaced straight grooves which break light into a spectrum by diffraction. Astronomical diffraction gratings are usually reflection gratings, giving several series or orders of spectra symmetrically arranged on either side of the reflected beam, becoming progressively fainter and more dispersed with distance from the central beam. The finer the spacing of the lines, the wider the spectrum produced. A typical diffraction grating has 6000 lines/cm. If the grooves are cut asymmetrically, in a sawtooth fashion, more of the diffracted light will be sent into one order of spectrum and the grating is said to be blazed.

Diffuse nebula A cloud of interstellar gas and dust which glows because of the effect on it of ultraviolet radiation from nearby stars. The term H II region is now preferred. The description 'diffuse' dates from earlier times when nebulae were defined according to their optical appearance. A diffuse nebula was one which retained its fuzzy appearance under magnification through a large telescope, as opposed to one which could be resolved into stars.

Digitizer A device that converts analog data to a discrete, digital representation. Any digitizer combines two operations:

sampling the input data, and

quantizing the samples.

Dimensions The fundamental measurables of a unit system in physics—those which are defined through operational definitions. All other measurable quantities in physics are defined through mathematical relations to the fundamental quantities. Therefore any physical measurable may be expressed as a mathematical combination of the dimensions. See: operational definitions.

Example: In the MKSA (meter-kilogram-second-ampere) system of units, length, mass, time and current are the fundamental measurables, symbolically represented by L, M, T, and I. Therefore we say that velocity has the dimensions LT-1. Energy has the dimensions ML2T-2.

Diogenite A class of calcium-poor achondrite meteorite; also known as hypersthene achondrites.They are named after the Greek philosopher Diogenes of Apollonia (fl. c. 440 BC), who recognized the cosmic origin of meteorites. Diogenites are composed almost entirely of the mineral bronzite with lesser amounts of the other minerals present in ordinary chondrites. The diogenites probably formed by partial melting of a parent body with a chondritic composition, followed by slow cooling within the crust of that parent body..

Dione The fourth-largest satellite of Saturn, 1120 km in diameter; also known as Saturn IV. Dione orbits 377 400 km from Saturn in 2.737 days, sharing its orbit with the much smaller Helene. Its axial rotation is the same as its orbital period. Dione was discovered in 1684 by G. D. Cassini. Parts of its surface are densely covered with impact craters, but there are also smooth plains with lower crater densities, and some major troughs around 8 km wide and up to 400 km long. Dione's density is 1.4 g/cm^3, suggesting that it consists of ice with a rocky core.

Dioptric system An optical system that uses only lenses to focus light. A refracting telescope is an example.

Diphda The star Beta Ceti, magnitude 2.0. It is a K0 giant, 53 L.Y away. An alternative name is Deneb Kaitos.

Dipole 1. A system of two equal and opposite closely spaced electric charges or magnetic poles (magnetic dipole).

2. A dipole antenna.

3. Any system having a bipolar structure about an axis of symmetry (e.g. a dipole field, or dipole anisotropy).

Dipole antenna An antenna consisting of a straight metal rod or wire broken in the centre to make two terminals. The terminals are connected to a feed which carries the signal to the receiver. The maximum sensitivity is in the direction perpendicular to the rods.

Dirac cosmology A cosmological theory built around the so-called large numbers hypothesis, which relates the fundamental constants of subatomic physics to large-scale properties of the Universe such as its age and mean density. It is due to the British mathematical physicist Paul Adrien Maurice Dirac.Dirac's theory is not widely accepted, but it introduced ideas related to the anthropic principle.

Direct Rotation or orbital motion in a counterclockwise direction when viewed looking down from above the north pole of the primary (i.e. in the same sense to most satellites); the opposite of retrograde. The north pole is the one on the same side of the ecliptic as the Earth's north pole. (The word "prograde" is sometimes used to mean "direct" in this sense.)

Direct motion Movement of revolution or rotation of a celestial object in the same sense as that of the Earth. For example, the Sun moves across the sky each day from east to west, an effect of the Earth's rotation on its axis. But, against the background of stars the Sun actually moves from west to east over the course of a year, the effect of Earth's orbit around the Sun. This west to east movement is direct motion. The Moon behave in exactly the same way. The planets also exhibit this west to east motion most of the time but owing to our perspective of the planets they occasionally appear to move in the reverse sense, known as retrograde motion.

Directivity The ratio of the maximum sensitivity of an antenna to its average sensitivity. Directivity is proportional to the effective aperture of the antenna and inversely proportional to the square of the wavelength. Where there are no losses in the antenna, the directivity is equal to the gain over an isotropic antenna.

Dirty beam In radio interferometry, simply a beam, but computed with precisely the same operations as those used to compute some companion dirty image (i.e., with the same *u-v* coverage, the same manner of gridding convolution, the same *u-v* weight function and taper, etc.). In deconvolving a dirty image, only the companion dirty beam should be used.

Dirty snowball A model of a comet's nucleus, proposed by F. L. Whipple in 1949, and confirmed by the Giotto space probe to Halley's Comet in 1986. According to this model, cometary nuclei consist of a relatively solid conglomerate of frozen ices and dust, rather than a loose 'flying sandbank' as suggested by an alternative model.

Disappearing solar filament (DSF) The sudden (timescale of minutes to hours) disappearance of a solar filament (prominence).

Disaster Literally "bad stars"; particularly apt in reference to a major asteroid impact.

Disconnection event A separation in the gas tail of a comet, resulting from changes in the strength and direction of the local interplanetary magnetic field carried by the solar wind. The existing gas tail separates from the coma and is swept away downwind, to be replaced by a new tail which develops in a slightly different direction governed by the prevailing IMF. The new gas tail can, in turn, be subject to disconnection. Several such events were recorded in the tail of Halley's Comet at its 1910 return, which coincided with a period of high solar activity.

Discovery program A NASA series of relatively simple, low-cost planetary missions capable of being developed and built within three years. The first was the Near Earth Asteroid Rendezvous launched in 1996.

Discrepancy (1) Any deviation or departure from the expected.

(2) A difference between two measurements or results.

(3) A difference between an experimental determination of a quantity and its standard or accepted value, usually called the *experimental discrepancy*.

Disk Disk is a general term that is used to describe flattened circular objects. The most common use of the word disk in astronomy is in describing round objects like the moon or planets. There is also the accretion disk which formed the solar system. Others uses of the term disk include disk-shaped galaxies and a cloud of stars in a spiral galaxy that surrounds the central bulge.

Disk galaxy A type of galaxy whose main structure is a thin disk of stars in approximately circular orbits about the centre, and whose light output typically falls off exponentially with radius. The term applies to all types of galaxy other than ellipticals, dwarf spheroidals, and some peculiar galaxies. The disk in lenticular galaxies (type SO) contains very little interstellar material, while the disks of spiral and irregular galaxies contain considerable amounts of gas and dust in addition to the stars.

Distributed object Generically, an object that may be accessed over a network connections. Thus the methods of the object may be executed by a simple syntax and without regard to the location of the object.

Diurnal tide A diurnal tide (also called a single day tide) is a tide with only one high water and one low water occurring during a lunar (tidal) day.

Dog star Popular name for Sirius, the most brilliant star in the sky, which lies in the constellation Canis Major, the Great Dog.

Dome 1. A steep-sided extrusion of viscous lava forming a bulbous mass over a volcanic vent; also known as a volcanic dome. An example is the dome that formed within the Mount St Helens crater, Washington state, after the dramatic eruption in 1980; other examples occur on Venus.

2. A low, circular raised area above a lunar mare, often with a pit in the centre; also known as a lunar dome. Slopes of lunar domes are usually only a few degrees. They are thought to be volcanic vents, in many cases the source of the fluid lavas that formed the maria, and are thus similar to shield volcanoes on Earth. Examples of lunar domes are found near the crater Hortensius.

Doppler correction The correction of an observed astronomical spectrum to a standard reference velocity system.

The observer rotates with the surface of the earth, which moves around the moon-earth barycenter, which moves around the solar system barycenter, which moves with respect to local stars, which revolve around the galactic center, etc. For standardization, apparent radial velocities of features in astronomical spectra are referred to one of a small number of defined velocity frames. These include the heliocentric velocity system, and the local standard of rest (LSR).

Doppler shift The Doppler shift (or Doppler Effect) is an increase or decrease in wavelength as the object emitting the wave moves relative to the observer. For example, a train whistle seems to be higher in pitch when the train is approaching you (the waves are compressed, shortening the wavelength), and lower in pitch when it is traveling away from you (the waves are elongated, lengthening the wavelength). The same thing happens with light waves when the light source is coming or going relative to us. For example, when a star is travelling away from Earth, its light appears redder (the light waves are elongated, lengthening the wavelength); this is called the red shift. The expansion of the universe was discovered when E. Hubble observed that the light from almost all other galaxies was red-shifted. The Doppler effect was named for Johann Christian Doppler (November 29, 1803-March 17, 1853), who first realized that it existed (1842).

Dorsum A ridge on a planetary surface; pl. dorsa. The name is not a geological term, but is used in the nomenclature of individual features, for example Schiaparelli Dorsum on Mercury.

Double star Two stars that appear close to one another on the sky. Such pairs may be divided into two classes: optical doubles, where the components are not gravitationally bound; and physical doubles, in which the stars are orbiting their common barycentre. The term 'double star' is often restricted to the former group, and the term binary star is used for the latter. In fact, optical doubles are relatively uncommon, and the majority of doubles are actually true binary systems.

Draco Draco (from the dragon in Greek mythology) is a constellation in the Northern Hemisphere. It is the radiant of the Quadrantids and Draconids (meteor showers). Draco is near the northern celestial pole; the tail of Draco is between the Big Dipper and the Little Dipper. The brightest star in the Draco is Thuban; it is located in the dragon's tail. (Thuban means "dragon or serpent" in Arabic. Thuban was the Earth's pole star about 5,000 years ago.) The second-brightest star in the Draco is Rastaban (which means "head of the dragon or serpent").

Draconic month The mean interval of time between successive passages of the Moon through the ascending node of its orbit, the point at which it passes from south to north of the ecliptic; also known as a nodical month. The Moon's orbit is inclined at about 5 to the ecliptic but it is not fixed in space. It has a slow variation which causes the two nodes to regress around the ecliptic every 18.61 years. Consequently the draconic month (27.2122 days) is about 2.5 hours shorter than the sidereal month.

Drake equation The Drake equation is a formula for estimating the possible number of intelligent civilizations in our galaxy. Most of the terms in this equations are unknown. In the equation, N is the number of civilizations, R is the average rate of star formation in the galaxy (about 20 stars per year), fs is the fraction of stars that are suitable (about 0.1), fp is the fraction of stars with planets (about 0.5), ne is the mean number of planets that are located in a "habitable zone," where water exists in liquid form, fl is the fraction of these planets on which any life form evolves, fi is the fraction of places where some the life becomes intelligent, fc is the is the fraction of intelligent species who could communicate with us, and L is the lifetime (in years) of a civilization (this is quite uncertain). The equation was formulated by radio astronomer Frank Drake in 1961.

Dubhe The star Alpha Ursae Majoris, magnitude 1.8. It is a K0 giant lying 100.L.Y away. It is a close double star, with an As dwarf companion of magnitude 4.8 that orbits it in a period of 45 years. It is one of the two Pointers that indicate the direction of the Pole Star, Polaris.

Dumbbell nebula An 8th-magnitude planetary nebula in Vulpecula, also known as M27 and NGC 6853. It is the most prominent planetary nebula for small telescopes. The Dumbbell Nebula is so named because of its double-lobed appearance, like an hourglass, in large apertures or on photographs. The Dumbbell lies about 1000 L.Y away.

Dust Small particles of solid matter found in space: in the Solar System, around stars (circumstellar dust), and between the stars (interstellar dust). The individual particles are usually called dust grains, and are of size 10 nm and upwards. Dust comprises about 10% by mass of interstellar matter. It extinguishes and reddens starlight, and can also be detected by its absorption and emission of infrared radiation and by its polarizing effect on starlight. The exact composition of interstellar dust is uncertain, but infrared absorption measurements indicate that a significant fraction of the material is organic, i.e. it contains carbon. Dust may account for the diffuse interstellar bands in the optical spectra of stars. Dust is produced in the cool outer envelopes of red-giant stars, in novae, and in supernovae. Dust is also present in the Solar System as interplanetary dust and cometary dust. Some dust is driven out of the Solar System by the radiation pressure of the Sun's light, while new dust is generated

Dwarf galaxy A small, low-luminosity galaxy. Dwarf ellipticals and dwarf irregulars are probably just fainter examples of this general type of galaxy. A separate type, dwarf spheroidals are similar to ellipticals in that they contain little or no interstellar material, but they are not so centrally concentrated, have low density and low surface brightness, and show a decrease of light output with radius which implies that they are more closely related to small disk galaxies. Most of the galaxies in the Local Group are dwarf spheroidals, and dwarfs in general are probably the most common galaxies in the Universe. There are no dwarf spirals, since no spirals are known with an absolute blue magnitude below -17 (about 109 solar luminosities).

Dwarf nova A form of cataclysmic binary, more specifically known as a U Geminorum star, that undergoes repeated outbursts. These occur when thermal instabilities trigger a sudden brightening of the accretion disk around the white-dwarf primary. The individual maxima have a general resemblance to the light-curve of a nova, although with a smaller amplitude (2-6 mag.), and a shorter duration (a few days). The intervals between outbursts are far shorter than those in a recurrent nova, and although irregular and unpredictable, show a mean cycle length for each individual star. Accretion on to the primary will probably eventually produce a nova outburst.

Dwarf star A star on the main sequence of the Hertzsprung-Russell diagram. Dwarf stars are of luminosity class V. Most stars are of this type, as is the Sun. They range in mass from about 0.1 to 100 solar masses. The term is used because stars on the main sequence are smaller than those of the same mass which have evolved into giants. White dwarfs, black dwarfs, and brown dwarfs are not dwarfs in this sense of being main-sequence stars.

Dynamical equinox The intersection of the celestial equator with the ecliptic, as determined from observations of the Sun or other members of the Solar System. The offset of the dynamical equinox from the catalogue equinox is derived by measuring the positions of Solar System objects relative to the catalogue positions of stars.

Dyson sphere A Dyson Sphere is an enormous, theoretical structure proposed in 1959 by physicist Freeman Dyson (1925-). This hollow sphere would be built surrounding a planet and its sun in order to collect all of the sun's emitted energy. The sphere would be assembled from many separate planetoids and planetary bits and pieces, put into orbit around the star. The shell material would come from disassembled planets and existing planetoids. Dyson calculated that in our solar system, there is enough matter to create a shell at least three meters thick around the Earth and Sun.

Dyson, Frank Watson English astronomer. He became the ninth Astronomer Royal in 1910. His main concern was with the accurate measurement of star positions and proper motions, which revealed the distribution of stars in our part of the Galaxy. From observations at several eclipses he identified elements in the solar chromosphere from lines in its spectrum; this success motivated the Greenwich eclipse expedition of 1919 which confirmed Einstein's prediction that a gravitational field bends starlight. He introduced the broadcasting of radio time signals from Greenwich in the 1920s.

E

E corona That part of the light from the Sun's corona that is due to emission lines from hot gas (the 'E' stands for emission). These emission lines include the so-called forbidden lines of highly ionized atoms of iron, calcium, and some other elements. The E corona is much fainter than the K and F coronae.

E layer A component of the Earth's ionosphere, at an altitude of 110 km; also known as the Heaviside layer after the English physicist Oliver Heaviside.It is used as a reflective surface by radio operators. Electron densities in the E layer show diurnal variation, being greater during daytime. Absorption of radio waves by the denser daytime E layer can be a problem. Localized patches of sporadic E ionization which commonly appear at the same atmospheric level during the summer months can also disrupt short-wave radio communication.

E line A Fraunhofer line at 527 nm wavelength in the spectrum of the Sun and other cool stars, due to a combination of absorption by iron and calcium.

E region A daytime layer of the earth's ionosphere roughly between the altitudes of 85 and 140 km.

Eagle nebula The Eagle Nebula (M16) consists of enormous columns of cool interstellar hydrogen gas and dust that are about 7,000 light-years from Earth (in the constellation Serpens). This star-forming cloud is illuminated by ultraviolet light that is emitted from newborn stars. The columns are called EGG's (Evaporating Gaseous Globules) because the ultraviolet light boils off some of the hydrogen gas (H_2), shaping the nebula in a process called photoevaporation. The biggest column is about 1 light-year tall.

Earth The third planet from the Sun. At perihelion each January it is 147 099 590 km from the Sun, as against 152096150 km at aphelion in July. Seen from far off in space it has a strong blue coloration, caused by its atmosphere. The Earth is slightly ellipsoidal in shape (equatorial diameter 12756 km, polar diameter 12714 km).

The Earth's atmosphere is composed (by volume) of 78% nitrogen, 21% oxygen, and 0.9% argon, plus carbon dioxide, hydrogen, and other gases in much smaller quantities. Water vapour is also present in variable quantities. White clouds of condensed water vapour may obscure a quarter of the Earth's surface at a time, belts of cloud being common around the equator and at temperate and polar latitudes. The pressure of the atmosphere at sea level varies around 1000 mbar.

The average atmospheric temperature at the surface is 15 °C, but ranges from -50 °C average in winter in Siberia up to +40 °C in the Sahara in summer.

Liquid water covers 71% of the Earth's surface.

Eclipse The passage of the shadow of one astronomical body over the surface of another. In any one year, the maximum total of solar and lunar eclipses visible from Earth is seven. The minimum number is two, both of which must be solar. Eclipses frequently come in pairs, with a lunar eclipse preceded or followed within a fortnight or so by a solar eclipse at the opposite node of the Moon's orbit.

Solar eclipses occur when the new Moon lies close to the node of its orbit, and at the same longitude as the Sun. The Moon obscures at least part of the Sun's disk as seen from a comparatively limited ground track where the cone of shadow cast by the Moon falls on the Earth's surface. Along the centre of this track a total eclipse may be seen; eclipses occurring around lunar apogee may be annular eclipses. A partial eclipse is seen either side of the narrow (no broader than 270 km) central path of totality or annularity, which sweeps eastwards at around 3200 km/h.

Eclipse, lunar The passage of the Moon through Earth's shadow. Lunar eclipses may be either total or partial. Totality may last up to one and three quarter hours although the period of totality is on average shorter.

Eclipse, solar The passage of the Moon in front of the Sun so that the Moon is directly in front of the Sun. Totality may last for a little over 7 minutes under favourable conditions. Partial eclipses occur when the Sun is incompletely covered. Annular eclipses occur when the Moon it near the farthest part of its orbit and hence appears smaller. In this case a bright ring of sunlight is seen around the dark body of the Moon. Technically speaking, a solar eclipse could be termed as an occultation of the Sun by the Moon.

Eclipse year The interval of time between successive passages of the Sun through a node of the Moon's orbit, when eclipses can occur. It lasts 346.62 days, the length of two eclipse seasons. An eclipse year is shorter than a sidereal year because the nodes of the Moon's orbit regress westwards around the ecliptic by about 19° per year. There are almost exactly 19 eclipse years in a Saros.

Edwin hubble Edwin Hubble was born in Marshfield, Missouri on November 20, 1889. He studied at the University of Chicago and at Oxford as a Rhodes Scholar. His interests turned to astronomy around 1914, but he delayed his entry into the field to enlist in the U.S. Army in 1917. He served in France during World War I and returned to the United States to begin work at Mount Wilson Observatory. There, he determined that other galaxies existed and were moving away from the Milky Way, proving that the universe was still expanding. He also discovered Cepheid variable stars in the Andromeda Galaxy as well as other galaxies. By comparing these variable stars to stars in the Milky Way, Hubble was able to determine that Cepheids were far beyond the boundaries of our galaxy. It was these compelling discoveries that proved the universe was far greater than imagined.

Effective temperature (symbol T eff or T e) The temperature of a black body that radiates the same total energy per unit area as a given object, such as a star. It is the most useful measure of a star's surface temperature. The effective temperature of the Sun, for example, is 5800 K. However, in other circumstances it may be misleading, such as for a planetary nebula which may have a kinetic temperature (obtained from the mean energies of the particles) of 10 000 K, but an effective temperature of only 50 K because of the low density of particles..

Egg nebula The Egg nebula (CRL2688) is a planetary nebula about 3,000 light-years from Earth. This huge cloud of dust and gas is expanding outwards at very high speeds. It was formed by a red giant star a few hundred years ago (this red giant was once a Sun-like star). Dust obscures our view of the star within.

Einstein coefficient One of three quantities, known as transition probabilities, used to describe the rate at which photons are absorbed or emitted from atoms or ions. The Einstein coefficients are widely used in the understanding of stellar spectra.

Einstein cross An example of a gravitational lens effect in which four images of a background object are formed. The first such example to be detected, the quasar G2237+0305, displays four distinct images arranged symmetrically around the image of a foreground galaxy, which is acting as the lens. The lensed quasar is about 8 billion L.Y away, while the galaxy lies 500 million L.Y from us. This object is sometimes also known as the Huchra Lens after its discoverer, the American astronomer John Peter Huchra.Similar objects subsequently discovered are now also termed Einstein crosses.

Einstein-de Sitter universe A type of universe in which the mean density of matter is precisely matched to the critical density. Such a model will not actually collapse, but will expand for ever with a continually decreasing expansion rate. This model lies on the dividing line between a closed Friedmann universe and an open Friedmann universe

Einstein ring A circular image of a distant source produced by a gravitational lens. Theoretically, such an image can be produced when a point mass lies exactly on the line of sight to a distant galaxy or quasar. In practice, nearcircular arcs have been detected in images of rich clusters of galaxies, and the radio emission from the quasar MG1654+1348 appears to have been lensed in this way.

Einstein, Albert German-Swiss-American theoretical physicist. His theories of relativity helped to shape 20th-century science and had profound implications for astronomy. The special theory of relativity (published in 1905) arose out of the failure to detect the ether, and built on the work of the Dutch physicist Hendrik Antoon Lorentz and the Irish physicist George Francis fitzgerald yields the relation $E = mc^2$ between mass and energy, which was the key to understanding energy generation in stars. The general theory of relativity (announced 1915, published in expanded form 1916), which encompasses gravitation, assumes great importance in very large-scale systems, and rapidly had an impact on cosmology. Astronomy has furnished observational evidence to support these theories. Einstein produced no subsequent work of great significance, searching unsuccessfully for a theory that would link relativity with electromagnetic forces.

Electromagnetic radiation Energy arising from the acceleration of electrically charged entities (e.g. electrons). Electromagnetic radiation can be considered to be composed of waves or particles, since it displays properties of both; this is referred to as the wave-particle duality. Electromagnetic waves are composed of oscillating electric and magnetic fields which lie at right angles to each other and to the direction of travel. They propagate through a vacuum at the speed of light, c; the speed is slower when travelling through a medium such as air, water, or glass.

Electromagnetic spectrum The complete range of electromagnetic radiation. From longest to shortest wavelengths, the electromagnetic spectrum consists of radio waves (10^5-10^{-3} m), infrared waves (10^{-3}-10^{-6} m), visible light ($4\text{-}7 \times 10^{-7}$ m), ultraviolet waves (10^{-7}-10^{-9} m), X-rays (10^{-9}-10^{-11} m), and gamma rays (10^{-11}-10^{-14} m). The speed of electromagnetic radiation, c, is constant in a given medium.Since c is equal to wavelength multiplied by frequency (expressed in appropriate units), for c to remain constant the frequency must become greater as the wavelength becomes shorter. For example, radio waves have long wavelength and low frequency, but gamma rays have high frequency and short wavelength. So both wavelength and frequency can take on a wide range of values. The visible region of the electromagnetic spectrum (i.e. light), detectable by the human eye, is subdivided into red, orange, yellow, green, blue, indigo, and violet.

Element An element is a chemically pure substance composed of atoms of a single type. The lightest element is hydrogen, which makes up almost 80% of the universe. It is followed by helium, which makes up a bit under 20% of the universe. Other common elements are oxygen, nitrogen, silicon, iron, and carbon. Hydrogen (deuterium) helium and some lithium were created by nucleosynthesis just after the Big Bang. The next heaviest elements (like carbon, nirogen, and oxygen) are formed inside stars via fusion. Most stars fuse hydrogen, forming helium. The heavier elements (like iron, copper and gold) are created as massive stars undergo supernova. Every atom around us and in our bodies was made in the big bang or in a star! A few man-made elements, like seaborgium, have been created in laboratories.

Elements, orbital Six quantities that describe the size, shape, and orientation of an orbit. They can be used to calculate the position of a body in its orbit at any given time. For planets orbiting the Sun, the ecliptic and the vernal equinox are used as a reference plane and reference direction respectively.The orbital elements are then: the longitude of the ascending node, O; the inclination, i; the argument of perihelion; the semimajor axis, a; the eccentricity, e; and a final number that gives the position of the planet in the orbit at a given time.This can be the time of perihelion passage, T or the longitude at the epoch, L. For a comet with a highly eccentric orbit, the semimajor axis is usually replaced by the perihelion distance, q.

If the orbiting body is the Moon, the orbital elements are modified; the argument of perihelion is replaced by argument of perigee, and time of perihelion passage by time of perigee passage.

Ellerman bomb A small, very bright point in solar active regions, visible in filtergrams tuned to the line wings of the Ha line. Ellerman bombs occur at the footpoints of arch filaments, often in the early stages of the development of major active regions. An alternative name is moustache, from the appearance of the Ha line, which has emission wings extending either side of a dark core. It is named after the American astronomer Ferdinand Ellerman.

Ellipse A closed curve like a flattened circle. The orbits of most celestial bodies are ellipses; the body being orbited lies at one focus of the ellipse. The longest diameter of an ellipse is termed its major axis, and the shortest diameter is the minor axis. Half the major axis is the semimajor axis, usually denoted by the letter a; the semiminor axis is usually denoted by b. The distance between the two foci of an ellipse divided by the length of the major axis is the ellipse's eccentricity, e. The eccentricity defines the shape of the ellipse, the ellipse being a circle when $e = 0$. When e is almost 1, the ellipse is long and narrow. For example, Halley's Comet follows an orbit for which $e = 0.967$.

Ellipsoid A body or surface in which every cross-section passing through the centre is a circle or ellipse. The Moon's shape conforms closely to an ellipsoid with three unequal axes, the shortest being the rotation axis and the longest aligned towards the Earth. Strictly speaking, the Earth is also ellipsoidal in shape since the equator is slightly elliptical, although the difference between its longest and shortest axes is less than 1 km.

Ellipsoidal variable A type of small-amplitude variable, consisting of a close, but not interacting, binary system; abbr. ELL. The two components are tidally distorted into ellipsoidal shapes. The inclination of the orbital plane to the line of sight is such that eclipses do not occur, but the orbital motion causes fluctuations in the visible area of the components and thus in the apparent magnitude of the system.

Elliptical galaxy A type of galaxy with a smooth, featureless circular or elliptical appearance, no spiral arms, and little or no interstellar gas or dust; symbol E. Elliptical galaxies range from about 107 solar masses and a few thousand light years in diameter for dwarf ellipticals to over 10^{12} solar masses and over 100 000 L.Y in diameter for giant ellipticals. Elliptical galaxies are classified from E0 to E7 according to their apparent shape, E0 appearing circular from our viewpoint and E7 the most elliptical. The degree of ellipticity is calculated from the ratios of the major (a) and minor (b) axes, using the formula 10(a - b)/a. Dwarf ellipticals are given the designation dE, while giant and supergiant ellipticals are known as D galaxies and cD galaxies respectively; such massive examples are usually found at the centres of clusters.

The stars in elliptical galaxies are mostly old (Population II), although some ellipticals also contain intermediate-age stars that formed more recently.

Elongation 1. The angle between the Sun and a body in the Solar System, as seen from Earth. Elongations are measured from 0° to 180° east or west of the Sun, in the plane passing through the Sun, Earth, and body. When a body is in conjunction with the Sun, its elongation is 0°; at opposition its elongation is 180°; and when at quadrature its elongation is 90° east or west. Greatest elongation is the maximum angular separation of Mercury or Venus from the Sun, either west or east. 2. The angle between a planet and a satellite, as seen from Earth. Elongations are measured from 0° east or west of the planet, in the plane passing through the planet, Earth, and satellite.

Emission The release of a photon from an atom when an electron in the atom jumps from a higher to a lower energy level. Emission from an astronomical object indicates that there must be a source of energy within it. A hot gas radiates because the motion of the atoms causes collisions which excite electrons in the atoms to higher energy levels. When the electrons return to lower energy levels, they emit the energy difference in the form of photons of specific wavelengths, giving rise to emission lines. If the electron is free and recombines with an atom, then the emission can occur at a continuous range of wavelengths, producing a continuous spectrum.

Emission line A bright line in a spectrum, marking a particular wavelength of radiation given out by hot or excited atoms. Emission lines can appear superimposed on a normal absorption spectrum, as caused by hot gas around a star, or they can appear on their own, as in the spectrum of a nebula excited by radiation from a nearby hot star. The lines allow the composition of the emitting gas to be determined.

Emission nebula A type of nebula that shines by emitting light when electrons recombine

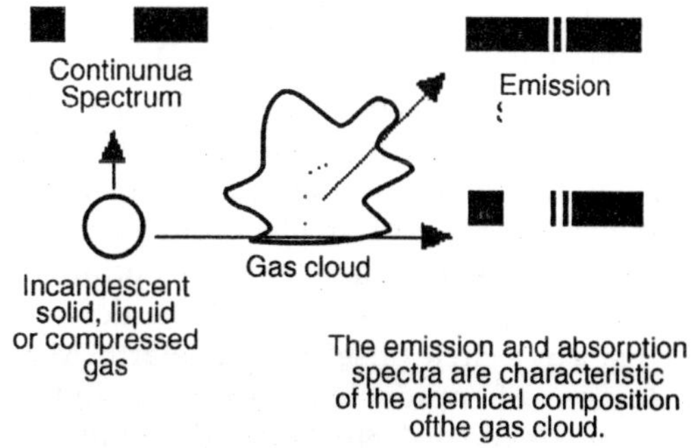

Fig. Emission spectrum

with protons to form hydrogen atoms. The electron frequently approaches the proton in steps emitting energy as light as it gets pulled in. In one of the most common "steps," the recombining electron emits a photon of red light. Since many atoms in the nebula do this all at once, the nebula appears red in color. This type of nebula is created when energetic ultraviolet light from a hot star shines on a cloud of hydrogen gas, stripping away electrons from the atoms (ionization). The free electrons can then begin the process of recombination. Plural of nebula: nebulae.

Enceladus The eighth-closest satellite of Saturn, distance 238 020 km, orbital period 1.370 days; also known as Saturn II. Its axial rotation period is the same as its orbital period. Enceladus was discovered in 1789 by F. W. Herschel. It is 500 km in diameter. Impact craters cover part of its surface, but over half the surface consists of smooth plains with almost no craters, suggesting that Enceladus may still be geologically active. These plains are separated from the cratered areas by long sinuous ridges. There are also grooves, folds, faults, and other signs of large-scale fracturing on the surface. Enceladus has an albedo of nearly 1.0, which means that it reflects nearly all the sunlight that falls on it. Hence it can absorb very little heat, and the surface temperature is -201 °C. Enceladus lies within Saturn's diffuse E Ring, and may be the source of the ring's particles.

Epicycle Epicycles are circular orbits within orbits that were used to (incorrectly) describe the orbits of objects in the Ptolemaic system (about A.D. 150). In Ptolemy's model of the solar system, an orbiting planet (or moon) moved in a series of circular orbits (epicycles) and the center of these epicycles orbited in another circular orbit (called the deferent) that was a circle offset from the Earth. Many layers of epicycles were needed to approximate real (elliptical) orbits with their retrograde motion. Copernicus also used epicycles in his heliocentric (sun-centered) solar system model (in the mid 1500's). Epicycles were used to predict orbits until Kepler discovered the elliptical nature of orbits (early in the 1600's).

Epoch The coordinates commonly used for the celestial sphere, which are analogous to latitude and longitude for the Earth's surface, are called right ascension and declination. The "prime meridian" of this system passes through the position of the Sun at the time of the vernal equinox. Thus its position changes slowly over the years, due to the precession of the equinoxes. The position of the celestial poles also changes with precession. Thus, to locate an object from its right ascension and declination, you must also know the date for which those coordinates are valid; that date is called the epoch of the coordinates.

Equator The intersection of the surface of a body with a plane perpendicular to the body's axis of rotation. The plane of the equator passes through the body's centre and divides the body into northern and southern hemispheres. All points on the equator are equidistant from the body's poles of rotation.

Equator, celestial The projection of the Earth's equator onto the celestial sphere.

Equinox Either of the two points at which the Sun's apparent yearly path intersects the celestial equator; or the dates on which this occurs—on or around March 21 and September 23. These points are also known as the first point of Aries and the first point of Libra. When the term 'equinox' is used without qualification, the vernal equinox is meant. Around the time of the equinoxes, night and day are equal in length the world over. The equinox is not a fixed point, but moves because of precession and nutation. If only precession is taken into account, the resulting point is known as the mean equinox of date. If, however, nutation is also included, the point is known as the true equinox.

Eratosthenes Eratosthenes (276-194 BC) was a Greek scholar who was the first person to determine the circumference of the Earth. He compared the midsummer's noon shadow in deep wells in Syene (now Aswan on the Nile in Egypt) and Alexandria. He properly assumed that the Sun's rays are virtually parallel (since the Sun is so far away). Knowing the distance between the two locations, he calculated the circumference of the Earth to be 250,000 stadia. Exactly how long a stadia is is unknown, so his accuracy is uncertain, but he was very close. He also accurately measured the tilt of the Earth's axis and the distance to the sun and moon, and devised a method for determining the prime numbers up to a given number (the Sieve of Eratosthenes). Eratosthenes made numerous contributions to the sciences and arts in many fields, including geography, mathematics, astronomy, chronography (calendars), music, and literature.

Ergosphere The region just outside the event horizon of a rotating black hole (i.e. a Kerr black hole), within which an observer is forced to rotate with the black hole, and cannot remain stationary with respect to the rest of the Universe. The outer boundary of the ergosphere is the static limit. In a non-rotating Schwarzschild black hole, the event horizon and stationary limit coincide. The ergosphere derives its name from the fact that energy can be extracted from the black hole in this region via the Penrose process.

Eridanus Eridanus (the River) is a southern constellation that is may have been named for the river Nile, the river Euphrates, or for the river of tears wept by the mythical Heliades. Eridanus is located near the constellations Fornax and Lepus. The brightest star in Eridanus, alpha Eri, is Achernar (meaning "end of the river"); it is the 9th brightest star in the sky (magnitude 0.46). The second-brightest, beta Eri, is Cursa (meaning "chair/footstool of Orion"). The third-brightest, gamma Eri, is Zaurak (meaning "boat"). The fourth-brightest, delta Eri, is Rana (meaning "frog"). The fifth-brightest, zeta Eri, is Zibal (part of the "Ostrich's Nest"). Eridanus was one of the original 48 constellations first noted by Ptolemy.

Eros Asteroid 433, discovered in 1898 by the German astronomer Gustav Witt and independently on the same day by the Frenchman Auguste Charlois.Eros was the first known asteroid with an orbit that brings it within the orbit of Mars; it is a member of the Amor group. Radar observations have shown that Eros is highly elongated with dimensions of 36 × 15 × 13 km. It is of S class, with a rotation period of 5.27 hours. Its orbit has a semimajor axis of 1.458 AU, period 1.76 years, perihelion 1.13 AU, aphelion 1.78 AU, and inclination 10°.8.

Error In colloquial usage, 'a mistake'. In technical usage error is a synonym for the experimental uncertainty in a measurement or result.

Escape velocity The minimum velocity required to escape from a gravitational field. The escaping object can be anything from a gas molecule to a spacecraft. A body's escape velocity is given by v(2GM/R), where G is the gravitational constant, M is the mass of the body, and R is the distance of the escaping object from the body's centre. An object moving at less than escape velocity enters an elliptical orbit; at exactly escape velocity it follows a parabolic trajectory; and if the object exceeds the escape velocity it moves away on a hyperbolic trajectory.

Eta carinae nebula The bright diffuse nebula NGC 3373 in Carina, about 2° across, containing numerous stars and clusters. It is divided into two parts by a dark V-shaped lane. At its heart is the variable star Eta Carinae, which is concealed by the smaller Homunculus Nebula. The Eta Carinae Nebula contains the Keyhole Nebula, a smaller dark nebula silhouetted against the brighter background near Eta Carinae. Its estimated distance is about 8000 L.Y

Ether A hypothetical medium once thought to permeate all space, through which electromagnetic radiation supposedly travelled; formerly spelt aether. On the basis of this supposition, the Earth should move with respect to the ether, and it was predicted that the speed of light would vary when measured in different directions. Experiments in the 19th century (e.g. the Michelson-Morley experiment) failed to detect any such variation in speed. The ether is now regarded as unnecessary, since it is recognized that electromagnetic radiation can propagate through empty space.

Eucrite A class of calcium-rich achondrite meteorite; also known as pigeonite-plagioclase achondrites. Two subclasses are recognized: the non-cumulate euctites, and the less-common cumulate eucrites. In their mineralogy, texture, and composition they closely resemble terrestrial and lunar basalts. The non-cumulate eucrites are essentially basalts that originated as surface lava flows, while the cumulate eucrites are intrusive rocks formed at shallow depth. Their parent body may have been the asteroid Vesta. Their name comes from the Greek eukritos, meaning 'easily discerned'.

Eudoxus of Cnidus Greek mathematician and astronomer, born in modern Turkey. He developed a model of planetary motion in which the Sun, Moon, and planets were carried around the Earth on a series of 27 Earthcentred spheres, with axes at different angles and rotating at different speeds. This theory of heavenly spheres, as modified first by Callippus and ultimately by Ptolemy, remained the orthodox view of the Universe for two thousand years. Eudoxus is also reputed to have introduced the constellation system from Egypt.

Eunomia family A small family of asteroids in the central region of the main asteroid belt, at a mean distance of 2.6-2.7 AU and with a mean orbital inclination of about 12°. The largest member is the S-class (15) Eunomia, diameter 260 km, while the two next-largest members, (85) Io and (141) Lumen, are of C class. Eunomia was discovered in 1851 by the Italian astronomer Annibale de Gasparis. Its orbit has a semimajor axis of 2.644 AU, period 4.30 years, perihelion 2.15 AU, aphelion 3.13 AU, and inclination 11°.8.

Europa The fourth-largest satellite of Jupiter and the second of the four Galilean satellites from the planet; also known as Jupiter II. Europa orbits in 3.551 days at a distance of 671 000 km. Its period of axial rotation is the same as its orbital period. Its diameter is 3138 km, making it only slightly smaller than our Moon. Europa's density of 2.97 g/cm^3 indicates that it is mainly composed of silicate rock, mixed with at least 5% water. It has a bright, icy surface of albedo 0.64, dominated by networks of dark, linear cracks, some over 1000 km long. Fewer than a dozen impact craters have been identified on Europa, indicating that the surface is very young and still geologically active.

European space agency An organization of European countries for cooperation in space research and technology, with its headquarters in Paris. ESA establishments include the European Space Research and Technology Centre at Noordwijk, Netherlands; the European Space Operations Centre Darmstadt, Germany, which controls European space missions; ESRIN, the centre for data from remote-sensing satellites at Frascati, Italy; and the European Astronaut Centre at Cologne, Germany. The ESA launch site is at Kourou, French Guiana, on the Atlantic coast of South America. ESA came into being in 1975 as successor to the European Space Research Organization and the European Launcher

Development Organization.Its member nations are: Austria, Belgium, Denmark, Finland, France, Germany, Ireland, Italy, the Netherlands, Norway, Spain, Sweden, Switzerland, and the United Kingdom. Canada is a Cooperating State.

Event horizon The surface of a black hole. For a non-rotating black hole, it is a spherical boundary at the black hole's Schwarzschild radius where the escape velocity becomes equal to the speed of light, so no events occurring within it can be seen from outside. However, the effects of the black hole's powerful gravitational field can still be felt outside the event horizon. For a rotating black hole, the event horizon is elliptical.

Evershed effect An outward flow of gas from the centre of a sunspot, proceeding from the umbra through the penumbra, and sometimes a little beyond it. The effect can be observed spectroscopically, particularly for spots near the solar limb, but may also be visible as outward-moving facular points, forming a so-called moat. The maximum outflow velocity is about 2 km/s. There is an inverse Evershed flow at higher altitudes, in which the flow (up to 20 km/s) is directed both inwards and downwards towards the sunspot. The effect is named after its discoverer, the English astronomer John Evershed.

Exobiology The scientific study of the possibility of life elsewhere in the Universe; also known as astrobiology. Exobiology embraces branches of biology (for example microbiology, biochemistry, and ecology) in addition to astronomy. Considerations include assessment of the likely conditions required for the origin and development of life, and the means of detecting such life. Investigations such as the direct sampling of Martian soil by the Viking landers in 1976, the study of organic molecules in meteorites, and the various SM programmes based on radio astronomy can be considered part of exobiology.

Exoplanet An exoplanet (short for extrasolar planet) is a planet that orbits a star other than the Sun. These exoplanets are detected by observing their star's "wobble" which the exoplanet's gravitational attraction causes.

Exosphere The outermost part of the Earth's atmosphere, lying at an altitude above 500 km, and petering out at greater heights into interplanetary space. At such levels the atmospheric particle densities are extremely low; the density at 1000 km altitude is estimated to be only 7.3×10^5 particles/ cm^3.Under such conditions, collisions between particles are rare.

Explorer An ongoing series of US scientific spacecraft. Explorer 1, launched on 1958 January 31, was the first successful US satellite; it discovered the Earth's Van Allen Belts. More recent Explorers have included the International Ultraviolet Explorer and the Cosmic Background Explorer. A concurrent Small Explorer Program consisting of simpler, cheaper satellites, began in 1992 with the Solar, Anomalous, and Magnetospheric Particle Explorer

Explosive eruption A dramatic volcanic eruption which throws debris high into the air for hundreds of miles; lava is low in silicate; can be very dangerous for people near by; an example is Mount St. Helens in 1980.

Extreme population I star A star belonging to the youngest stellar population, such as a T Tauri star, a star newly arrived on the zero-age main sequence, or a massive OB star with its associated H II region. Such stars have high metal abundances.They are found in localized regions of the galactic disk, notably the spiral arms, where star formation has occurred very recently.

Extreme ultraviolet A region of the spectrum between the ultraviolet and X-ray regions, covering the wavelengths 10-100 nm. The first EUV sources were detected by a telescope on the Apollo-Soyuz mission in 1975. The first survey of the EUV sky was carried out by Rosat in 1990/91. The Extreme Ultraviolet Explorer continued astronomy in this waveband.

Extreme ultraviolet explorer A NASA satellite to study the sky at extreme ultraviolet wavelengths, launched in 1992 June. First it made a six-month survey of the sky using three grazing-incidence telescopes, covering the wavelengths 6-70 nm. A fourth telescope pointing at right angles to the others surveyed a limited area of sky with deeper exposure. This telescope included a spectrometer that was used as the primary instrument after the initial sky survey.

Extremely low frequency (ELF) That portion of the radio frequency spectrum from 30 to 3000 hertz.

Extrinsic variable A variable star whose variations arise for some external reason, such as rotation, orbital motion, or obscuration, as distinct from intrinsic processes such as pulsation.

Extrusive Refers to igneous rocks erupted on the surface.

Eyepiece A lens or combination of lenses used to magnify the image formed by a telescope; also known as an ocular. The simplest form of eyepiece is a single converging lens of short focal length, but this has severe aberrations except at the very centre of its field of view. In practice, therefore, eyepieces usually have at least two elements. A field

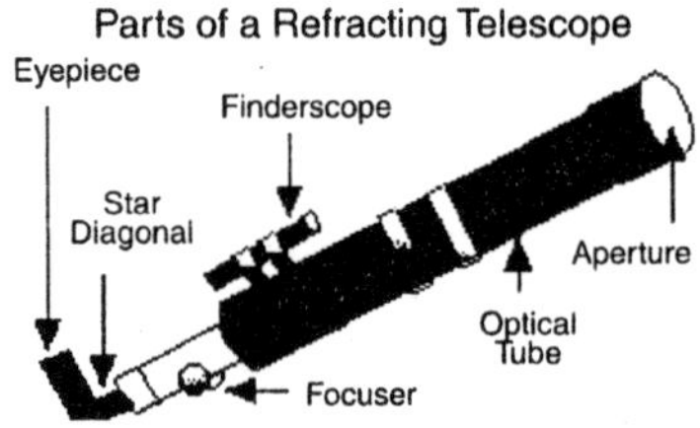

Fig. Eyepiece

lens, facing the objective or mirror, gathers light over a wider field than a single lens, while the eye lens, through which the observer looks, provides the magnification. This combination gives a good field of view while keeping aberrations under control. A field stop (a diaphragm located so that it is in focus as seen through the eye lens) provides a hard edge to the field of view. Many additional elements can be included to improve performance. Popular designs include the Erfle, Huygenian, Kellner, monocentric, Nagler, orthoscopic, Plössl, and Ramsden eyepieces.

F

F band A Fraunhofer line in the Sun's spectrum at 486.1 nm, due to absorption by hydrogen. It is also known as the Hß line in the Balmer series.

F corona The outer part of the Sun's corona, which is illuminated by sunlight scattered or reflected from solid dust particles. The same phenomenon also produces the zodiacal light much farther from the Sun. The dust grains are a few micrometres across, and are located in a disk extending outwards from about 1 solar radius (750000 km) from the Sun's surface. Unlike the electrons responsible for the K corona, the dust grains are relatively slow-moving. Thus the light scattered from them has the same spectrum as the photosphere, including its Fraunhofer lines.The F corona is the brightest part of the corona beyond 1.5 solar radii from the Sun's surface.

F star A star of spectral type F, slightly hotter and more massive than the Sun, and white in colour. On the main sequence, F-type stars have temperatures of 6100-7200 K, while supergiants are a few hundred degrees cooler. Masses are 1.2-1.6 solar masses, and luminosities 2-6.5 times the Sun's on the main sequence, but F-type supergiants have masses up to 12 solar masses and luminosities up to 32000 times the Sun's. The hydrogen Balmer lines weaken dramatically from spectral type F0 to F9, while the calcium H and K lines increase in strength and many other metal lines start to become visible. Polaris and Procyon are F-type stars.

Face on mars This photograph of the Cydonia Mense region of Mars was taken by NASA's Mars Global Surveyor in 1998. It is a coincidental alignment of rocks and other geologic formations that happens to look like a human face from this angle.

Facula 1. A brighter and hotter patch on the Sun's photosphere, visible in white light and best seen near the solar limb against the background of limb darkening. Faculae often appear shortly before a sunspot group forms, and remain visible for several days or weeks after the spots have vanished. High-latitude faculae also occur, well away from sunspots; these, unlike sunspot faculae, are most numerous on the rising part of the sunspot cycle. Faculae are slightly hotter, and coincide with bright patches in the chromosphere and the chromospheric network. Normally they can be resolved into small facular bright points, around 150 km wide, lasting 20 min or so. There is often a connection between faculae and structure within large sunspots, in particular light-bridges, which are ridges of bright material crossing a sunspot.

2. A bright spot on a planetary surface; pl. faculae.

Fall A meteorite that has actually been seen to fall and has been recovered at the place of impact. In many cases the meteorite is recovered immediately or very soon after falling, and is therefore relatively free of terrestrial contamination and weathering. Sometimes, however, the meteorite is not recovered until months or years after its observed fall. About 40% of known meteorites were seen to fall.

Falling star A falling star is not a star at all; it is a meteor (made of rock and/or iron) which is burning up in the Earth's atmosphere.

Family of constellations There are many families of constellations, constellations that are either close to one another or have some other relationship. Some constellation families include the Zodiac (the 12 constellations representing the signs of the Zodiac), the Ursa Major Family (10 constellations, including Ursa Major, Ursa Minor Canes Venatici, Bootes, Coma Berenice, Corona Borealis, etc.), the Perseus Family (9 constellations, including Cassiopeia, Cepheus, Andromeda, Perseus, etc.), the Hercules Family (19 constellations, including Hercules, Sagitta, Aquila, etc.), the Orion Family (5 constellations, including Orion, Canis Major, Canis Minor, Monoceros, and Lepus), the Heavenly Waters (9 constellations, including Delphinus, Columba, Equuleus, Vela, Puppis, etc.), the Bayer Group (11 constellations, including Hydrus, Dorado, Volans, Apus,etc.), the La Caille Family (13 constellations, including Norma, Fornax, Circinus, etc.).

Fanaroff-Riley class A category in a classification of extragalactic radio sources, typically radio galaxies or quasars, based on the separation between the brightest regions in their radio-emitting lobes. In Class I sources, the bright regions are separated by less than half the overall extent of the radio lobes. Hence these regions are closer to the nucleus than in Class II sources, where they are separated by more than half the total length of the source. Class II sources are in general considerably more powerful than Class I. The scheme is named after the South African astronomer Bernard Lewis Fanaroff and the British astronomer Julia Margaret Riley.

Fast nova A type of nova that declines by 3 magnitudes from maximum in 100 days or less; abbr. NA. An alternative classification is based on the time taken to decline by 2 magnitudes from maximum: very fast, less than 10 days; fast, 11-25 days; moderately fast, 26-80 days; slow, 81-150 days; very slow, 151-250 days.

Fermi acceleration In order to explain the origins of cosmic rays, Enrico Fermi (1949) introduced a mechanism of particle acceleration, whereby charged particles bounce off moving interstellar magnetic fields and either gain or lose energy, depending on whether the "magnetic mirror" is approaching or receding. In a typical environment, he argued, the probability of a head-on collision is greater than a head-tail collision, so particles would be accelerated on average. This random process is now called 2nd order Fermi acceleration, because the mean energy gain per "bounce" is dependent on the "mirror" velocity squared.

Bell (1978) and Blandford and Ostriker (1978) independently showed that Fermi acceleration by supernova remnant (SNR) shocks is particularly efficient, because the motions are not random. A charged particle ahead of the shock front can pass through the shock and then be scattered by magnetic inhomogeneities behind the shock. The particle gains energy from this "bounce" and flies back across the shock, where it can be scattered by magnetic inhomogeneities ahead of the shock. This enables the particle to bounce back and forth again and again, gaining energy each time. This process is now called 1st order Fermi acceleration, because the mean energy gain is dependent on the shock velocity only to the first power.

Fermion A type of elementary particle such as an electron, neutron, or proton having a +½ or -½ value of spin. A maximum of two electrons can therefore occupy the ground state, provided they have opposite spins. Hence elements other than hydrogen and helium must have some of their electrons in higher levels than the ground state. Fermions obey the Pauli exclusion principle. They are named after the Italian-American physicist Enrico Fermi.

Fibrille A feature in the low chromosphere of the Sun; also spelt fibril. Fibrilles are fine dark lines arranged in a near-spiral pattern around sunspots as seen in Ha light. They are particularly prominent around large, mature spots, from which they extend outwards for about 10 000 km, forming a superpenumbra. The whole pattern lasts as long as the active region containing the sunspots, but individual fibrilles live for only 20 min or so. Fibrilles are thought to delineate the magnetic field in the chromosphere around sunspots.

Filament 1. Along 'tongue' of relatively cool material (10 000 K) suspended in the much hotter solar corona (2 million 10.). Filaments appear dark when seen silhouetted against the Sun's disk in Ha light, but at the limb they appear as prominences. Quiescent filaments may show gradual changes, but portions of the filament may move more quickly, at speeds of a few kilometres per second. Loop filaments are sometimes seen near very large flares. The equivalent of an eruptive prominence is the disappearing filament, sometimes called a disparition brusque, from the French meaning 'sudden disappearance'. Disappearing or winking filaments may occur as a result of a Moreton wave.

2. A chain-like cluster or supercluster of galaxies. The large- scale structure of galaxy distribution in the Universe is dominated by such features, which may be tens of millions of light years long.

Filament channel a broad pattern of fibrils in the chromosphere, marking where a filament may soon form or where a filament recently disappeared.

Filamentary nebula A group of elongated clouds of gas and dust with a fine thread-like structure as seen from Earth. Many filamentary structures may actually be sheets or parts of shells seen edge-on, rather than threads. The best-known filamentary nebulae, such as the Veil Nebula, are supernova remnants. Although these filaments have temperatures of 10 000 K they are actually the coolest parts of the remnant, other parts of which remain at over 1 million K.

Filar micrometer A device for measuring the separations of two objects as seen through a telescope, such as the components of a double star. It consists of two fixed fine wires or webs in the field of view of an eyepiece, crossing at right angles, plus a third movable web, parallel to one of the fixed webs. The position of the movable web can be measured precisely against a scale. Many filar micrometers can also be rotated against an angular scale; such a device is known as a position micrometer. Both separation and position angle can then be measured.

Filigree A very fine structure on the Sun's photosphere, consisting of tiny bright dots arranged along the dark lanes separating granules. Sometimes there is a clustering of such dots to form crinkles. The smallest dots are about 150 km across and last for less than 30 min. The filigree structures are hotter than the surrounding photosphere in which they are located, and mark sites of strong magnetic field. They are visible in spectroheliograms made in weak Fraunhofer lines or the line wings of the Ha line.

Fireball A meteor whose apparent magnitude exceeds that of the planet Venus.Fireballs are comparatively rare, perhaps fewer than one per thousand meteors. During major annual meteor showers such as the Perseids or Geminids considerable numbers of fireballs may be seen. Fireballs are also associated with the atmospheric entry of large incoming bodies, which may result in meteorite falls. An exploding fireball is termed a bolide.

First animal and first dog in space Laika (meaning "barker" in Russian) was the first dog launched into Earth orbit (other dogs were launched earlier on sub-obital flights). She was part Siberian husky and was launched in Sputnik 2 on November 3, 1957. Laika survived in Earth orbit for four days. She died in space when the batteries to the cabin over-heated. Sputnik 2 fell back to Earth on April 14, 1958, burning up during re-entry.

First contact At a solar eclipse, the moment at which the leading limb of the Moon just touches the Sun's western limb, marking the beginning of the eclipse; or, at a lunar eclipse, the moment at which the Moon's eastern limb first enters the Earth's shadow.

First light First light is a term that refers to the earliest light emitted when the first galaxies were born. The first light occured long after the Big Bang happened. The Big Bang occured roughly 15 billion years ago, but the Universe was dark until the first light came millions of years after the Big Bang, after the smooth distribution of matter following the Big Bang coalesced into clumps, eventually forming galaxies.

First monkey in space The first monkey in sub-orbital space was a squirrel monkey called Gordo. Gordo was launched into space on December 13, 1958, in the nose cone of the US Army rocket Jupiter AM-13. Gordo did well in the flight, but drowned in the Atlantic Ocean when the flotation device on the nose cone failed and it sank. The next two monkeys in space, Able and Baker, were launched on May 28, 1959 (Jupiter AM-18). They were successfully retrieved after a 300 mile sub-orbital flight.

Flare, solar A sudden release of energy in the Sun's corona, lasting up to several hours or, exceptionally, more than a day. Flares emit radiation over the whole spectrum, from gamma rays to radio waves. They also throw out high-speed particles at speeds up to about 70% of the speed of light, which reach the Earth in 15 min or so, depending on their trajectory. Only the most energetic flares are visible in white light. Flares occur in active regions with complex magnetic fields, the largest flares being in the most complex regions. Most of the energy may be released in the first few minutes, with an impulsive stage that may last only a few seconds. The total energy released can be up to 10^{27} joules; there is no well-defined lower energy limit.

Flares are classified in two ways. By their appearance in Ha light, and by their soft X-ray emission. In Ha the term subflare is given to the smallest events, and the scale then runs from 1 to 4 with increasing area; a brightness code is added from faint via normal to bright.In soft X-rays (0-1-0.8 nm) flares are classified as C, M, or X according to increasing strength, with subdivisions from 1 to 9.

In Ha, the flare may start with the disappearance of a filament (i.e. a prominence seen from above), bright areas developing into ribbons either side of the magnetic inversion line. There is generally a rapid expansion stage called the flash phase. There are often hard X-ray and microwave radio bursts forming an early impulsive stage, with soft X-rays rising more gradually to maximum a few minutes later, followed by a decline.Flares derive their energy from the energy stored in magnetic fields, although the exact mechanism is not known.

Flux (symbol F) A measure of energy or number of particles passing through a given area of surface in unit time, usually per second. Luminous flux is the rate of flow of energy in the form of photons, measured in lumens;

magnetic flux is a measure of the strength and extent of a magnetic field perpendicular to a surface, measured in webers; particle flux is the number of particles, for example in a stellar wind, passing through a unit area per second.

Focal length Focal length is the distance from the lens (or mirror) in a telescope to the point at which the object being observed is focused.

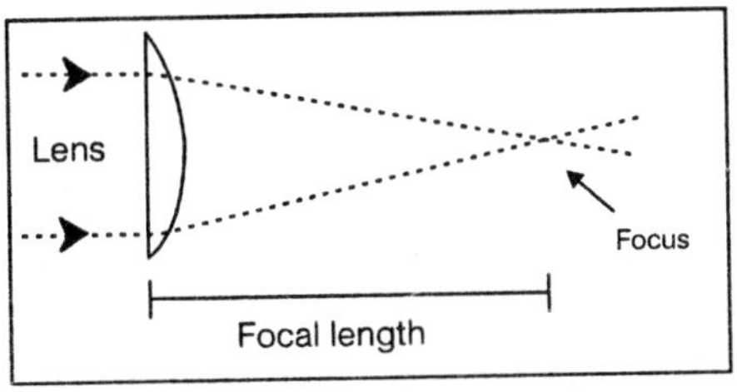

Fig. Focal length

Focal plane A flat surface at right angles to the optical axis of a lens or mirror which contains the image of a small, distant object and also the focal point. If an extended image lies not in a plane but on a curved surface because of curvature of field, it is said to be formed on the focal surface or Petzval surface.

Focus The point at which rays of light passing through a lens (or reflecting off a mirror) converge

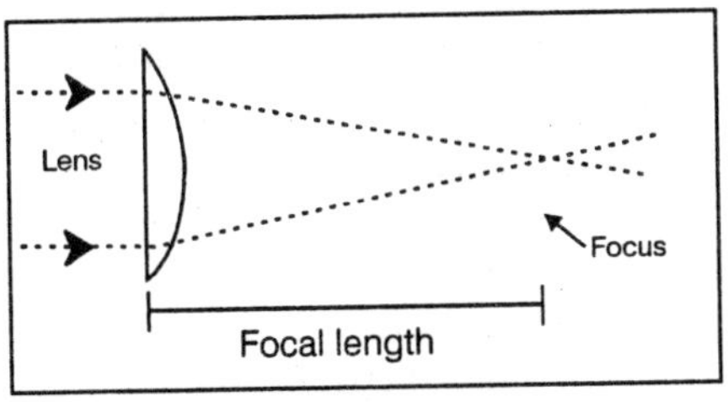

Fig. Focus

Focus, optical The point on the optical axis of an objective lens or mirror at which rays of light converge, and where an image is sharpest; also known as the focal point.

Focus, orbital One of the two points within an ellipse whose distance apart determines the eccentricity of the ellipse; pl. foci. The two foci lie on the major axis of the ellipse, at equal distances either side of its centre. In an elliptical orbit, one focus is occupied by the body being orbited; the other is referred to as the empty focus. If the orbit is a parabola or hyperbola, there is only one focus, occupied by the body being orbited.

Fomalhaut The star Alpha Piscis Austrini, magnitude 1.16. It is an A3 dwarf, 22 L.Y away.

Forbidden line An emission line in a spectrum that is emitted only by a lowdensity gas, as in a nebula and certain regions of active galactic nuclei. Such a line is said to be forbidden because it does not occur under normal conditions on Earth, where gases are denser. A forbidden line is produced when an electron jumps from an upper energy level, where it can remain for a long time, to a lower level; such a jump, or transition, is said to have a very low transition probability. On Earth, the excited atom would collide with other atoms and lose energy in the collision long before it could radiate the energy away. However, in the low densities of interstellar space and the regions around hot stars, collisions are extremely rare and there is time for the spontaneous decay to occur. Forbidden lines are denoted by square brackets, such as the [O III] lines of doubly ionized oxygen.

Fornax dwarf galaxy A dwarf spheroidal galaxy in the Local Group of galaxies, about 540000 L.Y away and nearly 6000 L.Y across. It has approximately 2.5×10^7 times the luminosity of the Sun and contains mostly old stars, but shows evidence of some younger stars about 3 billion years old. It was discovered in 1938 by H. Shapley.

Forward scattering Scattering in which photons emerge from the scattering medium travelling predominantly in the same direction as they entered. Forward scattering and backscattering always occur together, but the proportion of forward scattering increases with particle size. The haloes around the Sun and Moon in wet weather are caused by forward scattering by water droplets in the Earth's atmosphere.

Fossa A long, narrow, straight linear depression on a planetary surface; pl. fossae. The name, which means 'ditch' or 'trench', is not a geological term, but is used in the nomenclature of individual features, for example Erythraea Fossa on Mars, or Lakhmu Fossae on Ganymede.

Foucault pendulum A Foucault pendulum is a simple pendulum (a weight on a long string attached to a support) that tracks the rotation of the Earth. As the pendulum swings, the Earth rotates under the pendulum, so the pendulum seems to rotate. It was first demonstrated by Jean Bernard Léon Foucault, in 1851 at the Paris World's Fair.

Foucault test A method of determining the shape of a concave mirror; also known as the knife-edge test. A pinhole light source is placed at the mirror's radius of curvature, and a straight edge (e.g. a knife-edge) adjacent to the pinhole. The mirror is illuminated by light from the pinhole. If the mirror is spheroidal it will produce a sharp image of the pinhole at the knife-edge, and the tester will see the mirror's illumination abruptly cut off as the knife-edge is moved across the image. Any small irregularities in the figure show up clearly as humps or hollows. A paraboloid appears like a doughnut, deeper at the centre than at the edges. The test was devised by the French physicist Jean Bernard Léon.

Foucault, Jean Jean Bernard Léon Foucault (1819-1868) was a French physicist who was the first to demonstrate how a pendulum could track the rotation of the Earth (the Foucault pendulum) in 1851. He also invented the gyroscope (1852), showed that light travels more slowly in water than in air (1850), and improved the mirrors of reflecting telescopes (1858).

Fourier analysis A technique used to find the frequencies present in a complex signal; also known as frequency analysis. A series of measurements made at different times is broken down mathematically into the sum of simple oscillations of various frequencies. The lowest frequency present is called the fundamental frequency, and the higher frequencies are then harmonics of the fundamental frequency. The brightness fluctuations of some variable stars, for example, can be broken down into two or three simple sinusoidal variations by the use of Fourier analysis. The technique was invented by the French mathematician Joseph Fourier.

Fourier transform The mathematical breakdown of a non-repeating phenomenon into its component frequencies. It may be regarded as Fourier analysis applied to a function with an infinite period. The technique is widely used in image processing to reduce the effects of periodic and random errors. It is also needed to convert the outputs from aperture synthesis telescopes and Fourier transform spectrometers into images and spectra respectively.

Fourier transform spectrometer A spectrometer based on the principle of the Michelson interferometer, used for high-resolution infrared spectroscopy of bright stars. A Michelson interferometer splits the incoming starlight and sends it to two mirrors, one stationary, the other movable. When the light recombines, an interference pattern is produced. Since the incoming light consists of a range of wavelengths the interference pattern becomes very complicated as the mirror is moved. The input spectrum can, however, be recovered by taking the Fourier transform of the output signal. The Fourier transform can be thought of as breaking down the output signal into a sum of sine waves of different amplitudes. This technique is particularly useful in the infrared, where conventional spectroscopy is difficult.

Fourth contact At a solar eclipse, the moment at which the trailing limb of the Moon completely uncovers the Sun's eastern limb, marking the end of the eclipse; or, at a lunar eclipse, the moment at which the Moon's western limb exits the Earth's shadow; also known as last contact.

Fractional method A visual method of estimating the brightness of a variable star, using two comparison stars. The interval between the comparisons is mentally divided into a small number of steps. Estimates such as A V B or C V D result, where the figures in parentheses represent the number of brightness steps between the variable, V, and pairs of comparison stars, A, B and C, D. The actual magnitude may be calculated from the known magnitude of the comparison stars.

Fraunhofer lines Absorption lines and bands in the spectrum of the Sun that were first labelled by j. von Fraunhofer in 1814. In decreasing wavelength from the red, they are A, a, B, C, D, E, b, F, G, H, and K. The A and B bands are due to absorption by oxygen molecules in the Earth's atmosphere, and the a band is due to absorption by terrestrial water vapour, but the remainder result from absorption in the Sun's photosphere. Most prominent are the D lines of sodium, the H and K lines of calcium, and the G band caused by neutral iron and the CH molecule, all of which are features of stars of spectral types F, G, and K.

Fraunhofer, Joseph von German physicist and optician. In 1814 he made the first spectrometer, to measure the dispersive power of lenses, using a yellow flame as the light source. He compared the flame spectrum with a spectrum of the Sun produced by a prism, noting and recording the positions of the Sun's Fraunhofer lines. He later noted similar lines in the spectra of other stars. Fraunhofer also made achromatic lenses and the first diffraction grating, developed the equatorial mount, and constructed the 16-cm heliometer used by F. W. Bessel to measure the parallax of 61 Cygni.

Friedmann universe A model describing an expanding universe containing matter and radiation, but without a cosmological constant. Such a universe is both homogeneous and isotropic. There is, in fact, a family of such universes including those which expand for ever those that eventually collapse and the particular example of the Einstein-de Sitter universe which has a critical density of matter. The geometry of space-time in these universes is described by the Robertson-Walker metric and is, in the preceding examples, negatively curved, positively curved, and flat respectively.. The Friedmann models, originated by the Russian mathematician Alexander Alexandrovich Friedmann form the basis of the standard Big Bang theory.

Fringe 1. In optics generally, the successive light and dark patterns caused by the interference of waves that are in and out of phase at different positions, as in a Young's double slit experiment.

2. In radio astronomy, the oscillation in time of the output of a correlation-type interferometer before such oscillations are stopped.

Fringe pattern Another name for an interference pattern, used particularly in radio astronomy.

Fringe rotator In a correlating-type radio interferometer, a mechanism to introduce a time-varying phase shift into the local oscillator signal of a receiver, in order to reduce the frequency of the oscillations of the correlator output. Fringe rotation allows the correlator output (whose amplitude is proportional to visibility amplitude) to be sampled at a lower rate. The fringe rotation is chosen so that the fringe frequency for a point source located at the so-called fringe stopping center would be reduced to zero, or at least very close to zero. Usually the fringe stopping center and the delay tracking center coincide; both then are called the visibility phase tracking center.

Full moon A full moon appears as an entire circle in the sky; it is illuminated by the Sun and from the Earth, we see its entire daytime side. A full moon occurs once each lunar month, when the Moon is on the opposite side of the Earth from the Sun.

Fundamental star A star whose position and proper motion are known precisely enough to be included in a fundamental catalogue.

Fusion Nuclear fusion is an atomic reaction in which many nuclei (the centers of atoms) combine together to make a larger one (which is a different element). The result of this process is the release of a lot of energy (the resultant nucleus is smaller in mass than the sum of the ones that made it; the difference in mass is converted into energy by the equation $E=mc^2$). Suns are powered by nuclear fusion, mostly converting hydrogen into helium. In this process, 4 hydrogen nuclei combine to form a single helium nucleus; about 0.3% of the original mass is converted into energy.

Fusion crust The burned, often glassy surface layer of a meteorite. It is caused by intense heating during the meteorite's passage through the atmosphere and is usually black or blue-black in colour. Stony meteorites usually have a thicker fusion crust than do irons.

G

G band A broad Fraunhofer line in the Sun's spectrum at 430 nm, due to absorption by CH molecules and a closely spaced group of lines produced by neutral iron. The same feature is found in other stars, most notably of spectral types F to K.

G star A star of spectral type G, which includes the Sun. G-type stars on the main sequence have temperatures of 5300-6000 K, and therefore appear yellow. G-type giants are about 100-500 K cooler than main-sequence stars. G-type supergiants are about 4500-5500 K. The spectrum of the Sun is dominated by lines of singly ionized calcium and neutral metals. In cooler G stars the molecular bands of CH and CN become visible. Main-sequence G stars, and giants such as Capella, have masses of 0.8-1.1 solar masses, while supergiants are of 10-12 solar masses. The luminosities of G-type giants are about 30-60 times the Sun's, and for supergiants 10000-300000 times the Sun's.

Gagarin, Yuri Yuri Gagarin (1934-1968) was a Soviet cosmonaut and the first man to orbit the Earth. He piloted the Vostok 1 mission which launched April 12, 1961 and orbited the Earth. The flight lasted 108 minutes. The spacecraft was recovered later that day in the Saratov region of the Soviet Union. Gagarin died years later in a plane crash.

Gaia hypothesis Named for the Greek Earth goddess Gaea, holds that the Earth as a whole should be regarded as a living organism and that biological processes stabilize the environment. First advanced by British biologist James Lovelock in 1969.

Galactic centre The very central region or nucleus of our Galaxy, 26000 L.Y away in the constellation Sagittarius; coordinates RA 17h 46m, dec. -28° 56'. There is a clumpy ring of gas and dust about 5-15 L.Y from the exact centre. Inside this is a dense star cluster around the bright radio source Sagittarius A, which is believed to mark the very centre of the Galaxy and may contain a black hole of some millions of solar masses. Although highly obscured by dust from optical observation, the galactic centre has been extensively studied at infrared, radio, X-ray, and gamma-ray wavelengths.

Galactic co-ordinates A system of celestial co-ordinates based on the mean plane of the Galaxy, which is inclined about 63° to the celestial equator. Galactic latitude b is measured north (+) or south (-) from the Galactic equator, the great circle on the celestial sphere defined by the mean plane of the Galaxy. Galactic longitude l is measured eastward along the Galactic equator from the Galactic Center. Galactic co-ordinates are independent of precession.

Galactic disk The major structural component of spiral, lenticular, and some irregular galaxies, containing stars and gas and dust, orbiting the galaxy's centre. The thickness of the disk is small in relation to its diameter. In our own Galaxy the disk extends about 80000 L.Y from the galactic centre, while its total thickness is about 1500 L.Y as measured by the distribution of older stars, and only 600 L.Y for young stars, gas, and dust.

Galaxy A system of stars, often with interstellar gas and dust, bound together by gravity. Galaxies are the principal visible structures of the Universe. They range from dwarfs with less than one million stars to supergiants with over a million million stars, and in diameter from a few hundred to over 600000 L.Y Galaxies may be isolated, or in small groups such as our Local Group, or in large clusters such as the Virgo Cluster.

Galaxies are usually classified according to appearance.They come in two main shapes: spiral and elliptical Elliptical galaxies have a smooth, centrally concentrated distribution of stars, and very little interstellar gas or dust. Spiral galaxies are subdivided into ordinary spirals (S) and barred spirals (SB). Both types have disks containing interstellar material as well as stars. Lenticular galaxies (SO) show a clear disk, but with no spiral arms visible.

Irregular galaxies (Irr) have a rather amorphous, irregular structure, sometimes with evidence of a spiral arm or bar.

Galilean moons Jupiter's four largest moons: Io, Europa, Ganymede and Callisto; discovered independently by Galileo and Marius. (Galileo proposed that they be named the Medicean stars, in honor of his patron Cosimo II de Medici; the present names are due to Marius)

Galileo U. S. mission to study Jupiter's atmosphere, moons, and magnetosphere. The Galileo spacecraft was carried into space and released in 1989 by the shuttle Atlantis. It used its own rocket booster to leave Earth orbit. The Galileo spacecraft arrived at Jupiter in 1995 and completed its two-year primary mission before going on to explore Europa. The 14-year mission ended in 2003.

Galileo Galilei Italian astronomer, physicist, and mathematician. In 1609 he heard of the recent invention of the telescope, and began to make his own. He built several small refractors, up to 50 mm aperture, the highest magnification being about × 30. He published Siderius nuncius in 1610, in which he outlined his early telescopic discoveries, including mountains on the Moon, the four satellites of Jupiter and the innumerable stars of the Milky Way. Also in 1610 he observed the phases of Venus, and noted the unusual telescopic appearance of Saturn, although he did not recognize the rings' true nature. Galileo's discoveries added to his discontent with Aristotle's world-view, still widely believed, and he advocated the Copernican system, notably in his book Discourses and Mathematical Demonstrations on Two New Sciences. Conflict with the Church followed, and he was tried for heresy; he lived the rest of his life under house arrest.

Gamma-ray astronomy The study of electromagnetic radiation from space with the very shortest wavelengths and highest photon energies.. Gamma rays are produced in regions of extremely high temperature, density, and magnetic fields, sites of the most violent processes in the Universe.

Several dozen individual gamma-ray sources are known, plus the general gamma-ray background. Early experiments used balloons to carry instruments to altitudes where the atmospheric absorption of gamma rays is low. Exploratory observations were also made with spacecraft, including Ranger and Apollo missions, during the 1960s. The first sky surveys were made by the satellites SAS-2 and COS-B, launched in 1972 and 1974. In the late 1970s two High Energy Astrophysical Observatories carried gamma-ray experiments. The Granat satellite was launched in 1990, and the Compton Gamma Ray Observatory in 1991. Future missions planned include Spectrum X-gamma and Integral.

The large energy range necessitates several observational techniques.

Gamow, George Russian-American scientist born in Ukraine. He applied his early training in nuclear physics to astrophysics and cosmology. His version of the Big Bang theory was developed in the late 1940s, with the American physicists Ralph Asher Alpher and Robert Herman it proposed that the Universe underwent a rapid expansion from a hot, dense phase, full of radiation, building the nuclei of the elements from protons and neutrons. The theory predicted a Universe containing three-quarters hydrogen and one-quarter helium, roughly in accordance with present-day measurements. Alpher and Herman suggested that the present, cooled-down Universe should be pervaded by a cosmic background radiation with a temperature of about 5 K; their suggestion was forgotten until the radiation was detected in 1964.

Gas giants The gas giants are the large outer planets of our Solar System: Jupiter, Saturn, Uranus, and Neptune (but not tiny, rocky Pluto).

Gaspra Asteroid 951 Discovered in 1916 by the Russian astronomer Grigorii Nikolaevich Neujmin.It was the first asteroid to be seen in detail, when the Galileo probe flew within 1600 km of it in 1991. Gaspra, an S-class asteroid, is irregularly shaped with dimensions of 18.2 × 10.5 × 8.9 km, and is probably a fragment of a much larger body. As well as craters, Gaspra's surface has linear grooves similar to those on Mars's satellite Phobos. Gaspra's orbit has a semimajor axis of 2.210 AU, period 3.28 years, perihelion 1.83 All, aphelion 2.59 AU, and inclination 4°.1.

Gauss The gauss is a unit of magnetic induction (denoted B) in the cgs system (centimeter-gram-second). A changing magnetic field produces an electric current - this is called magnetic induction. The gauss was named for the German mathematician Johann Carl Friedrich Gauss (April 30, 1777-Feb. 23, 1855), who did work in magnetism.Magnetic flux is a measure of flux density (a Gauss is the magnetic flux per square centimeter).

Gauss, Carl Friedrich German mathematician. He devised a method for calculating the orbit of a body from just three observations, which allowed astronomers to recover the asteroid Ceres that had become lost behind the Sun after its discovery in 1801 by G. Piazzi. Gauss made a detailed study of celestial mechanics, developing a theory of perturbations which he applied to the determination of cometary and planetary orbits. He invented the method of least squares for use in correcting observational errors.

Gaussian gravitational constant (symbol k) A constant that appears in the precise form of Kepler's third law. Its value was calculated by C. F. Gauss, who found it to be 0.017 202 098 95. Subsequent measurements have given a slightly different value for k, but Gauss's value is still used when calculating the astronomical unit.

GBT (green bank telescope) The world's largest fully steerable single-dish radio telescope, under construction at the NRAO's facility in Green Bank, West Virginia. The GBT has a unique off-axis design intended to reduce sidelobe responses, and is due to be completed in 1997.

G-class asteroid A subclass of the C-class asteroids. Its members have low albedos (0.05-0.09), and are distinguished from class C by having a very strong ultraviolet absorption feature at wavelengths shorter than 0.4 μm caused by water of hydration in the surface material. The largest asteroid, Ceres, is now classified as G class. Other members include (176) Iduna, diameter 130 km, and (640) Brambilla, diameter 52 kin.

Gegenschein A faint, diffuse oval glow some 10 across, appearing in the midnight sky directly opposite the Sun; also known as the counterglow. The gegenschein is produced by the reflection of sunlight from small zodiacal dust particles, and has an extremely low surface brightness. It is visible only under the clearest and darkest sky conditions.

Geminga One of the brightest gamma-ray sources, lying in the constellation Gemini; the name is a contraction of 'Gemini gamma-ray source'. It was detected in 1972 by the satellite SAS-2, but not until 1988 was a faint optical counterpart of 25th magnitude identified. Geminga has since been found to be pulsating in X-rays and gamma rays every 0.237 s, suggesting that it is a spinning neutron star. In its gamma-ray, X-ray, and optical properties it is similar to the Vela Pulsar, although it emits no detectable radio waves. It is the nearest pulsar, 500 L.Y away, possibly the remnant of the supernova that formed the local bubble in interstellar space.

Gemini A constellation of the zodiac, representing the twins Castor and Pollux of Greek mythology. The Sun lies in Gemini for the last week of June and the first three weeks of July. The constellation's brightest stars are Castor and Pollux. Zeta Geminorum is a Cepheid that varies between magnitudes 3.7 and 4.2 in a period of 10.2 days. Eta Geminorum is a red-giant semiregular variable that ranges between Magnitudes 3.3 and 3.9. M35 is a 5th-magnitude open cluster; NGC 2392 is the Eskimo Nebula. The Geminid meteors radiate from the constellation every December.

Geminid meteors One of the three most active annual meteor showers, reaching a maximum ZHR of about 100 on December 13 from a radiant at RA 7h 28m, dec. +32°, near Castor. Activity is seen from December 7 to 15, and is produced by debris in a common orbit with the asteroid Phaethon. Geminid meteors are slow and often bright. The Geminids' apparent asteroidal origin differs from the cometary source of most meteors, and may account for their longer luminous duration in flight, and apparent dearth of persistent trains.

Geminid meteor shower The Geminids are a meteor shower that occur each year from Dec. 6-19, with a maximum on Dec. 13-14. This meteor shower occurs each year as the Earth passes through the orbit of the asteroid #3200 Phaeton, and dusty remnants of the asteroids burn up as they enter the Earth's atmosphere. The meteors in this shower seem to emanate from the constellation Gemini.

Gemini project A series of American manned space missions in the two-seater Gemini spacecraft. Astronauts practised the techniques of rendezvous and docking of spacecraft, and extravehicular activities in spacesuits, that would be needed for the subsequent Apollo Moon flights. The first manned Gemini flight was Gemini 3 in 1965 March; the series ended with Gemini 12 in 1966 November.

Gemini telescopes. Two identical 8-m telescopes for optical and near-infrared observations being built jointly by the USA, UK, Canada, Chile, Brazil, and Argentina. Gemini North, at an altitude of 4200 m on Mauna Kea, Hawaii, is due for completion in 1998; Gemini South, completion due in 2000, is on Cerro Pachón, Chile, at an altitude of 2715 m. The Gemini Telescopes project is managed by the Association of Universities for Research in Astronomy.

General catalogue of variable stars A listing of all known variable stars and their types, published by the Russian Academy of Sciences in Moscow. It first appeared in 1948. The fourth edition, containing 28 435 entries in three volumes, was published in 1985-7. A companion catalogue of suspected variables is also published, the most recent edition being the New Catalogue of Suspected Variable Stars (NSV), with 14 810 objects, in 1982.

General theory of relativity A theory announced by A. Einstein in 1915 that describes how space and time are affected by the gravitational fields of matter. The theory predicts that gravitational fields change the geometry of space and time, causing it to become curved. This curvature is apparent in a number of ways. First, light is bent in a gravitational field, a prediction that was confirmed by photographic measurements of the positions of stars near the limb of the Sun made during a total solar eclipse in 1919. The same effect manifests itself in a delay in radio signals from distant space probes as the signals pass the limb of the Sun. The curvature of space near the Sun also causes the perihelion point of Mercury's orbit to move forward, by 43 per century more than predicted by I. Newton's theory of gravity.In the orbits of pulsars in binary systems, the advance of perihelion can amount to several degrees per year.

Geocentric 1. With the Earth at the centre, as in for example the geocentric system of cosmology.

2. As seen from the centre of the Earth, as in for example geocentric coordinates.

Geocentric coordinates Any system of coordinates with their origin at the centre of the Earth. For bodies in the Solar System, geocentric coordinates differ slightly from the coordinates as actually measured by an observer on the surface of the Earth.. Various forms of geocentric coordinates are defined: geocentric equatorial coordinates, the most common, have the celestial equator as their reference plane, while geocentric ecliptic coordinates are referred to the ecliptic. Rectangular coordinates can also be geocentric in origin.

Geographos Asteroid 1620, a member of the Apollo group, discovered by R. L. B. Minkowski and the American astronomer Albert George Wilson in 1951 but not seen again until 1969. It is of S class. Radar observations show that Geographos is one of the most highly elongated objects in the Solar System, 5.1 × 1.8 kin, probably a splinter from a larger body. Its orbit has a semimajor axis of 1.246 AU, period 1.39 years, perihelion 0.83 AU, aphelion 1.66 AU, and inclination 13°.3.

Geometrical albedo A measure of the reflectivity of a surface, particularly that of a Solar System object such as a planet, a satellite, or an asteroid; also called physical albedo. The geometrical albedo is the ratio between the light or other radiation reflected from an object, as viewed from the direction of the Sun (i.e. at zero phase angle), and that which would be reflected by a hypothetical white, perfectly diffusely reflecting sphere (i.e. with an albedo of 1.0); this hypothetical sphere is assumed to have the same apparent size and be at the same distance as the real object. The wavelength or range of wavelengths at which the geometrical albedo applies must be defined. The bolometric geometrical albedo refers to reflectivity over all wavelengths.

Georgiaite A type of tektite found in Georgia, USA. The ablation age of georgiaites is 33-35 million years. This is very similar to the derived ages of the bediasites and the Martha's Vineyard tektite; together, they make up the oldest group of tektites.

Geostationary orbit A satellite is in geostationary orbit (GSO) when it remains over the same spot on the Earth (somewhere along the equator). The orbital period is 24 hours (equal to the earth's rotational period). A geostationary orbit is geosynchronous, circular and has zero inclination. There is only one geostationary orbit, a belt encircling the earth's equator at an altitude of about 22,236 miles (35,786 km). It is impossible to have a satellite in geostationary orbit over a point that is not along the equator. Many communication and weather satellites are in geostationary orbit, but their number is limited, sine they 're all in the same orbit and must be spaced enough to avoid interference.

Giant branch The region of the Hertzsprung-Russell diagram that extends diagonally to the upper right of the main sequence, containing giant stars. The giant branch ranges from stars with a surface temperature similar to the Sun's, but a luminosity some 30 times greater, to cooler but even more luminous stars. Giants greater than about 1 solar mass are separated from the main sequence by the Hertzsprung gap. Lower-mass giants are, however, linked to it by a subgiant branch.

Giant molecular cloud A massive cloud of interstellar gas and dust composed mainly of molecules. The typical diameter is 100 L.Y, and masses range from a few hundred thousand to 10 million solar masses. GMCs contain mostly hydrogen molecules (H_2, 73% by mass), helium atoms (He, 25%), dust particles (1%), neutral atomic hydrogen (H, less than 1%), and a rich cocktail of interstellar molecules (less than 0.1%). Our Galaxy contains over 3000 GMCS, the most massive of which lies near the radio source Sagittarius B2 at the galactic centre. They comprise half the mass of all interstellar matter, although they occupy less than 1% of its volume. The average gas density is a few thousand molecules per cubic centimetre. GMCs are found mainly in the spiral arms of disk galaxies, and they are the birthplace of massive stars. GMCs exist for some 30 million years, during which time only a small fraction of their mass is converted to new stars. The nearest GMC lies in Orion, and is associated with the Orion Nebula.

Giant planet A planet of much larger mass and diameter than the Earth, particularly one consisting mostly of gas; also known as a gas giant. In the Solar System the giant planets are Jupiter, Saturn, Uranus, and Neptune, but there is growing evidence of similar objects around other stars.

Giant star A star that has swollen in size towards the end of its life, having converted all the hydrogen in its core to helium. Such stars lie above the main sequence of the Hertzsprung-Russell diagram. A giant is brighter, larger, and cooler than a main-sequence star of the same mass. Giants have diameters 5-25 times that of the Sun, and luminosities from tens to hundreds of times the Sun's. They are of luminosity class II or III.

Gibbous moon A gibbous moon is between a full moon and a half moon, or between a half moon and a full moon.

Globular cluster A roughly spherical group of old stars in the halo of a galaxy. Globular clusters contain from tens of thousands to millions of stars, and have diameters of 100-300 L.Y At the centre of the cluster, where most stars are concentrated, the density may be over 250 stars per cubic light year. About 140 globular clusters are known in our Galaxy, travelling on highly elongated orbits around the galactic centre. They are very old, about 10^{10} years, having formed early in the history of the Galaxy. Stars in globular clusters are members of Population II, with a low content of heavy elements, although a few show values approaching that of the Sun. On the Hertzsprung-Russell diagram of a globular cluster, the main sequence and the giant branch are smoothly joined via a subgiant branch; also conspicuous is an asymptotic giant branch and a horizontal branch, punctuated by an instability strip containing RR Lyrae variables.

Glycolaldehyde Glycolaldehyde ($C_2H_40_2$) is a type of simple sugar that was recently found in a giant interstellar cloud of gas and dust [called Sagittarius B2 (North)], located near the center of the Milky Way Galaxy, about 26,000 light-years from Earth. This amazing find is the first sugar (and one of the most complex molecules) found in space. This molecule was detected by Jan M. Hollis (NASA/Goddard), Frank J. Lovas (University of Illinois), and Philip R. Jewell (NRAO/ Green Bank) using the National Science Foundation's 12 Meter Telescope, a radio telescope on Kitt Peak, Arizona. Hollis said, "Finding glycolaldehyde in one of these interstellar clouds means that such molecules can be formed even in very rarefied conditions. We don't yet understand how it could be formed there."

Gödel universe An unusual cosmological model which represents a rotating universe. This model possesses a number of strange mathematical features, including the fact that it allows time-travel to occur within it. It is due to the Austrian-American mathematician Kurt Gödel.

Gold, Thomas Austrian-American astronomer. In 1948, with H. Bondi and F. Hoyle, he proposed the steady-state theory of the Universe in which matter is continuously created. In the late 1960s he explained the signals from the recently discovered pulsars as synchrotron radiation beamed out by rapidly rotating neutron stars.

Goldstone The location in the Mojave Desert, California, of one of the three deep-space communications complexes that make up NASA's world-wide Deep Space Network.

Goodricke, John English amateur astronomer, born in Holland. Although both deaf and dumb, at the age of 17 he started to observe bright variable stars, encouraged by his neighbour, the astronomer Edward Pigott.In 1782 he announced that Algol's variability was regular, and suggested that its periodic dimmings were caused by a dark body orbiting it. He went on to observe other such eclipsing binaries. In 1784 he found that Beta Lyrae and Delta Cephei varied in brightness periodically, but did not realize that their variability had a different origin. In 1786, while observing Delta Cephei, he caught pneumonia and died, aged 21.

Googol A googol is the number 10^{100} (10 raised to the 100th power or 1 followed by 100 zeros= 10,000,000,000,000,000,000,000,000, 000,000,000,000,000,000,000,000,000, 000,000,000,000,000,000,000,000,000,000,000, 000,000,000,000,000).

Gould, Benjamin Apthorp American astronomer. In 1849 he founded the Astronomical Journal. In 1870 he moved to Argentina, founding the National Observatory at Córdoba, where he initiated the Córdoba Durchmusterung, a southern equivalent of the catalogue produced by F. W. A. Argelander for the northern stars. Gould's Belt is named after him.

Gould's belt A band of hot, bright stars (types O and B) forming a circle around the sky. It represents a local structure of young stars and interstellar material tilted at about 16° to the galactic plane. Among the most prominent components of the belt are the bright stars in Orion, Canis Major, Puppis, Carina, Centaurus, and Scorpius, including the Sco-Cen Association. The belt has a diameter of about 3000 L.Y (about one-tenth the radius of the Galaxy), and the Sun lies within it. Viewed from Earth, Gould's Belt projects below the plane of our Galaxy from the lower edge of the Orion Arm, and above the plane in the opposite direction. The belt is estimated to be about 50 million years old, but its origin is unknown. It is named after B. A. Gould, who established its existence in 1879.

Graben A strip of land, bounded by parallel faults, that has subsided between the faults; pl. graben. Many linear rilles on the Moon are graben, including the Rimae Sirsalis, about 3.5 km wide and 425 km long.

Grading function A complex function g(x,y) describing the amplitude and phase of the current distribution as a function of rectangular co-ordinates (x,y) over the exit or input aperture of an antenna. Also known as the illumination function. When normalized by the amplitude of the current at the point of maximum current density, also known as the taper function (compare u-v taper function for interferometry).

Gradual commencement The commencement of a geomagnetic storm that has no well-defined onset.

Grains, interplanetary Small particles of solid matter between the planets of the Solar System. They are produced mainly by collisions between asteroids and by the break-up of comets, but some enter the Solar System from interstellar space. There are trails of dust grains along the orbits of comets and asteroids, which are experienced as meteor streams if they cross the Earth's path. Interplanetary dust grains can also be detected when they scatter sunlight to form the zodiacal light, and by the small amount of infrared radiation they produce.

Granulation A mottling of the Sun's photosphere caused by numerous small light areas called granules. Individual granules are as much as 1000 km across, and may have polygonal shapes. They are separated from each other by darker intergranular lanes, about 400 K cooler than granules. Granule lifetimes are around 20 min, their birth being from the fragments of previous granules and their demise from fragmentation, fading, or sometimes explosion into small fragments. Granules are elongated near sunspots, and also occur within sunspot umbrae (umbral dots). Spectroscopic observations suggest that granules are convection cells of rising hot gas the intergranular lanes marking regions of descending cooler gas.

Graph A graph is a diagram that represents a series of points or lines.

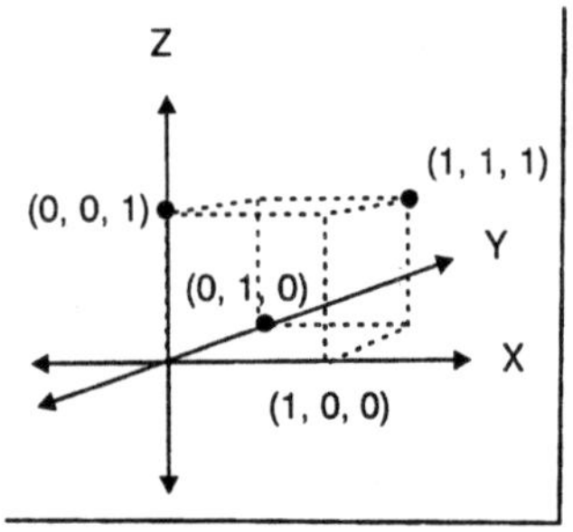

Fig. Graph

Gravistar A gravastar is an extremely dense, cold, dark, thick-shelled object that contains springy, oddly-behaving space inside it. A gravistar is the remnant of a dying star that has imploded; it has many similarities to a black hole, but emits far brighter X-rays than a black hole. Gravistars were theorized to exist in 2002 by Emil Mottola of Los Alamos National Laboratory, New Mexico, and Pawel Mazur of the University of South Carolina, Columbia. The existence of gravistars is not universally accepted.

Gravitation The force of attraction that operates between all bodies. The size of the attraction depends on the masses of the bodies and the distance between them; gravitational force diminishes with the square of the distance apart according to the inverse-square law. Gravitation is the weakest of the four fundamental forces in nature. I. Newton formulated the laws of gravitational attraction and showed that gravitationally a body behaves as though all its mass were concentrated at its centre of gravity. Hence a gravitational force acts along a line joining the centres of gravity of two masses. In the general theory of relativity, gravitation is interpreted as a distortion of space. Gravitational forces are significant between large masses such as stars, planets, and satellites, and it is this force which is responsible for holding together the major components of the Universe. However, on the atomic scale the gravitational force is about 1040 times weaker than the force of electromagnetic attraction.

Gravitational deflection The bending of the path of a beam of light or any other electromagnetic radiation by the gravitational field of a body. The amount of bending depends on the mass of the body and how closely the beam of radiation passes it. The effect was first measured for stars close to the limb of the Sun at the total eclipse of 1919 May. The maximum deflection at the Sun's limb is $1^{0}.75$ radially away from the

Sun. Gravitational deflection also results in highly distorted images of distant objects when a closer, massive object lies along the line of sight.

Gravitational energy The energy released by a body falling in a gravitational field. It is a form of potential energy. When an interstellar gas cloud collapses to form a protostar, the gravitational energy released heats its interior to the point at which nuclear reactions can commence. Gravitational energy also supplies the energies for stars condensing into white dwarfs, and plays a major role in the processes operating inside supernovae. Large amounts of gravitational energy are released by material falling into the accretion disks around black holes, and may provide the power source for quasars, Seyfert galaxies, and other active galactic nuclei.

Gravitational force The force of attraction, F, that exists between all bodies. For any two bodies, it is directly proportional to the product of the two masses and inversely proportional to the square of the distance between them. The constant of proportionality is G, the universal gravitational constant. Expressed in mathematical form, F = Gm 1 m 2 / d2, where m 1 and m 2 are the two masses and d is their distance apart.

Gravity The force of attraction between two or more masses. This force is dependant on both the masses themselves as well as the distance between them. Being a force, gravity is strictly speaking measured in Newton's but is commonly measured in Kilogrammes.

Great annihilator The Great Annihilator is a powerful, high-energy x-ray source located near the center of the Milky Way Galaxy. The Great Annihilator may be a black hole with a mass of roughly 10 to 1000 times that of the Sun. This object produces anti-matter (positrons, which are anti-electrons) which, when they collide with ordinary electrons, will annihilate the matter, resulting in the production of energy.

Great attractor A purported concentration of matter in the direction of the constellations Hydra and Centaurus which may be pulling surrounding galaxies, including ours, towards it. The Great Attractor is calculated to have a mass about 5×10^{16} solar masses and to lie about 150 million L.Y from our Galaxy. Its existence was originally inferred from studies of the motions of galaxies with respect to the Hubble flow. Although there clearly is a concentration of galaxies at the place where the Great Attractor supposedly lies, more recent studies indicate that the observed bulk flows of galaxies are probably due to the concerted pull of several distinct clusters.

Great bear Popular name for the constellation Ursa Major.

Great red spot A large oval marking in Jupiter's clouds at about 22° south latitude. Its dimensions are about 26000 km east-west and 14000 km north-south. It was first recorded by S. H. Schwabe in 1831 but attracted little attention until 1878-82, when it became a striking dark red. Since then it has varied greatly in size, colour, and intensity, sometimes being so faint that it is detectable only by the hollow it makes in the South Temperate Belt.A prominent, slightly smaller dark spot at a similar latitude to the Great Red Spot was seen in 1664 by the English scientist Robert Hooke and lasted until 1713. This was probably an earlier manifestation of a similar phenomenon. Space probes have shown the Great Red Spot to be a vast rotating vortex, equivalent to a storm or hurricane. Its top lies a few kilometres above the surrounding cloud deck, and its red colour may be due to compounds such as phosphine.

Great wall A roughly two-dimensional concentration of galaxies, at least 200 million by 600 million L.Y in extent, but less than 20 million L.Y thick. It contains many thousands of galaxies and has a mass of at least 10^{16} solar masses. The Great Wall extends more than 120° across the sky (8-16h RA) and is about 250 million L.Y away. This and other similar features, together with the many filamentary features in the distribution of galaxies, suggests that the large-scale structure of the Universe may be cellular in nature, with other Great Walls forming the faces of the cells and filaments of galaxies forming where these faces intersect.

Greatest brilliancy The greatest apparent magnitude of Venus at any particular apparition. It depends on both the planet's distance and its phase. Its brightness increases as it moves away from superior conjunction, because of the increasing apparent area of the illuminated disk, but eventually the rapidly thinning crescent counteracts the increase in apparent diameter, and the magnitude starts to fall again. Greatest brilliancy occurs about 36 days before and after inferior conjunction, when Venus is the brightest object in the sky after the Sun and Moon, reaching up to magnitude −4.7.

Greenhouse effect The elevation of a planet's surface temperature resulting from the absorption of long-wavelength radiation by gases in the atmosphere. For instance, carbon dioxide, methane, chlorofluorocarbons, and water vapour in the Earth's atmosphere are all transparent to incoming shortwavelength solar radiation, but absorb outgoing infrared. The Earth is, on average, about 35 K warmer than it would be if its atmosphere were completely transparent to outgoing infrared. On Venus, the greenhouse effect raises the surface temperature by some 500 K to around 730 K.

Greenwich mean astronomical time An obsolete form of Greenwich Mean Time, being the mean solar time on the Greenwich meridian but with the day beginning at noon rather than midnight. It is, therefore, the Greenwich hour angle of the mean sun. It was used by astronomers before 1925 to avoid the need to change date during the night.

Greenwich mean time The mean solar time on the Greenwich meridian, now more generally known astronomically as Universal Time.It is the Greenwich hour angle of the mean sun +12 hours, as the day starts at midnight.

Greenwich meridian The line of 0° longitude that passes through the Old Royal Observatory at Greenwich, London. It was adopted in 1884 as the prime meridian, and still serves as the reference point for the world's timekeeping and navigation systems.

Greenwich observatory The Royal Greenwich Observatory is athe national astronomical observatory of England. It is located in Greenwich, England (near London).

Greenwich sidereal date The number of sidereal days that have elapsed at Greenwich since a given starting date. It is analogous to Julian Date, and the initial epoch is chosen as the beginning of the sidereal day that was in progress at JD 0.0. The whole-number part of the Greenwich Sidereal Date is the number of sidereal days that have elapsed since that time (the Greenwich Sidereal Day Number), and the decimal part is the Greenwich Sidereal Time expressed as a fraction of a sidereal day.

Greenwich sidereal time The sidereal time on the Greenwich meridian; equivalently, the Greenwich hour angle of the equinox. The sidereal time will either be mean sidereal time or apparent sidereal time, depending on whether the mean equinox or true equinox is used.

Gregorian calendar The form of calendar that is now in almost world-wide civil use. It was devised with the help of the German Jesuit mathematics teacher Christopher Clavius and was introduced by Pope Gregory XIII in 1582 to replace the Julian calendar, which had got out of step with the seasons. In that year 10 days were omitted, Thursday October 4 being immediately followed by Friday

October 15. This change was made in order to re-establish March 21 as the date of the vernal equinox. Britain and America did not adopt the Gregorian calendar until 1752, by which time 11 days had to be omitted. Subsequently the calendar has been kept in step with the date of the vernal equinox by introducing a leap year every four years, as the Julian calendar had done, but suppressing the leap year in century years, unless the year is divisible by 400. Thus 1700, 1800, and 1900 were not leap years, but 1600 and 2000 are.

Grigg-Skjellerup, comet 26P A periodic comet discovered by the New Zealand amateur astronomer John Grigg in 1902, and independently by the Australian amateur John Francis ('Frank') Skjellerup in 1922. Its orbit is perturbed by Jupiter, which has led to an increase in its perihelion distance from 0.89 to 0.99 AU since 1922. Its orbital period is 5.1 years. Grigg-Skjellerup was visited by the Giotto space probe in 1992. The comet was found to have a much lower dust production than expected, and a gas production only about 1% of that of Halley's Comet at its 1986 perihelion. Debris from Comet Grigg-Skjellerup produces the periodic Pi Puppid meteor shower, seen around April 23 but only when the comet is close to perihelion. The comet's orbit has an eccentricity of 0.66 and inclination 21°.1.

Grubb, Howard Irish optician. In 1865 he joined the optical company established by his father, Thomas Grubb where he assisted with construction of a 48-inch (1.2-m) reflector for Melbourne, Australia, completed in 1867. Under his control, the company developed a reputation for large telescopes of high quality. In 1914 Grubb moved his business to England where in 1925, as the result of a merger, it became Sir Howard Grubb, Parsons & Company.

Grus A constellation of the southern sky, representing a crane (bird). Its brightest star is Al Na'ir (Alpha Gruis). Delta Gruis is a wide double of unrelated stars, magnitudes 4.0 and 4.1, as is Mu Gruis, magnitudes 4.8 and 5.1.

Guardians The stars Beta and Gamma Ursae Minoris.They lie in the bowl of the Little Dipper, Ursa Minor.

Guest star The ancient Chinese term for a star that newly appears in the night sky, and then later disappears. Later, the Europeans called this a nova.

Guide telescope A telescope designed for viewing a guide star when keeping a larger instrument correctly pointed as the Earth turns; also known as a guidescope. The actual guiding is done either by watching a guide star and keeping it precisely centred on cross-wires in the eyepiece, or using an autoguider. An alternative to a guide telescope is an off-axis guider.

Guider 'A device that allows a telescope to follow a guide star during photography or imaging. Guiding is necessary to compensate for small mechanical errors in the telescope and mounting, and changes in atmospheric refraction. There should be a means of monitoring a guide star and making small corrections to either the telescope's movement or the position of the camera. The monitoring may be carried out using a guide telescope or an off-axis guider, either visually or with an autoguider.

Gum nebula A large, almost circular emission nebula in the constellations Vela and Puppis, discovered by the Australian astronomer Colin Stanley Gum.The Gum Nebula is 36° across and is estimated to be 1300 L.Y away, which gives it a true diameter of about 800 L.Y It is expanding at 20 km/s, and is thought to have been formed by a supernova explosion over 1 million years ago. It contains the Vela OB2 association, the hot stars of which may cause it to shine. The Vela Supernova Remnant and the Vela Pulsar are embedded within the Gum Nebula, but are much younger.

Gutenberg discontinuity The Gutenberg discontinuity separates the outer core and the mantle of the Earth.

Gutenberg, Beno Beno Gutenberg (June 4, 1889 - 1960) was a German geophysicist. In 1913, he accurately determined the size of the core of the Earth. Gutenberg discovered that the Earth has a low-velocity zone in the upper mantle; this zone is now called the Gutenberg discontinuity. Gutenberg published a series of papers with Charles Richter (they were titled "On Seismic Waves" and published between 1931 and 1939) and *Seismicity of the Earth* (published in 1941).

Gyrofrequency The number of revolutions per second described by a charged particle circling in a magnetic field at a velocity much less than the speed of light; also known as cyclotron frequency. It is independent of the particle's velocity, but depends on its mass and charge and on the magnetic flux density.

Gyrosynchrotron radiation Electromagnetic radiation emitted by a charged particle moving in a magnetic field at an appreciable fraction of the speed of light. It is similar to cyclotron radiation, except that relativistic effects cause most of the energy to be emitted at higher multiples of the gyrofrequency. The slowly varying radio emission received from active regions on the Sun may consist of gyrosynchrotron radiation. Gyrosynchrotron radiation is believed to be responsible for the intense radio emission received from solar flares occurring in the magnetic fields above groups of sunspots.

H

H and K lines Two Fraunhofer lines in the ultraviolet part of the spectrum, at 396.8 and 393.4 nm respectively, due to singly ionized calcium.They are prominent in the spectrum of solar-type stars and cooler. Emission at the H and K lines is frequently found in stars where there is enhanced magnetic activity (e.g. flare stars and RS Canum Venaticorum stars).

H I region A region of interstellar hydrogen that is in the form of neutral, unionized atoms. The notation H I refers to the fact that the hydrogen atoms are not ionized.Each neutral hydrogen atom contributes just one particle to the gas. The density of H I regions is too low for hydrogen molecules to form, and starlight would dissociate any molecules that did, so the gas remains as atoms. Neutral hydrogen constitutes about half of all interstellar matter by mass and by volume, with an average density of 1 atom/cm^3. H I regions are cool (about 100 K) and emit no visible light, but they do give out the important radio spectral line at 21 cm wavelength.

H II region A region of interstellar hydrogen that is ionized. The notation H II refers to the fact that the hydrogen atoms are ionized.Each ionized hydrogen atom contributes two particles to the gas, namely a proton and an electron. H II regions are hot, with typical temperatures of 10000 K, and they are anything from 10 to 100000 times denser than H I regions. They are usually found around massive young O and B stars, the strong ultraviolet light from which ionizes the gas, causing it to shine. The Orion Nebula is a famous H II region. H II regions can be detected throughout the Galaxy by their strong radio and infrared emissions. The radio emission is bremsstrahlung from the ionized gas, and the infrared is thermal emission from dust.

H magnitude The magnitude in the infrared H band of the Johnson photometry system. The H-band filter has an effective wavelength of 1630 nm and a bandwidth of 307 nm. The importance of H is that it coincides with the minimum in the opacity of the H- ion, which is an important source of opacity in the atmospheres of cool stars.

Hadley circulation The movement of air from the equator towards the poles in a planet's atmosphere. Warm air near the equator rises and travels towards the poles, cooling as it does so until it becomes denser than the air below. It then sinks to the surface and returns to the equator, completing a cycle known as a Hadley cell. The tropical Hadley cells in the Earth's atmosphere extend to about 30°

north and south of the equator, but there are two further sets of Hadley cells, one between 30° and 60° latitude, and another between 60° latitude and the poles, which are less prominent and less permanent than the tropical ones. On Venus, there is a single Hadley cell extending from the equator to the poles. The effect is named after the English meteorologist George Hadley.

Hadley rille Hadley Rille is a long valley on the surface of the moon. This rille is 75 miles (125 km) long, 1300 feet (400 m) deep, and almost 1 mile (1500 m) wide at its widest point. It was formed by molten basaltic lava that carved out a steep channel along the base of the Apennine Front (which was explored by the Apollo 15 astronauts in 1971).

Hale, George ellery American solar astronomer. In 1889 he invented the spectroheliograph, which he used to study the Sun's prominences and surface features. In 1905 he found that sunspots were cooler than the surrounding photosphere, and in 1908 he showed from the Zeeman effect in their spectra that they had strong magnetic fields. In 1925 he found that the Sun's magnetic field reversed polarity in each successive sunspot cycle. Hale planned and organized funding for three telescopes, each of which was in turn the world's largest. The 40-inch (1-metre) refractor at Yerkes Observatory was completed in 1897; the 100-inch (2.5-metre) Hooker Telescope at Mount Wilson Observatory in 1917; and the 200-inch (5-metre) reflector (later named the Hale Telescope) at Palomar Observatory in 1948.

Hale-bopp comet A long-period comet discovered independently on 1995 July 23 by the American amateur astronomers Alan Hale and Thomas Joel Bopp At that stage it was still over 7 AU from the Sun. It was noted for a series of jets and shells of dust thrown off by the nucleus as it rotated in a period of about 11.5 hours. The comet reached pengee, 1.32 AU, on 1997 March 22, and perihelion, 0.91 AU, on 1997 April 1. Its maximum magnitude was about -0.5. The gas tail extended to 20° and the dust tail to 25°, corresponding to actual lengths of over 1 AU. Its orbit has a period of about 2400 years, eccentricity 0.995, and inclination 89°.4.

Half moon A half moon looks like half a circle. It is sometimes called a quarter moon (this moon has completed one quarter of an orbit around the Earth from its full or new position, and one quarter of the moon's surface is visible from Earth).

Halley, Edmond English scientist. From 1676 he spent two years observing the southern sky from St Helena, and in 1678 published a catalogue of 341 stars, the first southern star catalogue compiled from telescopic observations. He was the first to suggest that observations of transits of Venus could be used to measure the Sun's distance, which was first done long after his death. In 1683 he commenced a long series of lunar studies, discovering the Moon's secular acceleration in 1693. In 1684, having deduced the inverse-square law, he visited I. Newton and persuaded him to write the Principia. In 1705 Halley published the Synopsis of Cometary Astronomy in which he concluded that the comet he had observed in 1682 was the same as those of 1531 and 1607. He predicted it would return in 1758, which it did and was named Halley's Comet. In 1718 he concluded that the brightest stars had changed position since the time of Ptolemy Almagest, thus discovering proper motion.

Halley's comet The best-known periodic comet, returning to perihelion at average intervals of 76 years on a re-trograde orbit (the time between returns ranges from 74 to 79 years). The comet Halley has been observed at every return since 240 BC. It is named after E. Halley, who in 1705 showed that comets seen in 1531, 1607, and 1682 were identical, and successfully predicted the return of 1758. In 1910 Halley's Comet passed 0.15 AU from Earth, reaching magnitude 0 and showing a 100° tail; the Earth passed through the tail on May 20. At its most recent return, in 1986 it reached 3rd magnitude and had a 10° tail. On this occasion, five space probes were sent to investigate it: Giotto, Sakigake, Suisei, and Vega 1 and 2. It was found to have an irregular nucleus measuring 16 × 8 km with an albedo of 0.04. Halley's Comet is the parent of the Eta Aquarid and Orionid meteor showers. Its perihelion is 0.587 AU, aphelion 35.3 AU, eccentricity 0.967, inclination 162°.2.

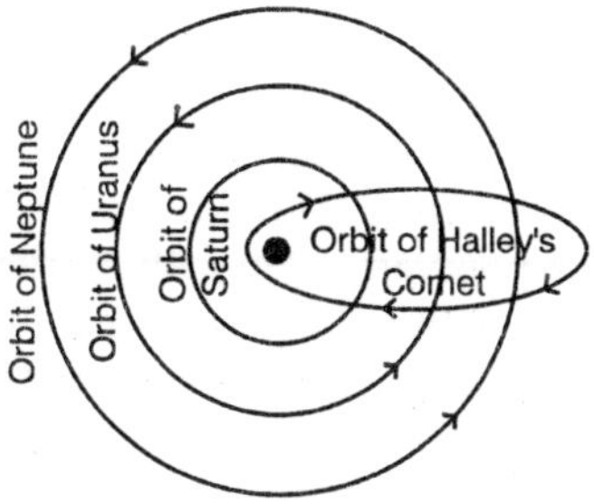

Fig. Halley's comet

Halo A halo is a luminous ring that is sometimes seen surrounding the sun or the moon. Some parts of the halo are very bright, others are not very bright. Sometimes, only a part of the ring is visible. The halo is produced as light is reflected and refracted through tiny, flat ice crystals in the atmosphere. Halos are always an angle of 22° away from the sun or moon, due to the hexagonal structure of the ice crystals. The diameter of the halo is about an eighth of the sky.

Harvard classification A system of classifying stars according to the characteristics of their spectra, introduced at Harvard College Observatory in 1890 by E. C. Pickering, which culminated in the Henry Draper Catalogue. At first, stars were ordered according to the strength of their hydrogen absorption lines from A (strongest) to P. Eventually several letters were merged or dropped, and the remaining spectral types were rearranged into the sequence O, B, A, F, G, K, M, in order of decreasing surface temperature. Early in the 20th century, spectral types R and N were applied to carbon-rich versions of types G, K, and and type S (M stars with heavymetal lines) was subsequently added.The Harvard system has since been superseded by the Morgan-Keenan classification system.

Harvard college observatory The astronomical observatory of Harvard University at Cambridge, Massachusetts, founded in 1839. HCO owns telescopes at Oak Ridge, Massachusetts, which are now operated by the Smithsonian Astrophysical Observatory.HCO is a partner in the Magellan Telescopes project. HCO and SAO jointly run the Harvard-Smithsonian Center for Astrophysics.

Harvard revised photometry A catalogue of stars brighter than visual magnitude 6.5, published by E. C. Pickering at Harvard College Observatory in 1908. In addition to 9096 stars, the catalogue contained nine novae or supernovae, four globular clusters, and the Andromeda Galaxy. This catalogue was the forerunner of the Bright Star Catalogue.

Harvard-Smithsonian center for astrophysics A research organization at Cambridge, Massachusetts, founded in 1973, which combines the research activities of Harvard College Observatory and the neighbouring Smithsonian Astrophysical Observatory.

Harvest moon The full Moon closest to the autumnal when the Moon rises around the time of sunset for several nights in succession. This is because the ecliptic, and hence the Moon's orbit, is then at its minimum angle to the horizon at the time of moonrise in mid- or high northern latitudes. The harvest Moon is so named because its light in the evening was of help to workers at harvest time.

Hasa A term used when a class has another class embedded within it (aggregation), or associated with it (a reference or "pointer" relationship). Another relationship is IsA.

Hat creek observatory The radio observatory of the University of California at Berkeley, sited at an altitude of 1040 m at Cassel, California. The observatory was founded in 1960. Its main instrument is currently the Berkeley-Illinois-Maryland Association Array, a millimetre-wave interferometer.

Hathor Asteroid 2340, a member of the Aten group, discovered in 1976 by the American astronomer Charles Thomas Kowal.Its diameter is about 200 m. Hathor can approach to within 0.006 AU (900 000 km) of the Earth's orbit. Its orbit has a semimajor axis of 0.844 AU, period 0.78 years, perihelion 0.46 AU, aphelion 1.22 AU, and inclination 5°.8.

Haute-provence observatory An observatory at an altitude of 660 m at St Michel, near Forcalquier in the Alpes de Haute Provence, southern France. It was founded in 1936 and is owned and operated by the Centre National de la Recherche Scientifique.Its main telescopes are a 1.9-m reflector, opened in 1958; a 1.5-m reflector, opened in 1967; a 1.2-m reflector, opened in 1943; and a 0.8-m reflector, originally opened at Forcalquier by the Paris Observatory in 1932 and moved to OHP in 1945.

Hawking radiation The emission of particles by a black hole. In the intense gravitational field around a black hole, pairs of virtual particles may live long enough for one member of the pair to be pulled towards the black hole, while the other moves outwards until they are too far apart to destroy each other. This gives the appearance that the black hole is emitting radiation like a black body, with a temperature inversely proportional to its mass. When the Hawking radiation exceeds the amount of matter and energy entering the black hole, it will start to evaporate. For a black hole of the Sun's mass, the temperature is only about 10^{-7} K, which is less than the temperature of the cosmic background radiation, so no evaporation would occur. However, mini black holes, with a mass of around 10^{12} kg and radius 10^{-15} m, would have a temperature of around 10^{11} K and would radiate strongly. The radiation was predicted to occur by S. W. Hawking.

Hawking temperature The temperature inferred for a black hole based on the Hawking radiation detected from it.

Hawking, Stephen William British theoretical physicist. Despite having suffered from a progressively disabling motor neurone disease for most of his life, he has established a powerful reputation. He has drawn on the general theory of relativity and quantum mechanics to study singularities in the Big Bang and black holes, showing that small black holes emit Hawking radiation. He turned to popularization with A Brief History of Time which sold more copies than any other book on science. Much of Hawking's work is directed towards a single theory that will unite gravity with quantum mechanics.

Hayashi contraction Hayashi contraction is a gravitational phemonenon in which a protostar becomes smaller, coalescing into a main seqence star.

Hayashi track The nearly vertical path on the Hertzsprung-Russell diagram that is followed by a fully convective star as it evolves. Forming stars descend such tracks on to the main sequence, becoming less luminous with time but maintaining a roughly constant surface temperature. The process occurs in reverse in old stars leaving the main sequence for the giant branch. It is named after the Japanese astrophysicist Chushiro Hayashi.

Haystack observatory A radio observatory at Westford, Massachusetts, founded in 1960. Its main instrument is a 36.6-m dish, housed within a radome (protective dome), opened in 1964 and originally used for radar observations of planets but now mostly used for radio astronomy. It is operated by the Northeast Radio Observatory Corporation a consortium of twelve educational institutions, on behalf of the Massachusetts

Institute of Technology, which owns the facility.

Hd number The HD (Henry Draper) number is an identifying number assigned to the strs in the Henry Draper catalog. In this system, every star is classified by its stellar spectrum. This sytem is named for the astronomer Henry Draper, but was cataloged by Annie J. Cannon (225,300 stars), and later extended by Margaret W. Mayall. For example, the star Vega is HD 172167 (the spectral type is not in the HD number).

Head, cometary The main part of a comet, consisting of the solid central nucleus and the surrounding coma of gas and dust While the head may sometimes appear to contain a star-like point Earth-based observations do not reveal the true nucleus itself.

Head-tail galaxy An elliptical galaxy in which strong radio emission in the nucleus is accompanied by an irregular tail of diffuse radio emission that may extend for hundreds of thousands of light years. The head and/ or tail may be double. The radio emission is synchrotron radiation from energetic electrons, as in double-lobed radio galaxies, but the motion of the galaxy through a tenuous intergalactic medium has apparently caused the plasma jets to be swept back in the wake of the galaxy.

HEAO (high energy astronomy observatory) A series of three satellites to observe the high-energy sky; HEAO-1 surveyed the 0.2 keV to 10 MeV sky; HEAO-2, renamed the Einstein observatory after launch, was the first fully imaging X-ray telescope which looked at the 0.2 - 20 keV sky during 1979-1981; HEAO-3 surveyed the 0.2 keV to 10 MeV sky during 1978-1981.

HEASARC (High Energy Astrophysics Science Archive Research Center) located at NASA's Goddard Space Flight Center. The HEASARC creates and maintains archives of data from ultraviolet, x-ray and gamma-ray satellites for use by astronomers around the world.

Heat Heat, like work, is a measure of the amount of energy transferred from one body to another because of the temperature difference between those bodies. Heat is not energy possessed by a body. We should not speak of the 'heat in a body.' The energy a body possesses due to its temperature is a different thing, called internal thermal energy. The misuse of this word probably dates back to the 18th century when it was still thought that bodies undergoing thermal processes exchanged a substance, called caloric or phlogiston, a substance later called heat. We now know that heat is not a substance.

Heat capacity The amount of heat required to increase the temperature of a material by one degree

Heat death of the Universe The long-term fate of the Universe, according to the laws of thermodynamics, in which all matter will eventually reach the same temperature. In this condition no energy is available for doing work and the entropy of the Universe is at a maximum. This outcome was predicted by the German physicist Rudolf Julius Emmanuel Clausius who introduced the concept of entropy.

Heavenly waters Heavenly Waters is a family of 9 constellations that includes Delphinus (the Dolphin), Columba (the Dove), Equuleus (the Little Horse), Carina (the Keel), Puppis (the Stern), Eridanus (the river), Pisces Austrinus (the Southern Fish), Vela (the Sails), and Pyxis (the Mariner's Compass).

Heaviside layer Another name for the E layer.

Heavy element Any element heavier than hydrogen and helium. In astronomy, the heavy elements are often termed metals.

Heavy-metal star A heavy-metal star is an unusual type of giant star. Heavy-metal stars include barium (Ba II) stars (a type of late giant star) and S stars (a type of red giant).

Hebe Asteroid 6, discovered in 1847 by the German amateur astronomer Karl Ludwig

Hencke 1793-1866). Hebe is an S-class asteroid of diameter 204 km. Its orbit has a semimajor axis of 2.423 AU, period 3.78 years, perihelion 1.94 AU, aphelion 2.91 AU, and inclination 14°.8:

Heisenberg's uncertainty principle Pairs of measurable quantities whose product has dimensions of energy×time are called *conjugate quantities* in quantum mechanics, and have a special relation to each other, expressed in Heisenberg's uncertainty principle. It says that the product of the uncertainties of the two quantities is no smaller than h/2. Thus if you improve the measurement precision of one quantity the precision of the other gets worse.

Hektor Asteroid 624, the largest known Trojan asteroid, discovered in 1907 by the German astronomer August Kopff It is a member of the group of Trojans at the L4 Lagrangian point 60° ahead of Jupiter. Hektor's brightness varies by a factor of three as it rotates with a period of 6.92 hours. The light curve indicates that Hektor is highly elongated, 150 × 300 km; alternatively, it might be dumbbell-shaped, two asteroids in contact, or a close binary. Hektor is a D-class asteroid. Its orbit has a semimajor axis of 5.172 AU, period 11.76 years, perihelion 5.05 AU, aphelion 5.29 AU, and inclination 18°.2.

Helene A satellite of Saturn, at a distance of 377 400 km; also known as Saturn XII. Helene's orbital period is 2.737 days. It shares the same orbit as the much larger satellite Dione, which it precedes by 60°. Helene is 36 × 32 × 30 km in size, and was discovered in 1980 by Voyager 1.

Heli-, helio Prefixes referring to the Sun.

Heliacal rising and setting The first visible rising of a celestial object in the morning sky after conjunction with the Sun; or the last visible setting of a celestial object in the evening sky before conjunction. The heliacal rising of Sirius was used by the Egyptians as a yearly marker for their calendar.

Heliocentric 1. With the Sun at the centre, as in for example the heliocentric system of cosmology.

2. As seen from the centre of the Sun, as in for example heliocentric coordinates.

Heliocentric coordinates Any system of coordinates with their origin at the centre of the Sun. They are often used for describing the positions of bodies in the Solar System. They can be either spherical coordinates or rectangular coordinates. Heliocentric spherical coordinates are given in terms of heliocentric latitude and heliocentric longitude. A variant system, barycentric coordinates, gives positions referred to the centre of mass of the Solar System which is slightly displaced from the centre of the Sun.

Heliocentric longitude (symbol l) A coordinate that gives the position of an object around the ecliptic as it would be seen from the centre of the Sun. It is measured in degrees from 0° to 360° clockwise along the ecliptic, starting at the vernal equinox.

Heliocentric parallax Another name for annual parallax.

Heliocentric Sun-centered solar system theories.

Heliocentric system In the heliocentric model of the solar system, all the planets orbit around the Sun.

Heliocentric velocity system Velocity measured relative to the *sun* taken as the focus of the earth's orbital motion (not relative to the solar system's barycenter). Heliocentric and barycentric velocities will differ by 0.012 km/s, so the distinction is sometimes ignored.

Heliographic coordinates The latitude and longitude of a feature on the Sun's surface. The heliographic longitude can be measured west or east of the central meridian of the Sun at any moment, or stated as the longitude with respect to the Carrington rotation system. The heliographic coordinates of sunspots and other features can be derived from latitude and longitude grids or mathematically. In either case the values of the position angle of the solar rotation axis and the latitude and longitude of the centre of the solar disk must be known; these are given in almanacs.

Heliometer An instrument formerly used for measuring the diameter of the Sun or the angular separation between close stars. A heliometer is a refracting telescope whose objective lens is split in half, so that the two halves can be moved along the dividing line. This allows two star images or opposite limbs of the Sun to be superimposed. The separation of the lenses and their position angle could be converted into angular measure. Heliometers were regularly used in the 19th century to measure the parallax of stars and angular diameters of planets, but have now been superseded.

Heliopause The heliopause is the boundary of the heliosphere of the Sunin which the solar wind's density decreases greatly (and its speed also declines). The location of this transition region is unknown, but must be at more than 50 AU from the Sun.

Helios probes Two German space probes, launched by NASA, to study the Sun and interplanetary space. Helios 1, launched in 1974 December, was put into an orbit that took it 45 million km from the Sun at perihelion, closer than any previous probe. Helios 2, launched in 1976 January, was about 43 million km from the Sun at closest approach.

Helioseismology The study of the Sun's interior by observing the oscillations of its surface. Solar oscillations occur on both a global and a local scale. Local oscillations are apparent through small Doppler shifts in Fraunhofer lines at different places on the Sun's disk. The period of the waves averages 5 min, with maximum velocities of about 0.5 km/s. Typically, a pattern of oscillations appears over a few thousand kilometres and persists for 30 min. The oscillations are due to the superposition of many standing sound waves, or p modes, which travel around the Sun between the surface and relatively shallow layers of its interior. Global oscillations are observed by the Doppler shifts in Fraunhofer lines in the Sun's integrated light. Such large-scale oscillations are due to p modes which travel from the Sun's surface to the deepest parts of the interior; the oscillation period averages 5 min, with a range of about 4-8 min.

Heliosphere The region of space around the Sun within which the solar wind flows. The heliosphere is thought to be about 100 AU in radius, and is bounded by the heliopause, beyond which interstellar gas exerts an equal pressure from outside. The shape of the heliosphere is unknown, but if there is a flow of interstellar material around it from a particular direction (an interstellar wind), the heliosphere may be like the Earth's magnetosphere: spherical on one side, but drawn out into a long tail on the other.

Heliospheric termination shock The heliospheric termination shock is the shock that occurs as the solar wind hits the heliopause and its speed slows greatly (down to about 20 km/s).

Heliostat A flat mirror on an equatorial mounting, driven to follow the Sun across the sky and reflect its light into a stationary telescope. If the telescope is oriented parallel with the Earth's axis, only one mirror is needed. However, the Sun's orientation as seen through the telescope will change during the day.

Helium Helium is an element with the atomic number 2. It has two protons and two neutrons in its nucleus which is orbited by two electrons. It is the second most abundant element in the universe. It is created from hydrogen atoms in the process of nuclear fusion that occurs within stars. The Sun is about 25% Helium. Helium was named after the Sun (called "Helios" in Greek) because it was first discovered on the Sun by Jules Janssen in 1868. Helium is plentiful on the Sun and rare on Earth.

Helium burning Helium burning is a stage in a star's life in which the star fuses helium into carbon and oxygen (through nuclear fusion). Every star that began with more than half a Solar mass will eventually burn helium.

Helium flash The explosive event that occurs when helium burning begins in the core of a low-mass star. The core must be dense enough for degeneracy effects to be important. A degenerate gas does not expand on heating, and is unable to cool through expansion in the way that a non-degenerate gas would. Thus, as helium begins to burn, the temperature rises rapidly, eventually becoming so high that the gas is no longer degenerate. At this point the core expands very rapidly, releasing its stored energy in less than 2 min.

Helium shell flash A cycle of unstable nuclear burning in asymptotic giant branch stars. It occurs when helium is being burnt in a thin shell around the star's core, and results from the strong dependence of the rate of energy generation on temperature. Any overheating causes considerable expansion followed by collapse, thus setting up large-scale pulsations.

Helium star The core of a formerly massive star, which has evolved and lost its hydrogen-rich envelope. Hydrogen loss can happen either through a strong stellar wind, as in Wolf-Rayet stars, or by mass transfer on to a companion, as in a close binary. A helium star is expected to evolve exactly as the core of a massive star would: by producing an iron core which collapses to cause a Type Ib or Ic supernova explosion, depending on the mass of the star. 'Helium star' is also an obsolete term for a normal B-type star.

Helix nebula A planetary nebula in Aquarius, also known as NGC 7293 or sometimes the Sunflower Nebula. It is the nearest planetary nebula, 450 L.Y away, and the largest in apparent size, nearly ¼°. It is ionized by a very hot 13th-magnitude central star of temperature 50 000 K.

Hellas planitia A circular lowland basin on Mars about 2500 km wide and 5 km deep, the site of an ancient impact, centred at -44° lat., 294° long. Through Earthbased telescopes it frequently appears bright due to mists and cloud. The great dust storms that sometimes obscure the surface of Mars after perihelion often start from Hellas.

Helmholtz contraction The Helmholtz contraction is the gravitational collapse of a protostellar cloud that is slowed by outward gas pressure and the limited rate at which radiation can escape.

Helmholtz, Hermann Von Hermann Ludwig Ferdinand von Helmholtz (1821-1894) was a German astrophysicist who studied solar energy production and star formation.

Helmholtz-kelvin contraction The contraction of a star under gravity, the potential energy thus lost being converted into heat and radiated away. The Helmholtz-Kelvin contraction time-scale is defined as the time it would take for a star to collapse if nuclear burning were switched off and it continued to radiate at the same luminosity. For the Sun this time would be about 30 million years. It is named after the German scientist Hermann Ludwig Ferdinand von Helmholtz and Lord Kelvin.

Hematite An iron-oxide mineral (Fe_2O_3) that may be the cause of the red color on Mars. Courser-grained gray hematite has the same chemical formula as the red variety, but a different crystalline structure. The deposits of gray hematite found in the Terra Meridiani region of Mars may suggest that water once circulated through the region's rock layers.

Hemisphere A hemisphere is half of a sphere.

Hemispherical albedo The fraction of light falling upon a non-luminous body, such as a planet, that is scattered by its surface, as a function of the angle of incidence. The body is assumed to be a sphere with a diffuse surface that reflects incoming parallel light in all directions.

Henry draper catalogue A catalogue listing the spectral types of 225 300 stars down to 8th magnitude, published in 1918-24. It was compiled at Harvard College Observatory by A. J. Cannon, who assigned spectral types on the basis of the Harvard classification system. The Henry Draper Extension containing 47000 spectra of fainter stars, was published in 1925-36, and a second supplement containing another 86 000 stars in 1949. The catalogue was named in honour of H. Draper. Although the Harvard classification system has been superseded, stars in the Henry Draper Catalogue and Extension are still widely known by their HD or HDE numbers.

Henyey track The roughly horizontal path on the Hertzsprung-Russell diagram followed by a pre-main-sequence star with a mass greater than about 0.4 solar masses. The track takes it on to the main sequence after it has descended the Hayashi track. Stars on Henyey tracks are becoming hotter, and therefore bluer, as they contract. The contraction ends once the core is hot enough for hydrogen burning to begin and the star joins the main sequence. Whereas stars on Hayashi tracks are fully convective, those on Henyey tracks are radiative. It is named after the American astrophysicist Louis George Henyey.

Hephaistos Asteroid 2212, a member of the Apollo group, discovered in 1978 by the Russian astronomer Ludmilla Ivanovna Chernykh.Hephaistos is 5.4 km in diameter. Its orbit has a semimajor axis of 2.163 AU, period 3.18 years, perihelion 0.36 AU, aphelion 3.97 AU, and inclination 11°.9. These orbital elements are similar to those of Comet Encke, and it has been speculated that Hephaistos and Comet Encke, as well as some other asteroids in similar orbits, may be remnants of a giant comet which split up within the past 20 000 years.

Heraclides of pontus Greek philosopher and astronomer, born in modern Turkey. He made the first known suggestion that the apparent rotation of the sky is caused by the axial rotation of the Earth. Often attributed to him is the idea that Venus revolves around the Sun, although there is no firm evidence for this idea appearing in ancient Greece until over a century after his death.

Herbig-haro object A small nebula with an emission-line spectrum, found in regions of star formation. HH objects have high velocities of several hundred kilometres per second relative to their surroundings. They are believed to be bow shocks formed when fast-flowing jets of material from a young star encounter interstellar matter. The emission lines result from the recombination of ions and electrons in the cooling gas behind the bow shock. They are named after the American George Howard Herbig and the Mexican Guillermo Haro.

Hercules cluster A loose, irregular cluster of galaxies some 500 million L.Y away in the constellation Hercules; also known as Abell 2151. The cluster is 1°.7 wide as seen from Earth, has a true diameter of 6 million L.Y, is flattened in shape, and is notable for the high proportion of spirals among its bright members. There are about 75 bright galaxies, and many fainter members.

Hercules Hercules is a Northern Hemisphere constellation that is the fifth largest in the sky. It is named for Hercules, the legendary hero of Greek mythology. The brightest of its rather dim stars is Ras Algethi (alpha Her), a variable red supergiant. The four stars of the central trapezoid within Hercules, epsilon Her, zeta Her, eta Her, and pi Her, form the asterism called Keystone. The globular star cluster M13 is located on the western part of the Keystone. The Tau Herculid meteor shower seems to radiate from Hercules.

Hercules x-1 A low-mass X-ray pulsar about 15000 L.Y away in the constellation Hercules. It is an eclipsing binary consisting of a visible star of about 13th magnitude accompanied by a neutron star. Material from the visible star is accreted by the neutron star, releasing its gravitational energy as X-rays. The optical counterpart, HZ Herculis, varies in spectral type from A to B during each orbit due to X-ray heating. The rapid spin of the neutron star produces X-ray pulses every 1.24 s, and the stars eclipse each other every 1.7 days. Hercules X-1 may also be a source of cosmic rays.

Herculina asteroid 532, discovered in 1904 by Max Wolf. Observations of a stellar occultation by Herculina in 1978 indicate the presence of a companion about 50 km wide and 1000 km away. Herculina is elliptical, with a mean diameter of 220 km, and is of S class. Its orbit has a semimajor axis of 2.773 AU, period 4.62 years, perihelion 2.28 AU, aphelion 3.26 AU, and inclination 16°.4.

Hermes Asteroid 1937 UB, a member of the Apollo group, discovered by the German astronomer Karl Reinmuth in 1937, when it approached within 0.006 AU (900000 km) of the Earth. It was then lost and remains so; hence it remains unnumbered. Hermes can approach to within 0.003 AU (450000 km) of the Earth's orbit. Its orbit has a semimajor axis of 1.639 AU, period 2.1 years, perihelion 0.62 AU, aphelion 2.66 AU, and inclination 6°.2.

Herschel, (Frederick) William German-born English astronomer and musician, originally Friedrich Wilhelm. He came to England in 1757, and became an organist. In 1773 William and his sister C. L. Herschel, who had moved to England the previous year, took up astronomy, observing with the first of many telescopes that he made himself. His discovery of Uranus on 1781 March 13 made him famous, and he was appointed private astronomer to King George III. With the king's patronage he was able to build a telescope with an aperture of 1.2 m, then the largest in the world. In 1787, Herschel discovered the two largest satellites of Uranus, and in 1789 the Saturnian satellites Mimas and Enceladus. During systematic surveys of the heavens he observed and catalogued many double stars and over 2000 nebulae and clusters; this work was continued by his son, J. F. W. Herschel. His observations of double stars revealed that many are in orbital motion around each other.

Herschelian telescope A reflecting telescope in which the primary mirror is tilted so that it focuses light to the top edge of the tube. The eyepiece is located at the top edge, looking down at the mirror. This design avoids light loss caused by a second reflection, which was particularly important when mirrors were made of speculum metal. But it results in astigmatism unless the focal ratio is large. The design was invented by F. W. Herschel, who used it for his giant 48-inch (1.2-m), but it is now of historical interest only.

Herschel-rigollet, comet 35P A periodic comet discovered in 1788 by C. L. Herschel. A comet discovered in 1939 by the French astronomer Roger Rigollet turned out to be a return of the same object. It is the periodic comet with the longest period, currently 155 years, to have been seen more than once. Its perihelion is 0.748 AU, eccentricity 0.974, and inclination 64°.2.

Hertz, Heinrich (1857 - 1894) A German physics professor who did the first experiments with generating and receiving electromagnetic waves, in particular radio waves. In his honor, the units associated with measuring the cycles per second of the waves (or the number of times the tip-tops of the waves pass a fixed point in space in 1 second of time) is called the hertz.

Hertzsprung gap An area of the Hertzsprung-Russell diagram containing no stars, between the top end of the main sequence and the giant branch. Stars cross this region very quickly once they leave the main sequence because their outer layers are expanding rapidly. While the star's luminosity remains roughly constant, the surface temperature goes down so that the star follows an approximately horizontal path from left to right on the HR diagram.

Hertzsprung, Ejnar Danish astronomer. In 1905 he showed how a star's luminosity was related to the width of lines in its spectrum, thus establishing spectroscopic parallax as a means of distance-finding. He inferred the distinction between red giant and dwarf stars from the fact that, although they were of the same spectral type, their spectra had different linewidths. Hertzsprung also proposed the modern definition of absolute magnitude. He went on to plot the first Hertzsprung-Russell diagram, for the stars of the Pleiades, in 1906. His work remained largely unknown, and in 1910 H. N. Russell independently developed the diagram in a slightly different form. In 1911 Hertzsprung discovered that Polaris is a Cepheid variable, and in 1913 estimated the distance to the Small Magellanic Cloud from the brightness of Cepheids within it.

Hertzsprung-russell diagram A graph on which a measure of the brightness of stars (usually their absolute magnitude) is plotted against a measure of their temperature. The diagram shows how the luminosities and surface temperatures of stars are linked. From a star's position on the diagram, astronomers can estimate its mass and the stage of its evolution.

Most stars lie on the main sequence, a strip which runs from the upper left to the lower right of the diagram. A star on the main sequence is burning hydrogen in its core, and during this phase of its life will remain at a point on the diagram that is determined by its mass. Other areas of the HR diagram are populated by stars that are not burning hydrogen in their cores, but may be burning hydrogen in a thin shell. The most prominent of these areas is the giant branch, consisting of stars which have exhausted the hydrogen fuel in their cores.

Hesperian epoch The geologic history of Mars has been divided into three broad time periods, or Epochs. From oldest to youngest, these are Noachian, Hesperian, and Amazonian (named after places on Mars). These Epochs are defined by the number of impact craters on the ground surface; older surfaces have more craters. The Hesperian Epoch extends from about 3.5 to about 1.8 billion years ago. The actual timing of the Epochs is not known because we have different models of the rates of meteorite falls on Mars.

Hess, Victor francis Austrian-American physicist, originally Viktor Franz. In 1912 he began to investigate the recently discovered ionization of the atmosphere. In a series of balloon ascents, he used electroscopes to show that radiation was always present above 150 m, and increased steadily with altitude. Its intensity did not depend on the time of day or night, and so could not come from the Sun—it had to be of cosmic origin. Hess's discovery of these cosmic rays was not recognized until after World War I; for it he shared the 1936 Nobel prize for physics.

Hevelius, Johannes German astronomer, born in modern Poland,. In the early 1640s he derived a reasonably accurate value for the Sun's rotation period, and described and named faculae. His Selenographia contains the first fairly detailed map of the Moon with named features, although few of his names survive. He also discovered the Moon's libration in longitude. In the 1650s he began an observational programme that resulted in an atlas, firmamentum Sobiescianum sive Uranographia, in which he introduced some new constellations, and an associated catalogue of 1564 stars, both published posthumously in 1690 by his second wife and assistant Elisabetha Hevelius, née Koopman.Hevelius was the last major astronomer to make positional measurements with naked-eye sighting instruments.

Hewish, Antony (1924-) English radio astronomer. In the 1950s at the Mullard Radio Astronomy Observatory, Cambridge, he collaborated with M. Ryle in surveys which led to the series of Cambridge catalogues of radio sources. In 1960 he and Ryle developed aperture synthesis. Hewish began to study fluctuating radio sources, and in 1967 his student S. J. Bell identified what proved to be the first signal from a pulsar. Hewish and Ryle shared the 1974 Nobel prize for physics for their work on pulsars and aperture synthesis.

Hexadecimal Numeration with a radix (base) of 16.

Hexahedrite A class of iron meteorite, containing not more than 4-6% nickel. Hexahedrites consist of large cubic crystals of kamacite over 50 mm across, and they can be cleaved in three perpendicular directions along the faces of a hexahedron, hence their name. They display a pattern of fine lines called Neumann lines. The hexa-octahedrite meteorites are transitional between hexahedrites and octahedrites. In these meteorites the thickness of the kamacite plates varies from 3 to 50 mm, most of the kamacite grains are rounded, and little or no taenite remains.

Hidalgo Asteroid 944, discovered in 1920 by W. Baade. It is of D class, and its diameter is 29 km. Hidalgo has a highly elliptical 14.15-year orbit that carries it from perihelion at 2.01 AU, within the main asteroid belt, out to aphelion at 9.69 AU, beyond the orbit of Saturn; the semimajor axis is 5.850 AU. Until the discovery of Chiron in 1977, it had the greatest aphelion distance and longest period of any known asteroid. Its orbit, which has the unusually high inclination of 42°.4, is similar to those of periodic comets, and it could be an extinct cometary nucleus.

Hierarchical cosmology A cosmological theory in which the Universe has structure on all scales, from individual galaxies, to clusters of galaxies, to superclusters, and so on without end. Each level of the hierarchy is supposed to look like an enlargement of the smaller scales. A hierarchical universe is not homogeneous on large scales and hence is not consistent with the cosmological principle. Observations suggest that our Universe is approximately hierarchical on scales up to perhaps 100 million L.Y, but thereafter becomes homogeneous.

High-energy astrophysics The study of X-rays, gamma rays, and cosmic rays coming from the Universe. The energy of the radiation is usually expressed in terms of electronvolts (eV). X-rays and gamma rays range in energy from 100 eV to over 500 000 eV. There is no sharp cutoff between the X-ray and gamma-ray regions of the spectrum. Cosmic rays are atomic and subatomic particles which move through space at velocities approaching that of light. Their energies range from about 10^8 to 10^{20} eV.

High energy astrophysical observatory A series of three NASA X-ray and gamma-ray astronomy satellites. HEAO-1, launched in 1977 August, surveyed the sky in the X-ray energy range 0.2-60 keV (0.02-6.2 nm) for discrete sources and studied the X-ray background. The satellite also studied low-energy gamma rays up to about 10 MeV. HEAO-2, also known as the Einstein Observatory, was launched in 1978 November and observed in the 0.1-4 keV (0.3-12.4 nm) range. HEAO-3, launched in 1979 September, was devoted to gamma-ray astronomy.

High frequency (HF) That portion of the radio frequency spectrum between between 3 and 30 MHz.

High latitudes With specific reference to zones of geomagnetic activity, "high latitudes" refers to 50° to 80° geomagnetic.

High mass X-ray binary (HMXB) A binary star system consisting of a compact object and a high mass companion (i.e. with a mass greater than about 10 times the Sun's mass)

High tide High tide is the time of high water. High tides occur when the gravitational attraction of the moon causes the oceans to bulge out in the direction of the moon.

Highlands Oldest exposed areas on the surface of the Moon; extensively cratered, and chemically distinct from the maria.

Highland "soil" Sediment on the surface of the lunar highlands; composed of broken rock and mineral fragments, and glass produced by impact.

Highlands, lunar Rough and heavily cratered bright terrain on the Moon. The highlands are generally 1-2 km higher than the dark lowland mare areas. The highlands are older, dating from the last stages of planetary accretion before about 4 billion years ago. The old, impact-shocked rocks of the highlands are chemically different from the maria, being enriched in calcium and aluminium. Mineralogically they consist mostly of feldspar.

High-speed stream A feature of the SOLAR WIND having velocities that are about double average solar wind values.

High-velocity cloud A cloud of neutral hydrogen travelling at a velocity far greater than can be explained by the normal rotation of the Galaxy. Some HVCs are debris from a tidal interaction between our Galaxy and the neighbouring Magellanic Clouds. The most spectacular example is the Magellanic Stream, which stretches nearly half-way across the sky, but is visible only to radio telescopes. The distances and origin of other HVCs are still unknown.

High-velocity star A star moving faster than 65 km/s relative to the average motion of the stars in the Sun's neighbourhood. Highvelocity stars are members of the galactic halo, moving in highly elliptical orbits around the galactic centre. Their high relative velocities result from the fact that they are passing through the galactic disk and do not share the rotation of the Sun and its neighbouring stars around the galactic centre. Such stars may have formed early in the Galaxy's history, or they may be remnants of smaller galaxies that have merged with ours.

HII region An area filled with clouds of ionized hydrogen; the ionization is usually caused by radiation from newborn stars

Hilda group A group of asteroids, often called the Hildas, marking the outer boundary of the main asteroid belt, concentrated close to the 3 : 2 resonance with Jupiter, at a mean distance of 4.0 AU from the Sun. In common with Trojan asteroids, Hildas may have more elongated shapes than main-belt asteroids of the same sizes. The group is named after (153) Hilda, a P-class asteroid of diameter 222 km, discovered in 1875 by the Austrian astronomer Johann Palisa in 1875. Hilda's orbit has a semimajor axis of 3.973 AU, period 7.92 years, perihelion 3.41 AU, aphelion 4.54 AU, and inclination 7°.8.

Hill, George William American mathematician and astronomer. Inspired by the work of L. Euler, he developed advanced mathematical methods for tackling problems in dynamical astronomy. His first major work was an application of the three-body problem: the orbits of Jupiter and Saturn and the perturbations exerted by these planets on the Moon. In his study of the Moon's motion he introduced a differential equation now known as Hill's equation.

Himalia Himalia is Jupiter's tenth moon. Himalia is 110 miles (170 km) in diameter and orbits 7,000,000 miles (11,480,000 km) from Jupiter. Himalia has a mass of 9.5 x

10^{18}kg. It orbits Jupiter in 250.5662 (Earth) days. Very little is known about Himalia. Himalia was discovered by C. Perrine in 1904.

Hind's crimson star The red-giant variable star R Leporis, named after the English astronomer John Russell Hind (1823-95), who remarked on its blood-red colour. It is a variable Mira Star, ranging from 6th to 12th magnitude in a period of about 14 months.

Hind's variable nebula A reflection nebula in Taurus, also known as NGC 1555, associated with the young, irregular variable star T Tauri. Both were discovered in 1852 by the English astronomer John Russell hind Major variations in the brightness, extent, and form of the nebula occur over decades. More rapid changes may be caused by variations in T Tauri itself and shadowing by dense clouds of material near the star.

Hinotori A Japanese satellite for solar X-ray studies, launched in 1981 February; before launch it was known as Astro-A. It carried high-resolution soft X-ray spectrometers and hard X-ray instruments which obtained the first images of solar flares at energies above 25 keV (wavelengths shorter than 0.05 nm).

Hipparchus Hipparchus (190-120 B.C.) was an ancient Greek astronomer who compiled first-known catalogue of stars and first map of the skies. He listed 850-1,000 stars, organized by constellation. He noted each star's position and brightness (he rated the brightness on a scale from 1 to 6, the brightest being 1). Hipparchus also devised the system of epicycles, an Earth-centered system in all celestial objects moved in perfect circles around the Earth. He also founded trigonometry.

Hipparchus of Nicaea Greek astronomer, geographer, and mathematician, born in modern Turkey. He put Greek astronomy on a more scientific footing, introducing arithmetic and early trigonometric methods. His many accurate astronomical observations resulted in a catalogue of 850 stars, giving their coordinates and dividing them into six magnitude classes. Ptolemy incorporated the catalogue and other findings by Hipparchus in the Almagest. Hipparchus made surprisingly accurate measurements of the precession of the equinoxes, the length of the year, and the Moon's distance. He may have been the inventor of the astrolabe.

Hipparcos An astrometry satellite launched by the European Space Agency in 1989 August. It measured the position, brightness, proper motion, and trigonometric parallax of 118 218 stars down to 12th magnitude to an accuracy of better than 0.002. It did so by measuring angular separations between pairs of stars as it scanned the sky continuously about a slowly varying rotation axis. In the simultaneous Tycho Survey, the satellite measured the magnitude and colour of over a million stars down to about 10th magnitude, plus their positions to about one-tenth the accuracy of the main survey. Observations ended in 1993. The final catalogue was published in 1997.

Hirayama family A group of asteroids whose members have very similar orbital characteristics and are believed to share a common origin. Hirayama families are thought to result from the break-up of larger asteroids several hundred kilometres in size following catastrophic collisions. They are named after the Japanese astronomer Kiyotsugu Hirayama who demonstrated their existence in 1928.

Histogram modification The histogram of pixel intensities in an image of a natural scene that has been linearly quantized is usually highly skewed toward darker levels. Detail is often imperceptible in the darker regions when such an image is viewed on a conventional display with only 8-bit pixels. Histogram modification can alleviate this problem by rescaling the original image so that the histogram of pixel intensities follows some preferred form. Histogram *equalization* flattens the output histogram, and may often be appropriate for astronomical images. Exponential and hyperbolic output histograms are also sometimes used, so that a final histogram "equalization" is performed (approximately) by the quasi-logarithmic response of the user's retina.

Hiten A Japanese Moon probe, known as Muses-A before its launch in 1990 January. It went into a highly elliptical orbit around the Earth from which it released a smaller craft, Hagoromo, into lunar orbit, although no results were received because of a transmitter failure. Hiten itself was put into orbit around the Moon in 1992 February. It was the first Moon probe not launched by the USA or the former Soviet Union.

Hoba west meteorite The world's largest known meteorite, which lies where it was found in 1920 at Hoba Farm, near Grootfontein, Namibia. It is an iron-nickel ataxite with an estimated mass of 59 tonnes and a maximum width of 2.8 m. It produced no crater, but remains partly embedded in the ground.

Hobby-eberly telescope An 11-m reflector at an altitude of 1980 m on Mount Fowlkes, Texas, opened in 1997. It is part of McDonald Observatory, and is jointly owned by the University of Texas, Pennsylvania State University, Stanford University, the Ludwig Maximilian University in Munich, and the Georg-August University in Göttingen. Its mirror consists of 91 hexagonal segments each 1 m wide. The telescope is designed primarily for spectroscopy rather than imaging. It is permanently angled at 55° altitude, but rotates freely in azimuth. It is named after William P. Hobby and Robert E. Eberly, supporters of public education in Texas and Pennsylvania respectively.

Hohmann orbit A trajectory along which a spacecraft moves from one orbit to another with the minimum expenditure of energy, first calculated by the German engineer Walter Hohmann in 1925. Such trajectories are used for changing the orbits of a satellite around the Earth, or for sending a probe from the Earth to another planet. A Hohmann orbit is elliptical, and just touches the orbits of origin and destination. It requires two firings of the spacecraft's rocket motor, one to break out of the original orbit and one to enter the destination orbit. The main disadvantage of the Hohmann orbit is the long flight times involved. From Earth to Mars the transfer time is 260 days; from Earth to Saturn would take about 6 years. For this reason gravity assists are often used to boost the spacecraft's speed and hence reduce journey times.

Holiosheath The interstellar gas presumed to lie between the heliopause and the shock front that may lie beyond it. If the heliosphere is analogous to the Earth's magnetosphere, the shock front would be expected to lie ahead of the heliosphere in the direction of interstellar gas flow.

Holmberg radius A measure of the size of a galaxy, based on its observed surface brightness. It is the radius at which the surface brightness is 26.5 magnitudes per square arc second in blue light, and represents about 1-2% of the brightness of the night sky. It is named after the Swedish astronomer Erik Holmberg.

Holography In radio astronomy, a method for refining the feed and panel alignment, thus for improving the aperture efficiency or antenna gain of a radio telescope. By scanning an antenna's beam over a raster around an unresolved radio source, and using another antenna pointing at the same source as a reference, information is obtained about the amplitude and phase distributions of the signal reflected from the antenna surface. These distributions are used to specify corrections (if needed) for the focus and alignment of the feeds or of the positions of individual panels in the reflector.

Homogeneity The property of being uniform throughout space. The assumption of homogeneity is part of the cosmological principle. The Universe is clearly not exactly homogeneous, so cosmologists define homogeneity on the large scale by saying that different parts of the Universe are, on average, identical when one looks at sufficiently large volumes.

Homologous flares Solar flares that occur repetitively in the same active region, with essentially the same position and with a common pattern of development.

Homology The use of the principle of *homologous deformation* in the design of antenna structures. Any massive structure must deflect under its own weight. As the reflector of an antenna in a radio telescope must move relative to the direction of the gravitational force, the deflections must change as the antenna moves. An antenna is said to obey the principle of homologous deformation if the entire structure above the elevation bearings is designed to maintain the reflector in the shape of a paraboloid of revolution when pointing to any position between the zenith and the horizon. This can be achieved if the focal length of the paraboloid is allowed to change, and if the electrical axis of the paraboloid is allowed to move in a slightly different way than the elevation axis rotates. (These deviations can be compensated by moving the feeds.)

Homunculus nebula A cloud of gas and dust that obscures the star Eta Carinae, with a shape that resembled a small human to early observers. The material was ejected by the star in 1843 and is now expanding at about 500 km/s. The nebula hides the star from view, and reradiates much of the absorbed energy at infrared wavelengths.

Hooker telescope The 100-inch (2.5-m) reflector at Mount Wilson Observatory, opened in 1917. It is named after the American businessman John D. Hooker who provided funds to buy the mirror. It was out of operation 1985-93, but was reopened when Mount Wilson Institute took over the running of the observatory.

Horizon The great circle formed by the intersection of the plane perpendicular to the observer's zenith with the celestial sphere; also called the astronomical horizon.

Horizontal branch A horizontal strip of stars in the Hertzsprung-Russell diagram of globular clusters, comprising stars somewhat bluer and fainter than those on the giant branch. These stars are thought to be burning helium in their core and hydrogen in a surrounding shell. They are of 0.6-0.8 solar masses, of which more than half is in the helium-burning core, and have probably lost substantial mass through stellar winds during their giant phase. Stars typically spend 50-100 million years on the horizontal branch.

Horizontal coordinates A system of coordinates that specifies the angular position of a celestial object relative to the observer's horizon at a given time. The coordinates used are altitude, the object's angular distance above the horizon, and azimuth, the bearing of the object clockwise from north measured parallel to the horizon. Sometimes the zenith distance is used instead of altitude. The coordinates change with the observer's position and as the Earth rotates.

Horizontal parallax The difference between the topocentric and geocentric positions of an object when it is on the astronomical horizon. The amount varies with latitude because the Earth is not exactly spherical, and is greatest at the Earth's equator.

Horizontal refraction The angular distance of an object below the horizon when it appears to lie on the horizon. The value 34' is used in calculating rising and setting times given in The Astronomical Almanac.

Horn antenna An antenna in the shape of a flared horn. It is often fed by a waveguide. A small horn is often used as the feed at the focus of a parabolic antenna. The cosmic background radiation was first detected with a large horn antenna.

Horologium An insignificant constellation of the southern sky, representing a pendulum clock. It contains little of interest. Its brightest star, Alpha Horologii, is of magnitude 3.9.

Horrocks, Jeremiah english astronomer. He refined J. Kepler Rudolphine Tables of planetary positions, measured the apparent diameters of the planets, and obtained a value of the solar parallax corresponding to an Earth-Sun distance of about 100 million km. In 1639 he observed a transit of Venus which he had predicted from his refinement of Kepler's tables, measuring the planet's diameter on a projected image. On the basis of Horrocks's lunar theory, J. Flamsteed later constructed tables of the Moon's motion which remained the best available until the mid-18th century.

Horsehead nebula A dark nebula in Orion shaped like the head of a horse, also known as Barnard 33. It is about 6' long and projects into the bright nebula IC 434 south of the star Zeta Orionis, but is well seen only on long-exposure photographs.

Horseshoe mounting An equatorial mounting in which the top end of a fork mounting is supported by a large circular bearing resembling a horseshoe. The design has great stability and has been used for large telescopes, such as the 200-inch (5-m) Hale Telescope at Palomar Observatory.

Horst A strip of land, bounded by parallel faults, that has been uplifted between them, the opposite movement to that in graben. Examples of horsts occur in the Ceraunius Fossae on Mars.

Hot big bang An alternative term for the standard Big Bang theory. The word 'hot' was initially used to distinguish it from a rival theory which had a cold initial phase. The existence of the cosmic background radiation requires that the Universe must have been hot in the past if the Big Bang picture is correct.

Hot dark matter A particular type of non-baryonic particle that, according to some theories, is created in the early stages of the Big Bang and survives to the present time in sufficient numbers to contribute significantly to the present density of the Universe. The term 'hot' signifies that these particles are fast-moving usually because they are of low mass. The favoured candidate for such a particle is a neutrino with a rest mass of around 10 eV, which is 1/500000 of the mass of the electron.

Hot spot 1. A compact, highly luminous region in a cataclysmic binary, located either where the stream of material hits the edge of the accretion disk, or at the inner edge of the disk. There may be small-scale flickering or larger variations in luminosity depending on fluctuations in the rate of mass transfer. The presence of a hot spot is often indicated by a hump in the orbital light-curve.

2. A small, bright region in a lobe of a radio galaxy. It is believed to be the site where high-velocity material in a jet collides with the leading boundary between the lobe and the surrounding intergalactic medium.

Hour angle The angle between the observer's meridian and the hour circle of a celestial object, measured anticlockwise along the celestial equator. The observer's meridian is therefore the meridian of zero hour angle. The hour angle of any object increases as the Earth rotates, and it can be expressed as a time by equating 360° with 24 hours. For example, an hour after an object crosses the meridian its hour angle is 1h, and so on. It is sometimes used as a coordinate in place of right ascension..

Hour circle A great circle on the celestial sphere passing through a celestial object and the celestial poles; also known as meridian of hour angle. It is perpendicular to the celestial equator.

Hourglass nebula The brightest central part of the Lagoon Nebula in Sagittarius, shaped like a figure-of-eight, near the 6th-magnitude star 9 Sagittarii.

Howardite A class of calcium-rich achondrite meteorite; also known as plagioclase-hypersthene achondrites. They are named after the English chemist Edward Charles Howard The howardites are breccias whose texture resembles that of the breccias found in the lunar regolith, and were apparently formed on the surface of their parent body by impact processes. Most of the fragments found in howardites resemble eucrite and diogenite material. The howardites are apparently related to the stony-iron mesosiderites.

Hoyle, Fred (1915) English astrophysicist and cosmologist. In 1948, with H. Bondi and T. Gold, he proposed the steady-state theory of the Universe in which matter is continuously created. Subsequently abandoned by most astronomers in favour of the Big Bang (so named from a dismissive remark by Hoyle), the steady-state theory nevertheless stimulated much important astrophysical research. Particularly significant was the work by Hoyle, W. A. Fowler, and G. R. and E. M. Burbidge on nucleosynthesis in stars. As well as his noted work on stellar evolution, Hoyle has propounded unorthodox ideas, suggesting for example that viruses and perhaps other life forms have been brought to Earth by comets. He has also been a popularizer of astronomy.

HPBW For an antenna beam in any one dimension, the full width in angle between the directions at which the received or transmitted power falls to half of its maximum value. (Synonymous with the FWHM of the antenna's power pattern in this dimension.)

H-R diagram The Hertzsprung -Russell (H-R) Diagram is a graph that plots stars color (spectral type or surface temperature) vs. its luminosity (intrinsic brightness or absolute magnitude). On it, astronomers plot stars' color, temperature, luminosity, spectral type, and evolutionary stage. This diagram shows that there are 3 very different types of stars:

Most stars, including the sun, are "main sequence stars," fueled by nuclear fusion converting hydrogen into helium. For these stars, the hotter they are, the brighter.

Giants and Supergiants are above the main sequence.

Faint white dwarfs (hot, dim stars) are below the main sequence stars.

HR number The number of a star in the Harvard Revised Photometry. The same numbers are used in the Bright Star Catalogue.

HST HST is an abbreviation for the Hubble Space Telescope, a powerful telescope in orbit around the Earth. HST takes pictures and spectra of objects in space without the interference of the atmosphere (which makes telescopic images from the ground have less detail); it transmits the pictures and spectra back to scientists on Earth. The telescope was launched into space in April, 1990, and was repaired in December, 1993. It was named for the American astronomer Edwin Hubble.

HSV (hue, saturation, value) A system for describing the physical perception of color, in terms of tint, perceived narrowness of the spectrum, and intensity (here substituted for by "value").

Hubble classification A widely used system for classifying galaxies according to their visual appearance, illustrated on the tuning-fork diagram. The sequence is based on three criteria: the relative sizes of the central bulge of stars and the flattened disk; the existence and character of spiral arms; and the resolution of the spiral arms and/or disk into stars and H II regions. The system was originated by E. P. Hubble.

The sequence starts with round elliptical galaxies showing no disks. Increasing flattening of a galaxy is indicated by a number which is calculated from 10(a - b)/a, where a and b are the major and minor axes as measured on the sky. No elliptical is known that is flatter than E7. Beyond this a clear disk is apparent in the lenticular or SO galaxies. The classification then splits into two parallel sequences of disk galaxies showing spiral structure: ordinary spirals, S, and barred spirals, SB. The spiral types are subdivided into Sa, Sb, Sc, Sd (SBa, SBb, SBc, SBd for barred spirals).

Hubble constant; h_0 (E.P. Hubble; 1925) The constant which determines the relationship between the distance to a galaxy and its velocity of recession due to the expansion of the Universe. After many years in which the Hubble constant was only known to be somewhere between 50 and 100 km/s/Mpc, it has been determined to be 70 km/s/Mpc ± 7 km/s/Mpc by the Hubble Space Telescope's Key Project team. (Advances in cosmology have shown that since the Universe is self gravitating, H_0 is not truly constant. Astronomers thus seek its present value.)

Hubble diagram A graph that plots either the redshift or velocity of recession of galaxies against their apparent magnitude or distance from us. The Hubble law appears in the form of a straight line on such a plot. The original diagram, presented by E. P. Hubble in 1929, was the first indication that the Universe is expanding. The Hubble diagram is now mainly used to test the geometry of the Universe, since at large distances any departures from the simple linear form of the Hubble law should show up as a curve.

Hubble flow The general outward motion of galaxies resulting from the uniform expansion of the Universe. All motions lie in a radial direction from the observer, and the velocities are proportional to the distance of the galaxies. The real pattern of galaxy motions is not exactly of this form, particularly close to us, because of the mutual gravitational interaction between galaxies; some nearby galaxies are even moving towards the Milky Way. At large distances, however, the discrepancies are small compared with the Hubble flow.

Hubble law The law that governs the expansion of the Universe. According to the law, the apparent recession velocity of galaxies is proportional to their distance from the observer. In mathematical terms, v = H0 r, where v is the velocity, r the distance, and H0 the Hubble constant. The law was put forward in 1929 by E. P. Hubble.

Hubble radius A distance defined as the ratio of the velocity of light, c, to the value of the Hubble constant, H0. This gives the distance from the observer at which the recession velocity of a galaxy would equal the speed of light. Roughly speaking, the Hubble radius is the radius of the observable Universe. Depending on the precise value of the Hubble constant, the Hubble radius lies between 9 and 18 billion L.Y

Hubble space telescope A 2.4-m reflector built by NASA and ESA, launched in 1990 April. It orbits at an altitude of about 600 km. Its original instruments included the Wide Field and Planetary Camera Faint Object Camera Faint Object Spectrograph Goddard High Resolution Spectrograph,and High Speed Photometer.During a servicing mission in 1993 December astronauts replaced the WF/PC with the improved WFPC2. They also installed the Corrective Optics Space Telescope Axial Replacement system in place of the High Speed Photometer to correct the telescope's spherical aberration, caused by a manufacturing error in the main mirror. On a second servicing mission, in 1997 February, astronauts installed the Space Telescope Imaging Spectrograph and the Near-Infrared Camera and Multi-Object Spectrometer in place of the GHRS and FOS.

Hubble time The time required for the Universe to expand to its present size, assuming that the Hubble constant has remained unchanged since the Big Bang. It is defined as the reciprocal of the Hubble constant, 1/ H0. Depending on the precise value of the Hubble constant, the Hubble time is between 9 and 18 billion years. In the standard Big Bang theory, the actual age of the Universe is always less than the Hubble time, because the expansion was faster in the past.

Hubble, Edwin powell American astronomer. He first studied nebulae, concluding in 1917 that the spiral-shaped ones were different in nature from diffuse nebulae, which he found to be gas clouds illuminated by stars. From 1923, using the 100-inch (2.5-m) telescope at Mount Wilson Observatory, he resolved the outer regions of the spiral nebulae M31 and M33 into stars, identifying over 30 Cepheid variables in them. This proved that such 'nebulae' were truly independent star systems like our own-other galaxies. In 1925 he devised the so-called tuning-fork diagram of galaxies, dividing them into ellipticals, spirals, and barred spirals, which he believed to indicate an evolutionary sequence. By 1929 Hubble had good distance measurements for over twenty galaxies, including members of the Virgo Cluster.

Hubble's variable nebula A reflection nebula in Monoceros, also known as NGC 2261, which is illuminated by the irregular variable R Monocerotis. The nebula's variability was detected by E. P. Hubble in 1916 and probably arises from changes in the star's luminosity and shadowing by dense clouds within the nebula. R Monocerotis is a young T Tauri star and a strong infrared source. It exhibits a vigorous bipolar flow, and is apparently surrounded by a protoplanetary disk—a disk of material from which planets may form.

Hubble-sandage variable An extremely massive, highly luminous hypergiant variable of a class originally identified in M31 and M33 by E. P. Hubble and A. R. Sandage in 1953. Such stars, currently classified as S Doradus stars, are among the most massive and luminous found in any galaxy.

Hue One of the three basic parameters (*hue*, *intensity*, and *saturation*) which may be used to describe the physical perception of the light that reaches one's eye. Hue, which is also termed *tint*, or simply *color*, refers to the dominant wavelength of the coloration, at a given location in an image or scene. The term also may be used to describe a multimodal color spectrum—e.g., one speaks of a purple hue. Different spectral distributions of light, of identical intensity and saturation, are capable of producing identical retinal responses; these unique responses comprise the set of perceptible hues.

Color matching tests have established that there are three basic types of human retinal receptors, whose peak responses are to red, green, and blue light. These are the three *primary hues* used in additive color mixing—e.g., in digital image display. They may be used to produce all, or virtually all, of the perceptible hues.

Huggins, William and margaret Sir William Huggins (February 7, 1824-May 12, 1910) was an amateur English astronomer who was the first person to use spectroscopy to determine the compositions of astronomical objects (in 1861). He determined that the Sun and the stars are composed mostly of the element hydrogen. He also examined the spectra of nebulae and comets. Huggins' wife (they were married in 1875), Margaret Lindsay Murray Huggins (1848-1915), was a self-taught astronomer who did extensive work in spectroscopy and photography. Margaret studied the Orion Nebula extensively. William and Margaret were the first people to realize that some nebulae, like the Orion Nebula, consisted of amorphous gases (and were not a congregation of stars, like the nebula Andromeda). A lunar crater, a Martian crater, and an asteroid (#2635 Huggins) have been named for William Huggins.

Hulst, Hendrik christoffel van de Dutch astronomer. His main field of study was interstellar matter. In 1944, while a student of J. H. Oort, he showed that neutral hydrogen should emit radio waves with a wavelength of 21 cm. His prediction was borne out in 1951 when the American physicists Edward Mills Purcell and Harold Irving Ewen detected 21-cm radio emission from interstellar hydrogen clouds. The same year, van de Hulst and Oort began to use Doppler shifts in hydrogen emission to examine the Galaxy's structure.

Humason, Milton Lasell American astronomer. He worked at the Mount Wilson (where he was employed originally as janitor) and Palomar Observatories. In the 1920s he participated in long-term projects with W. S. Adams, Alfred Harrison Joy and others to measure the absolute magnitudes and radial velocities of thousands of stars. From 1930, first with the 100-inch (2.5-m) telescope and later the 200-inch (5-m), he measured the radial velocities of hundreds of faint galaxies, finding progressively larger redshifts.

Hummocky terrain A type of terrain on a planetary surface consisting of irregular rounded hummocks in a chaotic pattern. The inner part of the ejecta blanket of a young impact crater is usually hummocky terrain, and it is also found in the furthest parts of large landslips.

Hungaria group A group of asteroids at a mean distance of about 1.95 AU from the Sun, near the inner edge of the main asteroid belt but separated from it by the Kirkwood gap at the 4:1 resonance with Jupiter. Hungaria group members have low orbital eccentricities, but high inclinations of 22-24°. The group has been separated from the asteroid belt by the gravitational effects of the major planets, notably Jupiter, and contains a broad representation of asteroid classes. All Hungaria objects are small. The group is named after (434) Hungaria, an E-class asteroid of diameter 11.4 km, discovered in 1898 by Max Wolf. Hungaria's orbit has a semimajor axis of 1.944 AU, period 2.71 years, perihelion 1.80 AU, aphelion 2.09 AU, and inclination 22°.5.

Hunter's moon The full Moon following the harvest Moon, usually occurring in October. As at harvest Moon, the full Moon rises around the time of sunset for several nights in succession in mid- or high-northern latitudes.

Huygenian eyepiece An eyepiece design invented by C. Huygens in 1703. It consists of two planoconvex lenses with their convex surfaces facing the objective lens. Because of its cheap and simple construction it is often used with small refractors, but will only perform well with large focal ratios. The Huygenian eyepiece suffers from aberrations if used on telescopes faster than about f/10. A variant known as the Mittenzwey eyepiece uses convex meniscus lenses and is usable down to f/8.

Huygens, Christiaan Dutch mathematician, physicist, and astronomer. In 1655 he discovered Titan, the largest satellite of Saturn. Soon after, he announced that the planet was surrounded by a 'thin flat ring'; others had observed the rings, but misinterpreted what they saw. In 1659 he discovered dark markings on Mars, notably the wedge-shaped Syrtis Major. With his brother, Constantijn Huygens (he constructed aerial telescopes (tubeless telescopes supported by cablesof very long focal length to overcome aberrations. He invented the Huygenian eyepiece. To physics he contributed the principle of conservation of momentum a theory of centrifugal force 1659), and the wave theory of light.He also made the first pendulum clock.

Huygens gap A division near the outer edge of Saturn's B Ring. It was discovered in 1981 on images from Voyager 2 and named after C. Huygens.

Hyades A large, V-shaped open cluster in Taurus, spanning over 5° of sky. It contains about 200 stars, and is estimated to be 660 million years old. The cluster lies 150 L.Y away, and its true diameter is about 15 L.Y

Comparison of the brightness of its stars with those in other clusters enables the distances to those other clusters to be calculated. The Hyades is thus an important yardstick in astronomical measurement.

Hyakutake, comet A long-period comet discovered on 1996 January 31 by the Japanese amateur astronomer Yuuji Hyakutake.It passed 0.10 AU from Earth on March 25, showing a bluish-green gas tail stretching up to 70°, while the head reached magnitude -1. Radar measurements revealed the comet's nucleus to be 1-3 km in diameter. Comet Hyakutake reached perihelion, 0.23 AU, on 1996 May 1. The comet's orbit has a period of about 20000 years, eccentricity 0.9997, and inclination 124°.9.

Hybrid eclipse A hybrid eclipse is an annular solar eclipse in which a brief total eclipse occurs in a small region along the eclipse's central path.

Hydra The largest constellation of all, representing a water-snake. Hydra meanders from the celestial equator into the southern celestial hemisphere, and is over 100° long. Its brightest star is Alphard.R Hydrae is a Mira star that varies between magnitudes 3 and 11 every 390 days, and U Hydrae is a semiregular variable that ranges between magnitudes 4.3 and 6.5. M48 is a 6th-magnitude open star cluster visible in binoculars, and M83 is an 8th-magnitude spiral galaxy. NGC 3242 is a planetary nebula known as the Ghost of Jupiter.

Hyder flare a filament-associated two-ribbon flare, often occurring in spotless regions. the flare presumably results from the impact on the chromosphere of infalling filament material.

Hydrazine A colorless liquid which burns rapidly and is used as a common rocket and missile fuel

Hydrocarbon A compound which is composed entirely of carbon and hydrogen atoms. Examples of hydrocarbons include methane, ethane, and propane. They are found in meteorites, comets, and nebulae.

Hydrogen (symbol H) The most abundant chemical element in the Universe, comprising some 73% of its mass. Hydrogen is found in several forms: molecular hydrogen, H_2, the most familiar form on Earth, occurs in interstellar gas clouds; atomic hydrogen, formed of individual hydrogen atoms, occurs in the more diffuse interstellar clouds and ionized hydrogen, in which each atom is broken down into a proton and an electron, and which is the most common form of all, being found in stars.

Hydrogen burning A series of nuclear reactions in which hydrogen nuclei (protons) fuse to form a helium nucleus, with the associated release of nuclear energy. In stars with core temperatures below about 18 million K. hydrogen burning occurs primarily through the proton-proton reaction. In hotter stars, the carbon-nitrogen cycle is primarily responsible.

Hydrologic cycle The Hydrologic Cycle (also known as the water cycle) is the journey water takes as it circulates from the Earth to the sky and back again.

Hydrogen emission region A region in interstellar space containing hot hydrogen gas which emits a characteristic red spectral line at 656.3 nm; such a region is the visible counterpart of an H II region. The 656.3 nm red line is the brightest line in the optical spectrum of hydrogen, and is known as the Ha line.It gives nebulae their characteristic red colour in photographs.

Hydrogen ion A hydrogen atom that has gained or lost an electron, and thus has an electric charge. The positive hydrogen ion, H+, has lost its only electron and therefore consists of a single proton. A nebula containing H+ ions is known as an H II region. The negative hydrogen ion, H-, has gained a second electron. H- ions are found in the outer layers of stars such as the Sun. Most of the light received from the Sun actually originates from the formation of H- ions.

Hydrogen maser A device that produces a radio signal directly from the transition between the upper and lower hyperfine levels in the ground state of atomic hydrogen. The most stable type of frequency standard or clock (for time scales of a few seconds to few hours) that is used in VLBI observations.

Hydrogen spectrum A distinctive pattern of spectral lines, either emission or absorption, produced by hydrogen, the simplest of all the atoms. The series of lines in the spectrum are labelled according to the lowest energy level involved in the transitions that give rise to the lines. If an electron jumps to or from the ground state (n = 1), it produces a line in the Lyman series; n = 2 corresponds to the Balmer series; n = 3 to the Paschen series; n = 4 to the Brackett series; and n = 5 to the Pfund series. The Balmer series was the first to be studied in the laboratory because all the lines are in the visible part of the spectrum. The Paschen, Brackett, and Pfund series lie in the infrared, whereas the Lyman series is entirely in the ultraviolet.

Hydroponics Hydroponics is a method of growing plants in water that contains dissolved nutrients (instead of in soil).

Hydrosphere The portion of Earth, or other planet, that is water, including liquid water, ice, and water vapor on the surface, underground, or in the atmosphere.

Hydrostatic equilibrium Hydrostatic equilibrium is a stable condition in a star in which the fluid matter within the star is at an equilibrium with respect to all forces, including the inward-pulling force of gravity, the out-ward pulling buoyancy due to pressure differentials, and the out-ward pulling forces of radiative pressure.

Hydroxyl (symbol OH) A molecule composed of an oxygen atom and a hydrogen atom; also called the hydroxyl radical. Hydroxyl is one of the commonest interstellar molecules. It was the first interstellar molecule discovered by radio astronomers, in 1963

Hydrus An unremarkable constellation near the south celestial pole, representing a small water-snake. Beta Hydri is its brightest star, magnitude 2.9.

Hygiea Asteroid 10, discovered in 1849 by the Italian astronomer Annibale de Gasparis.It is the fourth-largest main-belt asteroid, diameter 430 km, and is of C class. Its orbit has a semimajor axis of 3.134 AU, period 5.55 years, perihelion 2.76 AU, aphelion 3.51 AU, inclination 3°.8.

Hypatia Hypatia of Alexandria (AD 370-415) was a Greek mathematician, astronomer, teacher, and head of the Platonist school at Alexandria about AD 400. Hypatia wrote commentaries on the astronomical canon of Ptolemy and did work on conic sections. Her works are lost, but are referred to in the Suda lexicon. She was the daughter of the mathematician and philosopher Theon of Alexandria (he was also the last head of the Museum at Alexandria). A pagan, she was murdered in 415 by Christian monks in a religious/political struggle. The lunar Crater Hypatia and Rimae Hypatia were named for her.

Hyperbola A type of curve whose two 'arms' diverge and never rejoin, defined mathematically as a conic section with an eccentricity greater than 1. The orbit of one body moving past another without being captured has the shape of a hyperbola.

Hyperbolic comet A comet whose orbit around the Sun has an eccentricity greater than 1.0. Periodic comets may be injected into hyperbolic orbits following passage close to Jupiter, as with Comet Lexell in 1779. Comets on such trajectories are lost to the Solar System. Some comets making their first visit to the inner Solar System from the Oort Cloud may follow hyperbolic trajectories, never to return. The possibility that some hyperbolic comets come from outside the Solar System cannot be excluded.

Hyperbolic orbit An orbit having the shape of a hyperbola. A body in a hyperbolic orbit about another makes only one approach to the other body, in theory approaching from infinity and then receding to infinity. A spacecraft making a fly-by of a planet follows a hyperbolic trajectory.

Hyperbolic velocity The speed of an object that is moving faster than the escape velocity of a planet or other massive body. Hyperbolic velocity is capable of taking the object to an infinite distance from the planet or body.

Hyperboloid A surface or body obtained by rotating a hyperbola about its axis of symmetry. The secondary mirror in a Cassegrain telescope has a hyperboloidal surface.

Hypered film Film that has been treated, usually with gas, to enhance its response to low light levels

Hyperfine structure Very close splitting of spectral lines, due to the intrinsic spin of the atomic nucleus. It is a much smaller effect than fine structure and hence can be seen only at the very highest spectral resolutions.

Hypergalaxy A dominant galaxy with associated companion galaxies and intergalactic material. Many of the companion galaxies will be dwarfs. In the Local Group, both the Milky Way Galaxy and the Andromeda Galaxy are examples.

Hypergiant star A term usually now used to describe a star more massive than about 30 solar masses. The term was first used for the brightest objects observed in the Magellanic Clouds, because they appeared to be more luminous, and hence more massive, than any star in our own Galaxy. It is now known that these objects are not individual stars at all, but clusters of less massive stars.

Hyperion Hyperion is one of the 18 moons of Saturn. Hyperion is the largest-known irregular-shaped body in our Solar System. It has a variable rotational period. Craters mark the surface. Iapetus has a radius of about 180x140x112.5 km and is irregularly shaped. Hyperion orbits at about 1,481,100 km from Saturn. Its orbital period is 21.27 days. It was discovered by W. & G. Bond/W. Lassell in 1848.

Hypernove A hypernova is an enormous collapse and subsequent explosion of a supergiant star that eventually forms a black hole. This explosion is bigger than a supernova and is accompanied by a gamma-ray burst. The plural of hypernova is Hypernovae

Hyperon A baryon that is heavier than a proton or neutron. The term is becoming obsolete.

Hypersensitization A method of treating photographic emulsions to reduce the effects of reciprocity failure; also known as hypering. The most common modern technique involves soaking the film or plates in a chemically reducing gas-ideally, pure hydrogen. For amateur use, forming gas (8% hydrogen, 92% nitrogen) is preferred. Other methods include soaking the film or plates in water or ammonia; baking the emulsion for many hours at around 60 °C; and briefly preexposing the emulsion to light to achieve a background level on top of which light from the sky will be recorded more efficiently.

Hypersthene achondrite Another name for a diogenite.

Hypervelocity impact An impact of a particle into a surface at 1-2 km/s or more, a velocity at which solid rock behaves like a fluid. Hypervelocity impacts produce craters that are generally circular, as the crater is usually much larger than the irregular impacting body.

Hypothesis An untested statement about nature; a scientific conjecture, or educated guess. Formally, a hypothesis is made prior to doing experiments designed to test it. Compare: law and theory.

HZ herculis star A low-mass X-ray pulsar that is also an optically variable binary undergoing mass transfer. Such stars consist of a dwarf primary of spectral type B-F, and a neutron star which is an X-ray pulsar and possibly also an optical pulsar. There is a pronounced reflection effect caused by the X-rays incident on the primary. The X-ray emission switches periodically between a high state and a low state, and the optical luminosity is greatest when the X-ray source is in the high state. In the type star, also known as Hercules X-1, the amplitude is 2-3 mag.

HZ43 A hot white-dwarf star in the constellation Coma Berenices with a surface temperature of approximately 50 000 K. The SAS-3 satellite discovered that it was a strong emitter of soft X-rays, and it was also the first cosmic source of extreme ultraviolet radiation to be discovered. Because of its simple, well-known spectrum, it is used as a standard for calibrating EUV astronomy missions.

I

I magnitude The magnitude of a star measured at infrared wavelengths through an I filter. The wavelength of the I filter can have three separate values, depending on the system used. The wavelength and bandwidth on the Kron-Cousins RI photometry system are 797 and 149 nm; on the Johnson photometry system, 900 and 240 nm; and on the six-colour system, 10.3 μm and 180 nm. The first is now the most widely used, and is sometimes denoted by the subscript KC or C.

Iapetus The third-largest satellite of Saturn, diameter 1460 km; also known as Saturn VIII. It is the second-outermost satellite: orbiting at a distance of 3 561300 km. Iapetus orbits in 79.33 days, keeping the same face turned towards Saturn. Its leading hemisphere is far darker than the trailing hemisphere, the albedos being about 0.05 and 0.5 respectively, so that its mean opposition magnitude varies between 10.2 and 11.9. Iapetus was discovered in 1671 by G. D. Cassini. Impact craters cover its surface.

IAU Abbr. for International Astronomical Union.

IC Abbr. for Index Catalogue.

Icarus Asteroid 1566, a member of the Apollo group, discovered by W. Baade in 1949 during a close approach to Earth. In 1968, Icarus became the first minor planet to be observed by radar, when it came within 0.040 AU (6 million km) of Earth. Its diameter is about 1 km, and its rotation period 2.27 hours. Its orbit has a semimajor axis of 1.078 AU, period 1.12 years, perihelion 0.19 AU aphelion 1.97 AU, and inclination 22°.9.

Ice Used by planetary scientists to refer to water, methane, and ammonia which usually occur as solids in the outer solar system.

Ice age An ice age is a time lasting thousands of years during which the Earth is very cold and largely covered by ice and glaciers.

Ida Asteroid 243, discovered in 1884 by the Austrian astronomer Johann Palisa. It has a satellite, Dactyl. Ida, an S-class asteroid, measures 55 × 24 × 20 km and rotates in a period of 4.6 hours about its shortest axis. It is a member of the Koronis family. If its parent body was broken into several pieces, one of these fragments may have become its satellite. Its orbit has a semimajor axis of 2.861 AU, period 4.84 years, perihelion 2.74 AU, aphelion 2.98 AU, and inclination 1°.1.

Ideal-lens equation $1/p + 1/q = 1/f$, where p is the distance from object to lens, q is the distance from lens to image, and f is the focal length of the lens. This equation has important limitations, being only valid for *thin* lenses, and for *paraxial rays*. Thin lenses have thickness small compared to p, q, and f. Paraxial rays are those which make angles small enough with the optic axis that the approximation *(angle in radian measure)* = *sin(angle)* may be used.

Igneous A description applied to a rock formed by the cooling and solidification of a magma (molten rock), either below or on the surface of a planetary body. Igneous rocks are said to be extrusive if they solidify on the surface of a planet, and intrusive if they solidify at depth beneath the crust.

Igneous rock When molten rock cools, igneous rock is formed.

IGY Abbr. for International Geophysical Year.

Ikeya-seki, comet A long-period comet, discovered independently on 1965 September 18 by the Japanese amateur astronomers Kaoru Ikeya and Tsutomu Seki formerly designated 1965 VIII. It is a member of the Kreutz sungrazer group. Perihelion, 0.008 AU (1.2 million km), was on 1965 October 21, when its nucleus broke into three fragments. At perihelion the comet reached at least magnitude -10 and could be seen in broad daylight, while its tail reached about 60° in late October. Ikeya-Seki's orbit has a period of about 880 years, eccentricity 0.9999, and inclination 141°.9.

IKI The Space Research Institute in Russia. It is the equivalent of NASA in the U.S.

Illuminance (symbol E v) The energy in the form of visible radiation that falls on a surface in a given time. It is measured in lux.The energy incident at all wavelengths, not the visible, is termed irradiance.

Ilmenite Opaque mineral found in basalt; nearly pure iron-titanium oxide ($FeTiO_3$.)

Image 1. In computer graphics, two-dimensional data representing a scene. A digital image is composed of pixels arranged in a rectangular array with a certain height and width. Each pixel may consist of one or more bits of information, representing the brightness of the image at that point and possibly including color information encoded as RGB triples. Images are usually taken from the real world via a digital camera, frame grabber or scanner. See also *image formats*.

2. In radio astronomy, the above definition is generalized to mean a discrete representation of a segment of the sky brightness distribution as intensities in a finite-volume, linear, rectangular, or hyper-rectangular array of pixels. The term also is used (less technically) to refer to the *display* of such data, either on a computer graphics display or as a television picture.

3. In mathematical use, the image (or range) of a function is the set of values of that function applied to all elements of its domain.

Image co-ordinate system In AIPS++, an engine attached to an Image for turning positions within the Lattice to *Measures*, usually related to position on the sky and to Doppler velocity. An Image, for example a spectral line cube, may have more than one measure associated with its co-ordinate system.

Image enhancement Any process for improving the appearance, or suitability for a particular purpose, of an image using filtering, transformation or thresholding techniques, e.g. edge enhancement, histogram modification, median filtering, smoothing, sharpening, unsharp masking, false color or pseudo-color displays. In image enhancement there is no explicit effort to improve the fidelity of the reproduced image relative to some ideal form, as in the case of image reconstruction. Indeed, the goal is often to *emphasize* some feature of the image to draw a user's attention to that feature at the expense of fidelity to the natural "scene".

Image photon counting system An electronic system for recording faint images. It uses a four-stage image tube. The phosphor screen of the image tube is viewed by a sensitive television camera whose images can be stored and added. The IPCS can record photons of light striking the photocathode with a quantum efficiency of 20%. The system is less sensitive and more complex than a CCD which is more commonly used for routine observing, but it has the important advantage of being able to distinguish genuine photons from noise.

Imbrium basin An ancient impact basin on the Moon about 1300 km across, centred at about +33° lat., 16°W long. It is about 3.9 billion years old, and is the result of the largest impact on the Moon, which was nearly big enough to break the Moon apart. The Imbrium impact had far-reaching effects, sending ejected material over much of the Moon's surface and creating deep fractures within it. Following the impact, a long period of lava flooding occurred through these fractures, creating the Mare Imbrium and submerging much of the Imbrium Basin itself and a large area outside it.

Immersion An alternative name for the disappearance of a star at an occultation; or, the covering of the Moon by the Earth's shadow at a lunar eclipse.

Impact The forceful striking of one body, such as a meteorite, against another body such as a moon or planet.

Impact basin A impact basin is an impact crater that has a rim diameter greater than 185 miles (300 km). There are over 40 impact basins on the Moon, including Schrodinger (above). These catastrophic impacts produce faulting and other crust deformations. Material ejected from impact basins is distributed over wide areas.

Impact crater A depression caused by the high-speed collision of a meteoroid, asteroid, or comet with a solid surface. Impact craters are found on the Moon and all the terrestrial planets and satellites except Io. They range in size from tiny pits a few micrometres across up to vast impact basins thousands of kilometres wide. Young impact craters may be distinguished from volcanic craters by their circular shape, steep inner walls, and shallower outer slopes, as well as their ejecta blankets, secondary craters, and bright rays. Larger impact craters have terracing on their inner walls, flat floors, and central peaks. Old impact craters become progressively eroded and may lose some of these distinguishing features; others may be partially submerged or filled by lava flows or sediment.

Impact melt Impact melt is rock that melts during impact crater formation. When an objects strikes a planet or moon, forming a crater, the energy released at impact causes extremely high temperatures which can melt some surrounding rock, forming pools or sheets of lava at the bottom of the crater. Impact melt varies greatly in texture, but is uniform in its composition, made mostly of the planet's or moon's rocks, but also containing a bit of the impactor (asteroid, comet, or meteorite) material.

Impactite A substance composed of meteoritic material fused with melted rock, often in fragments smaller than a grain of rice. It results when a large meteorite strikes the Earth with sufficient energy to form a crater. impersonal astrolabe An alternative name for a Danjon astrolabe.

Implosion A violent inward collapse. An inward explosion.

In modern usage The measure of translational inertia is mass. Newton's first law of motion is sometimes called the 'Law of Inertia', a label which adds nothing to the meaning of the first law. Newton's first and second laws together are required for a full description of the consequences of a body's inertia.

The measure of a body's resistance to rotation is its *Moment of Inertia.*

Incandescent An incandescent material is so hot that it glows, producing light. Incandescent solids, liquids, and compressed gases produce a continuous spectrum; other gases produce a line or emission spectrum (only a few wavelengths are emitted).

Inclination (symbol i) 1. The angle between the orbital plane of a body and the reference plane centred on the object about which the body is revolving. For planets in the Solar System, the inclination of the orbit is usually given relative to the plane of the Earth's orbit, the ecliptic. For objects orbiting a planet, the inclination is usually given relative to the planet's equator. For double stars, the inclination is given relative to the plane of the sky. The inclination is one of the elements of an orbit.

2. The angle at which the rotation axis of a planet or satellite is tilted with respect to the perpendicular to some reference plane. The reference plane is usually the plane of the body's orbit.

Index catalogue Either of two supplements to the New General Catalogue compiled by J. L. E. Dreyer. The first Index Catalogue, containing 1529 newly discovered non-stellar objects, was published in 1895, and the second with 3857 objects, in 1908; both contained corrections to the NGC.

Indochinite A type of tektite found in Thailand and the area formerly known as Indo-China, comprising Laos, Cambodia, and Vietnam, one of many regions in Southeast Asia where tektites have been recovered.

Indus A constellation of the southern sky, representing a native American indian. Its brightest star is Alpha Indi, magnitude 3.1. Epsilon Indi is a K4 dwarf star 11.3 1.y. away.

Inequality A variation in the movement of a celestial object in its orbit about another which cannot be accounted for by their mutual gravitational attraction. Inequalities usually arise because of the perturbing forces of one or more other massive objects in the system. For example, the so-called great inequality in the orbital movements of the planets Jupiter and Saturn about the Sun is an oscillation in their heliocentric longitude with a period of some 900 years caused by their mutual perturbations and also by the nearly 2 : 5 resonance in their mean motions.

Inertia A descriptive term for that property of a body which resists change in its motion. Two kinds of changes of motion are recognized: changes in translational motion, and changes in rotational motion.

Inertial coordinate system The system of coordinates that is used in an inertial reference frame, normally in the special theory of relativity. The three space coordinates are usually Cartesian coordinates and the time coordinate is the time as measured by an observer at rest in the coordinate system. In astrometry, an inertial coordinate system is a reference frame formed by assigning coordinates to specific observable objects, such as the positions and proper motions of stars in a fundamental catalogue.

Inertial frame A non-accelerating coordinate system. One in which $F = ma$ holds, where F is the sum of all real forces acting on a body of mass m whose acceleration is a. In classical mechanics, the *real forces* on a body are those which are due to the influence of another body. [Or, forces on a part of a body due to other parts of that body.] Contact forces, gravitational, electric, and magnetic forces are real. *Fictitious forces* are those which arise solely from formulating a problem in a non-inertial system, in which $ma = F +$ *(fictitious force terms)*

Inertial mass A measure of a body's resistance to a change in its velocity or state of rest. Inertia is a direct property of the mass of a body: the greater the mass, the greater its inertia. Although mass is formally defined in terms of its inertia, it is usually measured by gravitation

Inertial reference frame A frame of reference in which any body initially at rest will remain at rest indefinitely, or in which a moving body moves in a straight line with constant speed indefinitely; in other words, it is free from any inertial forces. It can be defined as a frame of reference in which Newton's laws of motion apply exactly. The frame of reference fixed with respect to the Earth's surface is not inertial, because there are forces due to the Earth's rotation. The concept of an inertial frame of reference is particularly useful in the special theory of relativity, in which two different inertial frames have a fixed relative velocity but no relative acceleration. Inertial frames do not exist in general relativity, however, since, although a freely falling frame will appear inertial locally, differential gravitational effects prevent it from being globally inertial.

Inferior planets Mercury and Venus are called inferior planets because they are closer to the Sun than Earth. [Planets that are farther from the Sun than Earth are called superior planets.]

Inferometry If one combines two or more telescope's light, the resulting image would have the resolution of a telescope the size of the distance separating them

Inflation A brief and extraordinarily rapid period of expansion a fraction of a second after the Big Bang

Inflationary universe A variation on the Big Bang theory that includes a period of accelerated expansion in its early stages. In this model, energy is released during a so-called phase transition about 10^{-35} s after the Big Bang, similar to the release of latent heat as a liquid freezes. The energy released acts in the same way as a cosmological constant term and drives the Universe to expand much faster than in the standard Friedmann models. The ultra-rapid expansion stretches out any primordial 'wrinkles' in the curvature of space-time, rendering the Universe almost smooth and isotropic on the scales we can observe. Another feature of this theory is that it produces minute fluctuations in the density of the Universe which may be the seeds of galaxy formation.

Infrared astronomy The study of the Universe in the infrared part of the spectrum, at wavelengths of 1-300 μm. Infrared astronomy is hampered by the Earth's atmosphere, which is opaque and bright throughout much of the infrared band due mainly to water vapour and carbon dioxide. Another source of interference is warmth from a telescope's surroundings, including the telescope itself, which peaks around 10 μm. Ground-based infrared astronomy is restricted to the few infrared windows in the Earth's atmosphere, especially in the nearinfrared 1-5 μm region, and around 10 μm. Even then, infrared telescopes are placed on high, dry mountain tops. High-altitude balloons and aircraft have also been used, notably the Kuiper Airborne Observatory (KAO). But unimpeded viewing of the infrared sky requires telescopes in space, such as the Infrared Astronomical Satellite and the Infrared Space Observatory.

Prominent infrared sources include red giants and supergiants with dust shells, H II regions, the galactic centre.

Infrared astronomical satellite A joint US-Dutch-British satellite, launched in 1983 January, which made an infrared survey of the sky at wavelengths of 12, 25, 60, and 100 μm. It carried a 0.6-m telescope and an array of detectors cooled by liquid helium. In its 10-month lifetime it surveyed 95% of the sky, detecting 250 000 infrared sources. Among its discoveries were bands of cometary dust in the Solar System; several new comets and asteroids; infrared cirrus; possible forming planetary systems around certain stars; and a large number of galaxies emitting most of their energy in the infrared, including starburst galaxies.

Infrared citrus Wispy clouds of infrared emission first detected by the Infrared Astronomical Satellite and prominent at 60 and 100 μm. Most infrared cirrus is believed to be emission from dust grains in hydrogen clouds, heated by ultraviolet radiation to 20-30 K. Some of the denser cirrus clouds have cooler cores. The clouds are brighter at 12 and 25 μm than expected, possibly due either to a component of very small dust grains or to line emission from complex molecules.

Infrared excess Infrared emission from a star in excess of that expected from a black body with a temperature corresponding to the spectral type of the star. An infrared excess normally indicates that the star is surrounded by a shell or disk of warm dust heated by the star, for example a dusty shell around a red supergiant or a planetary system forming around a young star.

Infrared photometry Photometry in the wavelength range 1.22-21 μm, that is, in the near- and mid-infrared parts of the spectrum. The detectors are cooled with liquid nitrogen or helium to reduce noise. It is usual to chop back and forth between the object and a neighbouring patch of sky to subtract the strong background signal from the telescope, its surroundings, and the sky itself.Further, this part of the spectrum contains many strong bands of water vapour and other molecules in the Earth's atmosphere, so filters must be used to select the windows between them. Observations free from this interference have been made from satellites such as the Infrared Astronomical Satellite and the Infrared Space Observatory.

Inner planet The inner planets are those planets that orbit close to the sun. They are: Mercury, Venus, Earth, and Mars. They are relatively small, composed mostly of rock, and have few or no moons.

Innes, robert thorburn ayton Scottish astronomer, who worked mostly in South Africa. He introduced the orbital parameters for double stars now known as Thiele-Innes constants in conjunction with the Danish astronomer Thorvald Nicolai Thiele.He measured stellar proper motions, and introduced the blink comparator, which he used in 1915 to discover Proxima Centauri. His Southern Double Star Catalogue included over 1600 doubles he discovered.

Instrumental polarization Any contamination of a polarization measurement by an instrument's response to an undesired polarization state. In radio interferometry, the instrumental polarization arises mainly from imperfections in the feeds and from plumbing leaks between the feeds and the receiver front-ends. One tries to remove the instrumental polarization by applying corrections derived from observations of calibration sources whose polarization properties are known.

Integral A planned ESA gamma-ray astronomy satellite due for launch in 2001; the name is a contraction of International Gamma-Ray Laboratory. Integral will carry two main instruments spanning the energy range from 15 keV to 30 MeV (4×10^{-5} to 0.08 nm). The imaging system will be provided with a coded-mask telescope of high resolution. The spectroscopic instrument is likely to be an array of germanium detectors behind a lower-resolution coded mask. An X-ray monitor and optical burst monitor may also be included.

Integrated magnitude The apparent magnitude an extended object would have if all of its light were concentrated into a point source. This is a difficult quantity to measure for galaxies, nebulae, or clusters which cover a large area of sky. Integrated magnitudes may be measured with standard filters such as B or V, or bolometrically.

Integration time In data analysis, the interval over which data are averaged to reduce background noise and increase the signal-to-noise ratio.

Intensity 1. In visual perception, one of the three basic parameters (*hue, intensity,* and *saturation*) which may be used to describe the physical perception of color. Intensity is a measure of the energy of the spectral distribution, at a given point in an image or scene, weighted by the spectral response of the visual system. *Luminance* is the energy of the physical spectrum, but not weighted by the visual response. *Brightness* sometimes is used synonymously with either term. See *chromaticity*.

2. In photometry, a measure, sometimes called *specific intensity* of the amount of radiant energy received per unit solid angle per unit time per unit area of a surface element orthogonal to the direction of propagation of the radiation.

3. In radio astronomy, a term often used much more loosely to mean either the flux density of an unresolved radio source or the surface brightness of an extended source. The normalization and the units must be determined from the context because of the looser usage.

Intensity interferometer An optical interferometer for measuring the angular diameters of bright stars. An intensity interferometer operated from 1965 to 1972 at Narrabri in New South Wales, Australia, and consisted of two 6.5-m flux collectors on a circular railway track 188 m in diameter.

Intensive variable A measurable property of a thermodynamic system is intensive if when two identical systems are combined into one, the variable of the combined system is the same as the original value in each system. Examples: temperature, pressure. See: extensive variable, and specific.

Interacting binary A close binary in which there is mass transfer from the secondary to the primary component. In a detached binary, the primary may accrete material from the secondary's stellar wind.In very close, semidetached binary systems, a gas stream may impact on the primary or, at greater separations, an accretion disk may form When the primary has a strong magnetic field, it may gain material via an accretion column.

Interacting galaxy A galaxy whose structure is being altered by gravitational or other interaction with one or more nearby galaxies, for example the Whirlpool Galaxy.

Interamnia Asteroid 704, discovered in 1910 by the Italian astronomer Vincenzo Cerulli. It is the fifth-largest main-belt asteroid, diameter 338 km. Interamnia is an F-class asteroid, with a rotation period of 8.73 hours. Its orbit has a semimajor axis of 3.061 AU, period 5.36 years, perihelion 2.61 AU, aphelion 3.52 AU, and inclination 17°.3.

Interference 1. A phenomenon in which two or more sets of electromagnetic waves may combine so as to increase their total amplitude at a given point, or decrease it thereby forming an interference pattern. For interference to occur, the waves must be coherent (i.e. the crests and troughs must be in step with each other). Constructive interference occurs where the crests coincide with each other, and destructive interference where crests coincide with troughs.

2. Unwanted radio signals received by a radio telescope that originate in human activity rather than natural phenomena.

Interference filter A type of filter that uses the effect of interference of light to limit the wavelengths that pass through it. Ordinary filters are coloured, but interference filters consist of very thin multiple layers of alternating high and low refractive index (e.g. zirconium dioxide, magnesium fluoride) coated on glass. The incident light reflects backwards and forwards between the layers, thereby interfering with itself. By appropriate choice of thicknesses, the range of wavelengths that the filter transmits (the passband) can be adjusted. The antireflection coatings on camera and binocular lenses are simple interference filters.

Interference fringes A wave-like pattern resulting from the successful combination of two beams of light which amplifies the light

Interference pattern A pattern of alternating light and dark bands caused by the interference of two or more beams of light; also known as a fringe pattern. By extension, the term is used in radio astronomy to describe the interference of radio signals received by two or more aerials of an interferometer.

Interference, wave A wave is something that moves along and has high points (crests) and low points (troughs). If two (or more) different wave trains pass over one another the crests and troughs can add together to make bigger crests and troughs, and a crest and a trough can add together to produce zero. So if light is a wave phenomenon, then two light sources produce waves that in some places produce large amplitudes and other places produce zero. When to point sources of light are projected onto a screen this wave interference effect produces alternating light and dark spots. This demonstrates the wavelike nature of light.

Interferometer An instrument designed to achieve high angular resolution by connecting two or more telescopes or antennas to operate as a single instrument. The signals from each element are combined to form an interference pattern. The maximum resolution of the array is then determined not by the size of the individual elements, but by their maximum separation. Interferometers are widely used in radio astronomy, and also in infrared and optical astronomy. At optical and infrared wavelengths, the incoming beam is split and then recombined with itself to form an interference pattern. This can give high resolution of spectral lines, as in a Fabry-Perot interferometer.

Interferometric fringes A wave-like pattern resulting from the successful combination of two beams of light which amplifies the light

Intergalactic absorption The absorption of light in the space between galaxies. It is too weak to detect between our own Galaxy and its near neighbours. However, high-redshift (and hence distant) quasars frequently reveal many Lyman-a absorption lines at redshifts smaller than that of the parent quasar. This is the Lyman-a forest, which arises from absorption by intervening clouds in intergalactic space.

Intergalactic magnetic field The weak magnetic field of less than 10-12 tesla believed to be present in the space between galaxies. The intergalactic field may have arisen spontaneously in the very early Universe, or from the effect of atomic particles and magnetic fields from cosmic radio sources. It could be ten times stronger in clusters of galaxies.

International atomic time (The time-scale that has been used by international agreement since 1972. It is ultimately based upon atomic oscillations, rather than astronomical principles, and is the most accurate timescale available today. Its fundamental unit is the SI second, defined in terms of properties of the caesium atom.TAI differs from Coordinated Universal Time by an exact number of seconds, because leap seconds are introduced in UTC to keep it in step with the changeable rotation rate of the Earth. TAI was defined so as to be in agreement with UTI on 1958 January 1.

International cometary explorer A NASA spacecraft, formerly called International Sun-Earth Explorer-3 (ISEE-3), which became the first spacecraft to be sent to a comet. ISEE-3 was launched in 1978 August to study the solar wind. In 1983 December it was retargeted to Comet Giacobini-Zinner and renamed International Cometary Explorer. In 1985 September it passed through the tail of Comet Giacobini-Zinner, 7800 km from the nucleus. In 1986 March, ICE passed 28 million km from Halley's Comet.

International date line The imaginary line on the Earth's surface stretching from pole to pole through the Pacific Ocean which, by international agreement in 1884, marks the beginning or ending of a day. The line approximately follows the meridian of longitude 180°, bending where necessary to avoid cutting through land. To the east of the line, the date is one day behind the west of the line. On crossing this line in a westerly direction the calendar must be put forward one day, while if it is crossed in an easterly direction the calendar is put one day back.

International geophysical year An 18-month period, from 1957 July to the end of 1958, in which world-wide collaborative observations were made to study the connection between solar flares, geomagnetic disturbances, radio fadeouts, and particle emission from the Sun. The IGY was designed to coincide with maximum sunspot activity, which turned out to be the largest recorded since the invention of the telescope, and resulted in a greatly improved picture of solar-terrestrial relations. Sputnik 1, the first artificial satellite, was launched during the IGY by the Soviet Union.

International space The International Space Station is an orbiting structure where people can live in space. Construction (in orbit) on the new International Space station began in 1998.

International space station A joint venture by NASA, Russia, ESA, Canada, and Japan to build a space station in orbit around the Earth. The first section, the Functional Cargo Block will be launched by Russia in summer 1998, followed by the first US section, Node 1. The first occupants took up residence in 1999.

International sun-earth explorer A joint NASA-ESA series of probes to study the Earth's magnetosphere and the Sun's effect on it. The ISEE-1 and ISEE-2 spacecraft were both launched in 1977 October into orbits that carried them together into the magnetosphere. The instruments on board included magnetometers and particle detectors. ISEE-3 was launched in 1978 August into an orbit completely outside the magnetosphere to measure the solar wind flow independent of any influence from the Earth. ISEE-3 was later sent to intercept Comet Giacobini-Zinner, and was renamed the International Cometary Explorer.

Interstellar absorption The absorption of starlight by dust and gas in the space between the stars; also known as interstellar extinction. The dust and gas are strongly concentrated towards the plane of our Galaxy. Interstellar absorption increases towards the shorter-wavelength end of the spectrum, and hence makes stars appear redder, so that the terms interstellar absorption and interstellar reddening are used almost interchangeably. As well as continuous absorption there are diffuse interstellar bands, such as that at 443 nm, whose origin remains obscure. There are also atomic absorption lines such as the calcium H and K lines at 393 and 397 nm, the sodium D lines at 589.0 and 589.6 nm, and other weaker lines.

Interstellar dust Interstellar dust is composed of mictoscopic bits (on the order of a micron in diameter) of carbon and/or silicates. The origin of interstellar dust in unknown, but it seems to be associated with young stars. Interstellar dust is not at all like the dust we have in our houses (which is mostly bits of organic debris and lint).

Interstellar extinction As light from a star travels through interstellar space it encounters some amount of dust. This dust scatters some of the light, causing the total intensity of the light to diminish. The more dust, the dimmer the star will appear. It is important to take this effect into account when measuring the apparent magnitude of stars. The dark bands running across portions of the Milky Way in the sky are due to extinction by copious amounts of dust in the plane of our galaxy.

Interstellar maser A maser in a star-forming region, in the dense clouds of gas and dust that surround young stars or protostars. They are the brightest maser sources in our Galaxy. Some interstellar masers are associated with violent ejections of matter from young stars, while others are located in disks of material around young stars from which planetary systems might be forming.

Interstellar matter The material between the stars also called the interstellar medium. The ISM in our Galaxy consists of 99% gas and 1% fine dust particles by mass. There are also a few very energetic cosmic-ray particles. The ISM is spread very thinly, with an average density of only 1 particle/cm^3. However, there are enormous variations about this average. The densest parts, which include maser regions containing 10^{10} hydrogen molecules/cm^3, are also the coldest, with temperatures close to absolute zero. The hottest material has temperatures of 10^8 K, and densities of 1 particle/m^3 or less. The ISM is the reservoir of matter from which new stars form.

Interstellar medium The gas and dust between stars, which fills the plane of the Galaxy much like air fills the world we live in. For centuries, scientists believed that the space between the stars was empty. It wasn't until the eighteenth century, when William Herschel observed nebulous patches of sky through his telescope, that serious consideration was given to the notion that interstellar space was something to study. It was only in the last century that observations of interstellar material suggested that it was not even uniformly distributed through space, but that it had a unique structure.

Interstellar molecule Any molecule that occurs naturally in clouds of gas and dust in space. More than a hundred such molecules have already been identified, chiefly by their emission or absorption of radio waves at particular wavelengths. The known molecules include simple groupings such as ammonia and water simple organic molecules such as ethyl alcohol formaldehyde and acetic acid and a variety of ions and radicals which are unstable on Earth, such as hydroxyl (OH), sulphur monoxide and the formyl ion Many isotopic variants are also found, for example HDO, a form of water in which one hydrogen atom has been replaced by an atom of deuterium, a heavy form of hydrogen. The largest molecule yet identified is cyanooctatetra-yne (HC_9N). More complicated molecules could be responsible for spectral lines that have been detected but not identified. Interstellar molecules regulate the temperature of the gas clouds by radiating away energy at radio wavelengths.

Interstellar reddening As light from a star travels through interstellar space it encounters some amount of dust. This dust scatters some of the light, preferentially the short wavelength (blue) components. The spectrum of the light that remains is increasingly dominated by the long wavelength (red) end of the spectrum, hence the light is "reddened" as it travels through space. It is important to take this effect into account when measuring the color indices of stars.

Interstellar scintillation The fluctuation in brightness of a radio source due to the scattering of radio waves by irregularities in interstellar matter.Much of the fluctuation is caused by the relative motion of the Earth and the source rather than of the ISM, and occurs on a time-scale of minutes to hours. Measurements of scintillation rate can be used to determine the velocities of pulsars across the line of sight. Interstellar scintillation of pulsar signals is closely related to pulse broadening.

Interstellar wind The stream of interstellar gas and dust that appears to be moving past the Solar System because of the motion of the Sun around the Galaxy. It starts at the heliopause, where the solar wind gives way to the interstellar wind. The Pioneer and Voyager probes now travelling out of the Solar System are expected to encounter the interstellar wind early in the 21st century.

Intrinsic colour index The colour index a star would have in the absence of interstellar absorption. It is assumed that all stars of the same spectral type and luminosity class have the same intrinsic colour index.

Intrinsic variable A variable star whose fluctuations in brightness are caused by actual changes in the luminosity of the star itself, not by external processes such as rotation or eclipses.The majority of variable stars fall into this broad category.

Intrusive Refers to igneous rocks formed underground.

Invariable plane A fixed plane passing through the centre of mass of the Solar System, and oriented perpendicular to the axis about which the angular momentum of the Solar System is measured. The invariable plane is inclined at 1°.58 to the ecliptic, between the orbital planes of Jupiter and Saturn, but its precise location is not known because the masses of all the objects in the Solar System and their positions and velocities at a given time are not yet exactly known. Being fixed, it provides a permanent reference plane, whereas the ecliptic alters with time because of planetary perturbations.

Inverse compton effect The gain in energy of a photon when it is struck by a fast-moving electron; also known as inverse Compton scattering. The electron passes on a small proportion of its energy to the photon, and the photon's wavelength decreases. The electron has to suffer a large number of collisions before it loses an appreciable fraction of its energy. It is the opposite of the Compton effect.

Inverse-square law A law that describes how the strength of a force (e.g. gravitation) or a flow of energy (e.g. light) weakens with distance from the source. According to the inverse-square law, the magnitude of a given quantity is inversely proportional to the square of the distance from the source. For example, if an object's distance were doubled, the strength of the light or gravity from it would fall by four times; if its distance were tripled, the strength of its light or gravity would be nine times less.

Invisible distribution In the context of radio interferometry, a function *f* (or a generalized function-or distribution) whose Fourier transform *F* vanishes everywhere that the interferometer pairs have sampled the *u-v* plane. This term was introduced by R. N. Bracewell and J. A. Roberts (*Aerial smoothing in radio astronomy*, Austr. J. Phys., 7 (1954) 615-640). Also see *principal solution*.

For an actual interferometer, there exist fewer physically plausible invisible distributions than for an idealized interferometer. This is because each visibility sample is not a point sample of the Fourier transform *F*, but rather some kind of local average. By the Paley-Wiener theorem, if *F* is nontrivial and vanishes in some open neighborhood, then *f* cannot be of compact support, and hence it may be considered implausible.

Io The third-largest satellite of Jupiter, diameter 3630 km, and the innermost of the four Galilean satellites; also known as Jupiter I. It orbits in 1.769 days at a distance of 422000 km, keeping the same face turned towards Jupiter. Its geometric albedo is 0.6 and its opposition magnitude 5.0. The Voyager spacecraft in 1979 revealed that Io has explosively erupting volcanoes which eject plumes of debris 70-280 km high, some of the material landing up to 500 km from the eruption site. The sources of the eruptions are volcanic calderas or fissures, of which there are over 300. Huge lava flows radiate from many of the volcanoes, and the whole surface is yellowish in colour due to deposits of sulphur or sulphur oxides. There are extensive plains and mountainous regions on Io but no impact craters, indicating that its surface is geologically very young. Io's density, 3.57 g/cm^3, suggests that it has an iron-sulphur core about 1500 km across and a silicate mantle.

Io torus The Io Torus is a doughnut-shaped plasma cloud around Jupiter near Io's orbit (also known as the "Io plasma torus") This torus is caused by Jupiter's strong magnetic field, which strips ions from Io as it rotates; Io acts like an electrical generator.

Ion An atom or molecule that has lost or gained one or more electrons. An ion is positively charged if it has lost one or more electrons, and negatively charged if it has gained one or more electrons. In the notation for ions, the chemical symbol is followed by a superscript. Thus doubly ionized iron is represented as Fe2+. Alternatively, the symbol is followed by a Roman numeral which is one more than the number of electrons lost (e.g. Fe III for doubly ionized iron, Fe I for neutral iron).

Ion engine An ion engine is an engine that ionizes a gas in order to propel a spacecraft. A electrical charge is used to ionize xenon gas (an inert gas like neon or helium, but heavier). In an electrical field the ions are accelerated to a speed of about 30 km/second. The high-speed xenon ions are expelled as exhaust, pushing the spacecraft in the other direction (since momentum is conserved). This efficient type of engine delivers 10 times the thrust of conventional engines for the same mass of fuel. This engine is only used in outer space. The ion drive is used in NASA's Deep Space 1; it carries 181 pounds of xenon.

Ionic (or ionized) gas Gas whose atoms have lost or gained electrons, causing them to be electrically charged. In astronomy, this term is most often used to describe the gas around hot stars where the high temperature causes atoms to lose electrons.

Ionization The process by which an atom or molecule loses or gains electrons. Atoms that have lost or gained one electron are said to be singly ionized; if they have lost or gained two electrons they are doubly ionized; and so on. Ionization occurs at high temperatures as in a star, or by the impact of high-energy atomic particles (e.g. electrons, protons, alpha particles) or short-wavelength radiation (ultraviolet, X-rays, gamma rays).

Ionization equilibrium A balance achieved in a hot gas when the number of ionizations of a particular ion equals the number of recombinations into that ion. An example of a gas in ionization equilibrium is the Sun's corona. Thus, in ionization equilibrium, the number of ionizations of the ion $Fe9^+$, which emits the red coronal line, to the $Fe10^+$ ion equals the number of recombinations from $Fe10^+$ to give $Fe9^+$.

Ionization front A boundary separating electrically neutral gas from ionized gas. There are several astronomical situations in which light or ultraviolet radiation falls on neutral gas and ionizes it. For example, when a massive young star first begins to shine it ionizes the surrounding gas, first the nearest gas and then gas progressively farther away. The ionization front is the moving boundary of the ionized region.

Ionization potential (symbol I) The energy required to remove the least strongly bound electron from an atom or ion. It is usually measured in electronvolts (eV). The energy required increases as the level of ionization increases. Thus the ionization potential of neutron iron is 7.9 eV, that of singly ionized iron is 16.2 eV, doubly ionized iron, 30.7 eV, and so on.

Ionization temperature The temperature of a gas or plasma obtained from the relative numbers of neutral atoms and ions. It is the temperature at which the Saha ionization equation most nearly gives the observed levels of ionization in a star.

Ionized gas A gas that has been heated to a state where it contains ions and free-floating electrons; also known as a plasma

Ionopause The upper boundary of the ionosphere, lying at an altitude of around 250 km at the top of the F layer.

Ionospher The ionosphere is one of the highest layer of the Earth's atmosphere. The ionosphere starts at about 43-50 miles (70-80 km) high and continues for hundreds of miles (about 400 miles = 640 km). It contains many ions and free electrons (plasma). The ions are created when sunlight hits atoms and tears off some electrons. The ionosphere is located between the mesosphere and the exosphere (and is part of the thermosphere). Auroras occur in the ionosphere.

Ionospheric scintillation The fluctuation in brightness of a radio source due to the scattering of radio waves by irregularities in the Earth's ionosphere.

Ionospheric storm A disturbance in the f region of the ionosphere, which occurs in connection with geomagnetic activity.

Iota aquarid meteors A comparatively weak meteor shower, with ill-defined activity limits during July and August. The most prominent maximum falls on August 6, when the ZHR may reach 10. Iota Aquarids tend to be faint and fairly swift. The meteor stream, like many others that lie close to the ecliptic, is split into at least two components. At maximum, the Southern Iota Aquarids have a radiant at RA 22h 10m, dec. −15°, while the Northern Iota Aquarids emanate from RA 22h 04m, dec. −06°.

IRAS-Araki-Alcock comet A long-period comet detected by the Infrared Astronomical Satellite on 1983 April 25, and found independently on May 3 by the Japanese amateur astronomer Genichi Araki (1954-) and the English amateur George Eric Deacon Alcock formerly designated 1983 VII. It made an unusually close passage of 0.031 AU (4.6 million km) to Earth on 1983 May 11, when it was of 2nd magnitude and moving rapidly across the northern sky. Because of its proximity to Earth, the comet appeared diffuse, with a diameter of 2°. Perihelion, 0.99 AU from the Sun, was on 1983 May 21. Its orbit has a period of about 1000 years, eccentricity 0.990, and inclination 73°.3.

Iridium anomaly Iridium is an element that is rare on the Earth's surface, but abundant on chondritic meteors. The presence of excess Iridium at the K-T layer (the Iridium anomaly) supports the Alvarez asteroid theory.

Iris Asteroid 7, discovered in 1847 by the English astronomer John Russell Hind. Iris is an S-class asteroid, of diameter 208 km. Its mean magnitude at opposition is 8.4; of the main-belt asteroids, only Vesta, Ceres, and Pallas can become brighter. Its orbit has a semimajor axis of 2.386 AU, period 3.69 years, perihelion 1.84 AU, aphelion 2.93 AU, and inclination 5°.5.

Iron An element with atomic number 26; sumbol: Fe. Iron is one of the most abundant elements in the rocky planets. It is the most abundant element in the metallic cores of the inner planets.

Iron meteorite A meteorite composed primarily of nickel-iron; also known as a siderite. Some iron meteorites are probably material from the cores of asteroidsized parent bodies, while others may have originated at shallower depths. Irons are the easiest meteorites to identify, because they are heavy, look metallic, are strongly magnetic, and are usually covered with pits, depressions, and grooves from their passage through the atmosphere. On impact they have a blue-black fusion crust, with traces of metal melted during the flight, but after exposure on Earth they rust, turning brown. Iron meteorites comprise only 4% of observed falls, but resist weathering better than stony meteorites and so are found more easily. Iron meteorites are divided into three main subgroups on the basis of the structure of their nickel-iron alloy: hexahedrites, octahedrites and the ataxites. An iron meteorite which cannot be included in one of these groups is termed anomalous.

Iron peak Elements whose nuclei contain a total of 56 protons and neutrons (i.e. they have a mass number of 56), for example 56Fe, 56Ni, or 56Co. Their formation by nuclear fusion inside a star marks the end-point of the star's energy output. Further fusion to create elements with higher mass numbers would absorb energy rather than liberate it. Nuclei of 56Fe have the maximum binding energy.

Irradiance (symbol E e) The energy at all wavelengths that falls on a surface in a given time. It is measured in watts per square metre. The energy input at visual wavelengths alone is termed illuminance.

Irregular eruptive Variables are divided into three broad groups. The poorly studied variables, designated I, are subdivided into IA for early spectral types and IB for those with intermediate to late spectra.Stars associated with nebulosity may vary by several magnitudes with rapid changes (up to 1 magnitude in 1-10 days), and are divided by spectra into types INA and INB. Finally, there are IS stars with rapid variations (0.5-1 mag. in hours or days), subdivided into types ISA and ISB, although this group is poorly defined. Stars with specific emission features are classified as T Tauri stars (INT) or in rare cases when nebulosity is not present, IT. When absorption indicates the infall of material, the star is a YY Orionis star, type IN(YY).

Slow, irregular, pulsating giants or supergiants are designated type L. Many are poorly studied, and may subsequently be found to be semiregular variables or other types. All have late spectral types (K-M, C, and S).

Irregular galaxy A type of galaxy with ill-defined structure; Hubble type Irr or Ir. There are two main types in the Hubble classification. Irr I galaxies are not as massive as large ellipticals or spirals, often have a high gas content, and are undergoing star formation. The categories Irr+ and Irr- are sometimes used to denote the degree of resolution of the structure of the galaxy into star clusters, H II regions, and other features, with Irr+ showing higher resolution than Irr. The high gas content of Irr I galaxies implies that they may not have undergone much evolution since their formation. A subdivision of Irr I, the so-called Magellanic irregulars, symbol Im, resemble the Magellanic Clouds in the Local Group. The Hubble type Irr II denotes galaxies of unusual appearance that are classified as irregular simply because they do not fit into any other class, and may in many cases represent interacting or merging systems.

Irregular galaxy cluster An irregular galaxy cluster is a group of any type of galaxies (not just irregular galaxies) that forms an irregular (non-spherical) shape. The local group is an irregular galaxy cluster.

Irregular variable A star that exhibits an irregular light-curve. There are two types, with very different characteristics: eruptive variables near the main sequence; and evolved pulsating variables.

Isaac newton group (ING) A group of telescopes at the Roque de los Muchachos Observatory on La Palma in the Canary Islands, jointly owned and operated by the UK and the Netherlands. As well as the 4.2-m William Herschel Telescope and the 2.5-m Isaac Newton Telescope, the group includes the 1-m Jacobus Kapteyn Telescope,opened in 1984. The headquarters of the group is at Santa Cruz on La Palma island.

Isochron A straight line on a graph from which the age of a rock can be derived. The graph plots the abundances of parent isotopes against the abundance of their daughter isotopes for several minerals from the same rock, or several rocks formed from the same source at the same time. The points lie on a straight line whose slope can be used to calculate the age of the rock. A steeper slope means an older rock, less of the parent isotope remaining.

Isotropy A characteristic of a substance or body in which physical properties are the same in all directions, such as the cosmic background radiation and the very large-scale distribution of matter throughout the Universe. An object such as a star that emits radiation equally in all directions is known as an isotropic radiator. An ideal antenna that is equally sensitive in all directions is known as an isotropic antenna.

IUE(International Ultraviolet Explorer) An ultraviolet space observatory launch in 1978. Originally designed for a 3 year mission, IUE exceeded all expectations and functioned for over 18 years, finally ceasing operation in September 1996.

Izar The star Epsilon Boötis, a close double consisting of a K0 giant and an A0 dwarf of magnitudes 2.7 and 5.1, appearing orange and blue-green. Their colours give rise to the alternative name Pulcherrima, meaning 'most beautiful'. The pair lie 150 L.Y away.

J

J magnitude The magnitude of a star as measured through the infrared J filter in the Johnson photometry system. The J filter has an effective wavelength of 1220 nm and a bandwidth of 213 nm.

James Clerk Maxwell telescope A 15-m radio telescope opened in 1987 at an altitude of 4090 m at the Mauna Kea Observatories, Hawaii, owned by the UK, Canada, and the Netherlands, and operated by the Joint Astronomy Centre in Hawaii. It is designed to work at millimetre and submillimetre wavelengths. It is named after J. C. Maxwell.

Jansky The jansky (abbreviated Jy) is a unit of radio flux density (the rate of flow of radio waves); 1 Jy = 10^{-26} watts/m^2/Hz. A typical radio source has a spectral flux density of roughly 1 Jy. The jansky was named to honor Karl Gothe Jansky who developed radio astronomy in 1932.

Jansky, Karl Guthe American radio research engineer. In 1931, he began to study atmospheric 'static' that interfered with telecommunications. By the end of 1932, he had accounted for all but one persistent signal, which he concluded came from outside the Solar System in the direction of the constellation Sagittarius, towards the centre of our Galaxy. Thus was born radio astronomy, although Jansky himself never followed up his discovery. In radio astronomy, the unit of flux density is now named the jansky in his honour.

Janssen, Jules César French spectroscopist. In 1862 he discovered and named the telluric lines in the Sun's spectrum, originating in the Earth's atmosphere, and realized that similar lines in the spectra of planets would reveal the composition of their atmospheres. In 1868, independently of J. N. Lockyer, he observed the spectra of solar prominences at a total eclipse, and went on to invent the spectrohelioscope. He found a new line in the solar spectrum, which Lockyer attributed to a new element later named helium. In a celebrated adventure in 1870, during the Franco-Prussian War, Janssen escaped by balloon from a besieged Paris and flew to the Atlantic coast to observe an eclipse.

Janus Janus is one of the smaller of the 18 moons of Saturn. This heavily-cratered moon orbits at a mean distance of about 94,000 miles (151,472 km) and revolves around Saturn in about 18 hours. Janus has a diameter of about 185 miles. Janus shares an orbit with Epimetheus (they are only separated by about 50 kilometers (31 miles)). Once every four years they approach each other, exchange some momentum and switch orbits. This icy moon was discovered by the French astronomer Audouin Dollfus in 1966.

Jeans length The minimum size of a cloud of gas, of given temperature and density, which is capable of collapsing under its own gravity. Gas clouds smaller than this do not collapse because their gas pressure more than balances their gravitational force. The Jeans length is proportional to the square root of the temperature, and inversely proportiohal to the square root of the density. For the dense gas characteristic of star-forming clouds it is a few tenths of a light year. It is named after J. H. Jeans.

Jeans mass The mass contained in a sphere whose radius is the Jeans length, and thus the minimum mass that can collapse under its own gravity. For the dense gas characteristic of star-forming clouds, it is about a solar mass.

Jeans, James Hopwood English mathematician, physicist, and astronomer. By considering the stability of a rotating mass of fluid, and building on earlier work by J. H. Poincaré and others, he developed theories of the formation and evolution of stars, double-star systems, and spiral galaxies. In 1917 he proposed a theory, subsequently expanded on by the English mathematician Harold Jeffreys in which the planets condensed from filaments pulled out of the Sun by a passing star. In 1928 Jeans advanced a cosmological theory in which matter was continually being created—a cornerstone of the later steadystate theory. He then turned to popular science writing and broadcasting.

Jemison, Mae Mae C. Jemison (October 17, 1956 -) was the first African-American woman in space. Dr. Jemison is a medical doctor and a surgeon, with engineering experience. She was accepted into NASA's astronaut program in 1987. She flew on the space shuttle Endeavor (STS-47, Spacelab-J) as the Mission Specialist; the mission lifted off on September 12, 1992 and landed on September 20, 1992. Dr. Jemison founded the International Science Camp in Chicago in 1994, a program designed to interest children in science and space.

Jet A narrow, bright feature associated with several types of astronomical object. Jets are seen mainly at radio but occasionally at other wavelengths. They emerge from the core of certain active galaxies and may extend for many hundreds of kiloparsecs. In some sources, notably radio galaxies, jets are commonly seen on both sides of the nucleus while in others, such as quasars, only one jet is normally seen (single-sided jet). Some jets exhibit the phenomenon of superluminal velocity. Optical stellar jets are seen emerging from young stars such as T Tauri stars and FU Orionis stars, and are associated with Herbig-Haro objects. The X-ray binary system SS433 has two jets.

Jet propulsion laboratory A NASA establishment in Pasadena, California, founded in 1944 for missile development but which became part of NASA in 1958. It is operated for NASA by the California Institute of Technology.JPL tracks and controls space probes, as well as developing new missions. JPL manages the world-wide Deep Space Network, including its own antennas at Goldstone in the Mojave Desert. JPL also operates an observatory at Table Mountain, California, altitude 2290 m, equipped with a 1.2-m reflector and smaller instruments.

Jet stream The jet stream is a high-speed wind that is usually found at high altitudes, between 25,000 and 45,000 feet (just below the tropopause). It is called the jet stream because the first people who encountered the jet stream were pilots who described it as though the wind was being propelled from a jet engine.

Jets, radio Narrow, collimated beams of plasma that are producing radio (synchrotron) emission. These jets emerge from the cores of radio galaxies and can extend outward across regions of space larger than the size of the galaxy itself. These jets are believed to be powered and launched from an accretion disk orbiting a supermassive black hole at the galaxy's core.

Jewel box A 4th-magnitude open cluster in Crux, near the Coalsack dark nebula, also known as NGC 4755 or the Kappa Crucis Cluster. Most of the bright stars in the cluster are blue-white in colour, including Kappa Crucis, a magnitude 5.9 supergiant, but an 8th-magnitude red supergiant lies near the centre. The cluster is 7600 L.Y away. It was named by J. F. W. Herschel who compared it to a collection of jewels. The name Jewel Box is sometimes also applied to M6 in Scorpius, better known as the Butterfly Cluster.

Jilin meteorite The greatest recorded fall of stony meteorites, which occurred near Jilin in Manchuria, northeastern China, on 1976 March 8. About 4 tonnes of fragments of the H5 chondrite meteorite were collected. One fragment, weighing 1.77 tonnes, is the largest single piece of stony meteorite ever recovered.

Job's coffin A quadrilateral of stars in Delphinus, formed by Alpha, Beta, Gamma, and Delta Delphini.

Jodrell bank Jodrell Bank in northeastern Cheshire, England, is the location of a radio astronomy observatory, Nuffield Radio Astronomy Laboratories. The radio telescope's size is 250 ft (76 m).

Johnson photometry Photometry based on the work of the American astronomer Harold Lester. It falls into three natural parts: UBV, RI, and infrared photometry. UBV photometry was based on a photomultiplier sensitive to the wavelength range 300-600 nm. The U and B magnitudes were formed by splitting the old photographic magnitude into blue and ultraviolet parts. The V magnitude is similar to but more accurate than the old visual system. Johnson later extended this work into the red and infrared with two further filters designated R and I, using a photomultiplier sensitive to wavelengths 0.3-1.1 μm. The third innovation was the extension into infrared photometry with J, H, K, L, M, and N filters covering the wavelength range 1.25-10.4 μm.

Johnson space center A NASA establishment near Houston, Texas, founded in 1961 to design and develop manned spacecraft, to select and train astronauts, and to conduct their missions. JSC houses the Mission Control for Shuttle flights and also a separate control centre for the International Space Station. Lunar samples collected by the Apollo missions are stored at JSC.

Joint astronomy An establishment in Hilo, Hawaii, owned by the UK, which operates the United Kingdom Infrared Telescope and the James Clerk Maxwell Telescope, the latter jointly with Canada and the Netherlands. It will also operate the Gemini North telescope. The JAC was originally founded in 1978 as the headquarters for the UK Infrared Telescope, but its role was enlarged and it was renamed in 1988.

Julian calendar The form of calendar first introduced in 46 BC by the Roman emperor Julius Caesar, after whom it is named. It was prepared in consultation with the Greek astronomer Sosigenes.Each month was assigned the number of days it has today, and a normal year had 365 days. It was intended to be a solar calendar in which the date remained in step with the seasons. To preserve this link, every fourth year was a leap year with February having an additional day. The average length of a year is then 365.25 days, close to but not exactly equal to the tropical year. The Julian calendar was superseded by the more precise Gregorian calendar in 1582.

Julian date (JD) A calendar and timekeeping system introduced to give unambiguous dates and times of celestial events, unaffected by changes in the civil calendar. The system was begun in 1582 by the French scholar Joseph Justus Scaliger who named it to honour his father, the Italian-born French scholar Julius Caesar Scaliger.The starting date is chosen as January 1 of the year 4713

BC, sufficiently far in the past that it predates all known recorded astronomical observations. Time is measured from mean noon (12h UT) on that date by the number of days and fractions of a day elapsed. For example, an observation made on 1962 June 24 at 18h UT was made at JD 2 437 840.25. The integral part of the Julian date is called the Julian day number, and the fractional part is the universal time expressed as a fraction of a day.

Juliet The sixth-closest satellite of Uranus, distance 64350 km, orbital period 0.493 days; also known as Uranus XI. Its diameter is 84 km, and it was discovered in 1986 by the Voyager 2 spacecraft.

Juno Asteroid 3, the third asteroid to be discovered, by the German astronomer Karl Ludwig Harding in 1804. Its diameter is 248 km. Juno is of S class, with a rotation period of 7.21 hours. Its orbit has a semimajor axis of 2.668 AU, period 4.36 years, perihelion 1.98 AU, aphelion 3.36 AU, and inclination 13°.0.

Jupiter The fifth planet from the Sun, and the largest planet in the Solar System, with a mass two and a half times greater than all the other planets combined; Its maximum opposition magnitude is -2.9, making it normally the second-brightest planet after Venus. Jupiter is distinctly ellipsoidal in shape (equatorial diameter 142 984 km, polar diameter 133 708 km). The rotation period of the visible surface is about 9h 50m in equatorial regions and about 9h 56m for the rest of the planet The rotation of the solid surface, derived from radio observations, is thought to be 9h 55m 29s, known as System III.

Jupiter has a thick atmosphere composed of about 90% hydrogen and 10% helium plus traces of methane, ammonia, water, ethane, ethyne, phosphine, carbon monoxide, and germanium tetrahydride. Near the top of the atmosphere the temperature is around -143 °C. Internally, Jupiter is thought to have an iron-rich rocky and perhaps icy core about 10 times the mass of the Earth.

Jupiter comet family A group of periodic comets whose orbits have been modified by close passages to Jupiter. Members of Jupiter's comet family have aphelia near 5 AU (Jupiter's orbital distance) and orbital periods of 6-12 years. An example is Comet 16P/ Brooks 2, whose orbit was shortened from an initial period of 29 years to only 7 years after passing within 0.001 AU of Jupiter in 1886. The comet's perihelion distance was decreased from 5.48 to 1.95 AU. Tidal disruption by Jupiter's gravity split the nucleus of Comet Brooks 2 into several fragments.

Jy Jy stands for "jansky (named to honor Karl Gothe Jansky who developed radio astronomy in 1932)". The jansky (abbreviated Jy) is a unit of radio flux density (the rate of flow of radio waves); 1 Jy = 10^{-26} watts/m^2/ Hz. A typical radio source has a spectral flux density of roughly 1 Jy.

K

K corona The bright inner part of the Sun's corona, caused by sunlight scattered from electrons. It is the true corona, as opposed to the F corona which is due to light scattered by dust particles. Owing to the extremely large speeds of the free electrons (averaging about 10000 km/s for a coronal temperature of 2 million K), the Fraunhofer lines of the photospheric spectrum are smeared out so that the spectrum of the K corona is almost a pure continuum.The K corona is brighter than the F corona out to 1.5 solar radii from the Sun's surface.

K index A 3-hourly quasi-logarithmic local index of geomagnetic activity relative to an assumed quiet-day curve for the recording site.

Range is from 0 to 9. The K index measures the deviation of the most disturbed horizontal component.

K magnitude The magnitude of a star as measured through the infrared K filter of the Johnson photometry system. The K filter has an effective wavelength of 2190 nm and a bandwidth of 390 nm.

K star A star of spectral type K, somewhat cooler than the Sun and appearing orange in colour. K-type stars on the main sequence have surface temperatures in the range 3900-5200 K, while giants are about 100-400 K cooler, and supergiants a few hundred degrees cooler still. Masses of main-sequence K stars are 0.5-0.8 solar masses, and their luminosities 0.1-0.4 times the Sun's. Giants are of 1.1-1.2 solar masses and have luminosities 60-300 times the Sun's, while K supergiants can be up to 13 solar masses and have luminosities as much as 40 000 times the Sun's. The dominant spectral features of K-type stars are the neutral metal lines of iron and titanium, with calcium (both Ca I and Ca II) being particularly strong. The molecular bands due to cyanogen (CN) and titanium oxide (TiO) strengthen considerably from K0 to K9. The best-known stars of this type are Arcturus, a Kl giant, and Aldebaran, a K5 giant.

Kant, Immanuel German philosopher. He proposed a cosmogony, published in 1755, in which the Solar System was formed via a disk which condensed out of primordial material. The Solar System was part of a larger system and many of the nebulae seen by astronomers were in fact other galaxies, which he termed island universes. Kant was influenced by I. Newton's theories and also by the English philosopher Thomas Wright of Durham.

Kappa crucis cluster Another name for the Jewel Box cluster.

Kappa cygnid meteors A weak meteor shower, active August 17-26. Maximum (maximum ZHR 5) occurs on August 20 from a radiant at RA 19h 20m, dec. +55°. The Kappa Cygnids are reputedly rich in fireballs.

Kappa mechanism The process that drives the pulsations of many types of variable star, including Cepheid variables and RR Lyrae stars. It is named from the Greek letter kappa the symbol for the coefficient of absorption of stellar material. In an ionization zone within a star, any small, chance rise in density produces increased opacity and thus increased absorption of energy from the stellar interior. This causes heating and expansion of the layer, which overshoots its rest position, leading to a drop in pressure, density, and temperature.The effect then reverses, setting up an oscillation and causing the star's outer layers to pulsate.

Kapteyn selected areas Areas of sky approximately 1° × 1° square which were chosen by J. C. Kapteyn in 1906 for his study of the structure of the Galaxy. The 206 areas were uniformly spaced at about 15° intervals over the whole sky. In addition, 46 areas were chosen for their special significance in the Galaxy, such as the galactic pole regions. In each area the magnitudes, colours, spectral types, and proper motions of the stars were measured to as faint an apparent magnitude as possible.

Kapteyn, Jacobus Cornelius Dutch astronomer. He developed mathematical methods for reducing and handling large amounts of observational data on a statistical basis. He compiled the Cape Photographic Durchmusterung, a catalogue of southern stars from photographs taken in South Africa by D. Gill. A programme of measuring proper motions (which turned up Kapteyn's Star in 1897) revealed two preferred directions of motion a result of the rotation of the Galaxy. In 1906 he began his plan of measuring the positions and brightnesses of stars in 206 areas distributed regularly over the sky. With his compatriot Pieter Johannes van Rhijn Kapteyn attempted to estimate the size and shape of the Galaxy. However, he wrongly concluded that the Sun lay near the centre of the Galaxy, being unable to gauge correctly the effect of interstellar absorption which obscured the Galaxy's true extent.

Kapteyn's star A 9th-magnitude red dwarf, 12.7 L.Y away in Pictor. It has the second-largest proper motion of any star. It is named after J. C. Kapteyn, who discovered its fast motion in 1897.

Kaus australis The star Epsilon Sagittarii, magnitude 1.85. It is a bright giant of spectral type A0 lying 76 L.Y away.

K-correction An adjustment made to the photometric magnitudes and colours of distant galaxies; also known as K-term. It takes account of the effect of the redshift on a galaxy's spectrum.

Keck observatory The W. M. Keck Observatory is located on the top of the dormant volcano Mauna Kea in Hawaii. The Keck observatory has world's largest infrared and optical telescopes, called the Keck 1 and the Keck 2. Each telescope has a 10-meter (33 ft) primary mirror that is made up of 36 hexagons (each of which is 1.8 meters (6 feet) wide and weighs 880 pounds). The Keck 1 telescope opened in 1993; the Keck 2 telescope opened in 1996. Both telescopes are 8 stories tall. The Keck observatory is run by the California Institute of Technology, the University of California, and NASA.

Keck telescopes Two identical 10-m reflectors with mirrors composed of hexagonal segments at the W. M. Keck Observatory, Hawaii.

Keeler gap A division in Saturn's rings, very close to the outer edge of the A Ring. It lies 136 530 km from Saturn's centre and is 35 km wide. Its existence was confirmed by the Voyager probes; it is named after the American astronomer James Edward Keeler who saw it in 1888.

Keeler, James E. 1857-1900 American astrophysicist, probable discoverer of the dark narrow gap in the outer part of the A ring of Saturn, and the second director of Lick Observatory. Keeler was (probably accidentally) cheated of his rightful fame when the A ring gap became known as "Encke's Division". Encke had earlier seen a broad, poor contrast feature in the A-ring (now called the "Encke Minimum" by some amateurs) which is quite different from the sharp, distinct gap that Keeler recorded on the very first night of observing with the Lick 36-inch refractor. Images from Voyager and Cassini show a much smaller gap near the very outer edge of the A ring which has been named the Keeler Gap. Of course, Keeler never saw it. On the other hand, the "Encke" gap which was seen by Keeler may have been seen even earlier by Francesco De Vico, William Lassell and/or the Rev. William R. Dawes.

Kellner eyepiece An eyepiece design which is in effect an achromatic version of the Ramsden eyepiece, and is therefore sometimes known as the achromatic Ramsden. It has a planoconvex field lens and an achromatic eye lens with a flat surface facing the eye. Its field of view is fairly large, 45-50°, and is free from aberrations. For this reason the Kellner eyepiece is often used in binoculars. It is a good all-purpose eyepiece, giving acceptable results even with low f-numbers, although it is prone to ghost images caused by internal reflections. It was invented in 1849 by the German optician Carl Kellner.

Kelvin temperature scale Kelvin is a temperature scale designed so that 0K is defined as absolute zero and the size of one unit is the same as the size of one degree Celsius. Water freezes at 273.16K; water boils at 373.16K. This temperature scale was designed by Lord Kelvin (William Thompson, 1824 - 1907). [K = C + 273°, F = 9/5C + 32°].

Kelvin wave The Kelvin wave is a gentle but huge swell of warm water in the Pacific Ocean. This mass of water is a few degrees warmer than surrounding water, is only 5-10 cm high, but is hundreds of kilometers wide.

Kennedy space center A NASA establishment on Merritt Island at Cape Canaveral, Florida, responsible for launching manned and unmanned spacecraft. Launch facilities built for the Apollo Moon missions have been modified for use by the Space Shuttle. KSC also has a landing runway for Space Shuttles. KSC is adjacent to Cape Canaveral Air Force Station, where the launchpads for unmanned rockets are located. Cape Canaveral began operation as a missile test centre in 1950. KSC opened on its present site in 1965.

Kepler, Johannes German mathematician and astronomer. In 1600 he became Tycho Brahe's assistant in Prague where he undertook to complete the tables of planetary motion Tycho had begun. Kepler first calculated the orbit of Mars. He spent much time trying to reconcile Tycho's accurate observations of the planet with a circular orbit, but concluded that Mars moved instead in an elliptical orbit. Thus he established the first of his laws of planetary motion A theory that the Sun controlled the planets by a magnetic force led him to the second and third of his laws, which were published as part of his treatise on theoretical astronomy, Epitome astronomiae Copernicanae The Rudolphine Tables (named after Tycho's patron, the Holy Roman Emperor Rudolph II) of planetary motion appeared in 1627 and were still in use in the 18th century. Kepler also wrote De stella nova, on the supernova of 1604 and Dioptrice on optics and the theory of the telescope.

Kepler's laws Three laws governing the orbital motions of the planets, discovered by J. Kepler. The first law states that the orbit of a planet is an ellipse with the Sun at one focus of the ellipse. The second law states that the radius vector joining planet to Sun sweeps out equal areas in equal times. The third law states that the square of the orbital period of each planet in years is proportional to the cube of the semimajor axis of the planet's orbit. The first law gives the shape of the planet's orbit; the second describes how the planet must continuously vary its speed as it follows its orbit, moving fastest at perihelion and slowest at aphelion. The third law gives the relationship between the planets' average distances from the Sun and their periods of revolution. 1. Every planet orbits the Sun in an ellipse, with the Sun at one focus of that ellipse.

2. The radius vector, a line from the Sun to the planet, sweeps out equal areas within the ellipse in equal times.

3. A fixed ratio exists between the time taken to complete an orbit and the size of the orbit, and this ratio is the same for every planet.

Kepler's star A supernova in Ophiuchus, first observed on 1604 October 9, and described by J. Kepler in his book De stella nova. Observations from Europe and Korea show that it reached a maximum apparent magnitude of -3 in late October. The star remained visible for almost a year. The light-curve is that of a Type Ia supernova. The visible supernova remnant consists of a few faint filaments and brighter knots at a distance of approximately 30000 L.Y it is the radio source 3C 358.

Kernel 1. In computing, the essential part of an operating system, responsible for resource allocation, low-level hardware interfaces, security etc.

2. In programming, an essential subset of a programming language, in terms of which other constructs are (or could be) defined.

Kerr black hole A rotating black hole, as distinct from a non-rotating Schwarzschild black hole. Black holes are expected to rotate rapidly, since the stars that formed them would have been rotating; hence they will be Kerr black holes. Several consequences arise from the addition of rotation to a black hole. First, the event horizon becomes elliptical, and its surface area becomes less than that for a static black hole of the same mass. If the black hole were rotating sufficiently quickly, the area of the event horizon would reduce to zero, leaving the central singularity visible from outside.Second, there is a region around a rotating black hole, the ergosphere, in which objects are forced to spin around the black hole. The outer edge of the ergosphere is the static limit. Third, a new, inner event horizon forms, and it becomes possible to travel through the black hole, and emerge into a new universe or perhaps another part of our own Universe, through this second event horizon.

Kirchoff, Gustav Gustav Kirchoff (1824-1887) was a German physicist who realized that each element gave off a characteristic color of light when heated to incandescence. When separated by a prism, the light for each element had a specific pattern of wavelengths. Kirchoff, together with Bunsen, used his techniques to discover two new elements, cesium (1860) and rubidium (1861). Kirchoff found that when light shines through a gas, the gas absorbs some of the light, the same wavelengths of light that it would emit when heated. He applied his techniques to the Sun, explaining Fraunhofer lines. He also found that incandescent solids, liquids, and compressed gases emit a continuous spectrum.

Kirchhoff's first law An incandescent solid or gas under high pressure will produce a continuous spectrum.

Kirchhoff's second law A low-density gas will radiate an emission-line spectrum with an underlying emission continuum.

Kirchhoff's third law Continuous radiation viewed through a low-density gas will produce an absorption-line spectrum.

Kirkwood gaps Any of several narrow regions within the main asteroid belt where few bodies are found, as a result of gravitational perturbations by Jupiter. These gaps occur at mean distances from the Sun which correspond to zones of commensurability. In these regions of the belt, an asteroid orbits the Sun in a simple fraction of Jupiter's orbital period. These commensurabilities are generally written in the form of a ratio, for example 3: 2 for objects which orbit in exactly two-thirds of Jupiter's orbital period. A few zones of commensurability contain isolated concentrations of asteroids in stable orbits, but most are virtually devoid of asteroids. Examples are the gaps at the 2:1, 5:2, and 3:1 commensurabilities (mean distances of 3.28, 2.82, and 2.50 AU from the Sun). Such gaps were first noted by D. Kirkwood in 1857.

Kirkwood, Daniel American astronomer. He is best known for his discovery of the Kirkwood gaps. He also explained how the Cassini and Encke Divisions in Saturn's rings resulted from orbital resonances with the planet's larger satellites. He pointed out that certain groups of asteroids shared very similar orbital elements, such as (153) Hilda and other asteroids with periods two-thirds of Jupiter's.In 1880 Kirkwood was the first to suggest the existence of a group of sungrazing comets.

Knots Saturn's outermost ring, the subdivided "F" ring, has many visible knots or clumps of matter. These knots may be clumps of particulate ring material or tiny orbiting moons of Saturn.

Kochab The star Beta Ursae Minoris, magnitude 2.1. It is a K4 giant 83 L.Y away. With Pherkad, it forms the so-called Guardians of the Pole.

Kohoutek, comet A long-period comet, discovered by the Czech astronomer Luboš Kohoutek on 1973 March 18; formerly designated 1973 XII. When discovered, Comet Kohoutek lay beyond Jupiter's orbit and was unusually bright for an object so distant. This led to some over-estimates of the comet's likely magnitude close to perihelion, 0.14 AU, on 1973 December 28. Astronauts aboard the Skylab space station observed the comet with a coronagraph during its perihelion passage at magnitude -3. Comet Kohoutek emerged into the evening sky early in 1974 January. Although then of 4th magnitude with a tail up to 25° long and an antitail, it was much less spectacular than originally anticipated. Its orbit is calculated to be hyperbolic with an inclination of 14°.3.

Koronis family A well-defined Hirayama family of asteroids in the outer region of the main asteroid belt at a mean distance of 2.87 AU from the Sun. The members are mainly S-class asteroids with very similar colours, spectra, and albedos. They are believed to have originated in the break-up of a highly homogeneous parent body, presumably of S class. The largest family members are (208) Lacrimosa, diameter 48 km, and (167) Urda, diameter 44 km. Ida is also a member. The family is named after (158) Koronis, an S-class asteroid of diameter 36 km, discovered in 1876 by the German astronomer Victor Carl Knorre.The orbit of Koronis has a semimajor axis of 2.871 AU, period 4.86 years, perihelion 2.72 AU, aphelion 3.02 AU, and inclination 1°.0.

Kowal, Charles T. 1940- American astronomer; discovered Leda and the comet-like object 2060 Chiron (aka 95 P/Chiron).

Kp index A 3-hourly planetary geomagnetic index of activity generated in Gottingen, Germany, based on the K index from 12 or 13 stations distributed around the world.

Kramers opacity The main source of resistance to the flow of radiant energy in gases with temperatures about 10^4-10^6 K. The Kramers opacity is mostly due to energy transitions of free electrons, and declines as the temperature increases. It is the main source of opacity in the interiors of stars of up to about 1 solar mass. In stars much more massive than the Sun, the Kramers opacity is dominant only in the surface layers, electron scattering opacity being more important at deeper levels. It is named after the Dutch physicist Hendrik Anthony Kramers.

Kreep Crystalline rock from the highlands of the Moon, enriched in potassium rare earth elements, and phosphorus.The presence of KREEP rocks suggests that chemical separation has occurred within the Moon. They usually have high concentrations of uranium and thorium, which are radioactive and therefore produce heat, so KREEP rocks were important to the Moon's internal development.

Kreutz sungrazer A group of apparently related long-period comets with very small perihelion distances (less than 0.01 AU). They are named after the German astronomer Heinrich Carl Friedrich Kreutz.who studied them in 1888. Sungrazer orbital elements fall into two subgroups, travelling in retrograde orbits with periods of about 500-1000 years. It thus appears that the sungrazers are derived from the break-up of a single large progenitor and the further disintegration at subsequent returns of two or more of its fragments. Notable sungrazers include the Great September Comet of 1882 and Comet Ikeya-Seki in 1965. Numerous sungrazers have been discovered by coronagraphs aboard spacecraft.

Kron-cousins ri photometry The most commonly used red (i.e. 600-1100 nm) photometric system with two filters, called R and I, of effective wavelengths and bandwidths 638 and 138 nm, and 797 and 149 nm, respectively. R - I is more sensitive to temperature than B - V for cool stars because of the number of lines and molecular bands in the B and V passbands. It was originated by the American astronomer Gerald Edward Kron and modified by the South African Alan William James Cousins.

Kruskal diagram A diagram used to plot trajectories in space-time near a black hole. The vertical and horizontal coordinates, called Kruskal coordinates, are two complicated functions of time and distance from the black hole. Lines of constant time radiate from the origin of the diagram, with steeper slopes corresponding to later times. Photons always travel along diagonal lines at ±45° to the vertical. The trajectory of an object falling into the black hole is shown as a curving line moving upwards on the diagram at less than 45° to the vertical. The diagram is named after the American physicist Martin David Kruskal.

Kuiper airborne observatory A NASA Lockheed C-141 transport aircraft modified to carry a 0.91-m Cassegrain reflector for infrared astronomy. The KAO operated from 1975 to 1995 and flew at altitudes of up to 13.7 km. Its replacement is the Stratospheric Observatory for Infrared Astronomy.The KAO was named after G. P. Kuiper.

Kuiper belt A region of the outer Solar System containing an estimated 107-109 icy planetesimals, or comet nuclei. The Kuiper Belt is an inner, flattened extension of the Oort Cloud. It lies more or less in the same plane as the planets and extends outwards from around 30 AU (the orbit of Neptune) to perhaps 1000 AU. Such a vast reservoir of comets beyond Neptune was first proposed in 1951 by G. P. Kuiper. In 1992, the British-born American astronomer David Clifford Jewitt (1958-)and the Vietnamese-born American astronomer Jane Luu discovered the first Kuiper Belt object, 1992 QB1. This has a diameter of about 200 km, semimajor axis 44.0 AU, orbital period about 296 years, perihelion 40.9 AU, aphelion 47.1 AU, and inclination 2°.2. Since then, dozens more have been found. The Kuiper Belt could be the source of most periodic comets. Unusual objects such as Chiron and Pholus may have originated in the Kuiper Belt.

Kuiper, gerard peter Dutch-American

astronomer. In a search for new planetary satellites he discovered Miranda (in 1948, orbiting Uranus) and Nereid (in 1949, orbiting Neptune). His spectroscopic studies revealed methane in the atmospheres of Uranus and Neptune. He also found methane bands in Titan's spectrum, demonstrating that the satellite has an atmosphere. He suggested the existence of the Kuiper Belt as the source of short-period comets. Kuiper was advisor on many American lunar and planetary missions, and proposed the idea of flying infrared telescopes on board high-altitude aircraft, which led to the Kuiper Airborne Observatory.

L

L chondrite A chondrite (a stony meteorite containing small, round, silicate granules called chondrules) that has a low amount of iron

L magnitude The magnitude of a star as measured through the infrared L filter of the Johnson photometry system. The L filter has an effective wavelength of 3450 nm and a bandwidth of 472 nm. More recent observers have redefined the L filter to avoid the worst of the absorption bands in the Earth's atmosphere. The new filter has a longer central wavelength of 3800 nm, and is called L′. There is also an L magnitude in Walraven photometry, but the context should always make clear when this is intended.

La superba The star Y Canum Venaticorum. It is a semiregular variable red supergiant of type SRB ranging between 5th and 7th magnitudes with a period of just over 5 months. It was named by P. A. Secchi for its remarkable spectrum.

Labes A landslide on a planetary surface; pl. labes. The name is not a geological term, but is used in the nomenclature of individual features, for example Ophir Labes on Mars.

Labyrinthus A complex, intersecting network of linear depressions on a planetary surface; pl. labyrinthi. The name is not a geological term, but is used in the nomenclature of individual features, for example Noctis Labyrinthus on Mars.

Lacaille, Nicolas Louis De French astronomer. From the Cape of Good Hope he surveyed the southern skies in 1751-3. He discovered 24 new nebulae and clusters, and charted the positions of nearly 10 000 stars. While there he measured the position of the Moon simultaneously with J. J. de Lalande in Berlin to obtain an accurate value of its distance. Lacaille published a star chart which introduced 14 new southern constellations; his full southern star catalogue, Coelum australe stelliferum, was published posthumously in 1763.

Lacerta A faint constellation of the northern sky, representing a lizard. Its brightest star is Alpha Lacertae, magnitude 3.8. The constellation contains BL Lacertae, the prototype of a class of peculiar galaxies.

Lacertid An alternative name for a BL Lacertae object.

Lacus A small, irregular dark patch on a planetary surface; pl. lacus. The name is not a geological term, but is used in the nomenclature of individual features, for example Lacus Somniorum on the Moon and Solis Lacus on Mars. The word means 'lake', and was first applied to lunar features in the 17th century, when it was thought the dark patches might be bodies of water.

Lagoon nebula The diffuse nebula M8 in Sagittarius, also known as NGC 6523. It is about 1 ½° long and ½° wide. The Lagoon Nebula gets its name from a dark lane that divides it in two. The brightest part of the nebula is called the Hourglass. The Lagoon Nebula contains the 5th-magnitude star cluster NGC 6530. The main illumination of the nebula comes not from this cluster but from the 6th-magnitude blue supergiant 9 Sagittarii. The Lagoon Nebula lies 5200 L.Y away.

Lagrange Joseph Louis De French mathematician, born in Italy. In celestial mechanics he studied perturbations and stability in the Solar System. He examined the three-body problem for the Earth, Moon, and and the motion of Jupiter's satellites.In 1772 he found the particular solutions to the problem that give rise to the equilibrium positions now called Lagrangian points. Lagrange also studied the Moon's libration.

Lagrange, Joseph (1736 - 1813) A French mathematician of the eighteenth century. His work Mecanique Analytique (Analytical Mechanics; 1788) was a mathematical masterpiece. It contained clear, symmetrical notation and covered almost every area of pure mathematics. Lagrange developed the calculus of variations, established the theory of differential equations, and provided many new solutions and theorems in number theory. His classic Theorie des fonctions analytiques laid some of the foundations of group theory. Lagrange also invented the method of solving differential equations known as variation of parameters.

Laika Laika (meaning "barker" in Russian) was the first dog sent into orbit around the Earth. She rode on Sputnik 2, a Russian mission that blasted off on November 3, 1957 (Sputnik 2 was a metal sphere that weighed about 250 pounds = 113 kg). Laika was originally thought to have survived in Earth orbit for four days, dying in space when the batteries to the cabin over-heated. In 2002, it was revealed that Laika died roughly 5 to 7 hours into the flight, from overheating and stress. After orbiting the Earth 2,570 times, Sputnik 2 fell back to Earth on April 14, 1958, burning up during re-entry. Laika was a female mutt who was part Siberian Husky.

Laplace, Pierre Simon De French mathematician and astronomer. He established celestial mechanics as a mathematical discipline, applying Newton's theory of gravitation to the study of orbits, perturbations, and stability in the Solar System. He explained the small, long-term variations in the orbital speeds of Jupiter and Saturn and the Moon.His nebular hypothesis of the Solar System's origin,in which planets condense from different zones in a nebula, is similar to I. Kant's theory, of which he was apparently unaware. His Traité de mécanique céleste is a summary of 18th-century celestial mechanics.

Laplacian plane A plane about a planet upon which a satellite's orbital plane precesses because of perturbations. In the usual case where the satellite's orbit is perturbed both by the oblateness of the planet and by the Sun, the Laplacian plane lies between the planet's equatorial plane and the planet's orbital plane about the Sun. It is named after the French mathematician P. S. de Laplace.

Late-type galaxy A spiral or irregular galaxy. The name comes from the conventional position of these galaxies in the tuning-fork diagram of galaxy types. For similar reasons, an Sc or Sd spiral galaxy may be referred to as a latetype spiral, as opposed to an early-type Sa or Sb spiral.

Late-type star A star with a surface temperature cooler than the Sun, with a spectral type K, M, C, or S; G stars are often included as well. Late-type stars can be either of low mass, if they are on the main sequence, or more massive than the Sun if they are giants or supergiants. The designation 'late' derives from the time when it was wrongly thought that stars with K or M spectra were old and evolved.

Latitude In a planeto-centric co-ordinate system, the angle measured at the center of the planet from the equator to a point on the planet's surface, towards one of the poles.

Also used, by analogy, as one co-ordinate in the *ecliptic* and *galactic* systems.

Latitude variation The slight change in the geographical latitude of an observing site which results from motion of the Earth's poles due to the Chandler wobble of the Earth's rotational axis. This may be detected by observations of stars at the zenith, since the declination of the zenith point is equivalent to the latitude of the observing site.

Latitude, celestial The angular distance of a celestial object from the nearest point on the ecliptic.

Lava Fluid magma that flows onto the surface of a planet or moon; erupted from a volcano or fissure. Also, the rock formed by solidification of this material.

Law A statement, usually mathematical, which describes some physical phenomena. Compare: hypothesis and theory.

LBV Abbr. for luminous blue variable.

Le Verrier, Urbain Jean Joseph French celestial mechanician. From 1838 he studied the long-term stability of the Solar System, in particular the causes of perturbations, leading to a revision of the masses of the planets, the dimensions of the Solar System, and the velocity of light. In 1845 he suggested that a planet within Mercury's orbit would account for irregularities in Mercury's observed positions.The same year, D. F. J. Arago alerted him to similar irregularities with Uranus, and in 1846 Le Verrier supplied the position (calculated independently of J. C. Adams) that enabled J. G. Galle to locate Neptune.

Lemaître universe A model of the universe containing a cosmological constant term, named after G. É. Lemaître. In this model, space has a positive curvature but expands for ever. The Lemaître universe is both homogeneous and isotropic. The most interesting aspect of such a universe is that it undergoes a so-called coasting phase in which the cosmic scale factor is roughly constant with time.

Leo A constellation of the zodiac, representing a lion. The Sun passes through Leo from the second week of August to the third week of September. Leo's brightest star is Regulus Beta Leonis is Denebola; Gamma Leonis is a handsome double star. R Leonis is a deep-red Mira Star, ranging from 6th to 10th magnitude with a period of about 10 months. Leo contains two pairs of spiral galaxies visible in small telescopes: M65 and M66, and M95 and M96. Every November the Leonid meteors radiate from near Gamma Leonis. Leo also contains the third-closest star to the Sun, 7.8 L.Y away, an M6 dwarf called Wolf 359, magnitude 13.5.

Leo minor An insignificant constellation of the northern sky, representing a little lion. It contains little of interest. Its brightest star, 46 Leonis Minoris, is of magnitude 3.8.

Leonid meteors A meteor shower showing weak activity (maximum ZHR 15) in most years, although meteor storms sometimes occur when the parent comet, Tempel-Tuttle, returns to perihelion, at approximately 33-year intervals. Leonid storms were seen in 1799, 1833, 1866, and 1966. During the storm of 1966 November 17, the Leonid ZHR may have reached 100 000 for 40 min. Activity is usually substantial for several years either side of the storms. Leonids are seen from November 15 to 20, with maximum on November 17, when the radiant lies at RA 10h 08m, dec. +22°, in the Sickle of Leo. Leonids have the highest geocentric velocities of any meteor stream, 70 km/s, and a large proportion leave persistent trains.

Lepton Leptons are subatomic particles that are susceptible to the weak nuclear force but not the strong force (the force that binds an atomic nucleus together). There are six leptons: the electron, muon, tau, electron neutrino, muon neutrino, and tau neutrino.

Lepton era The interval, commencing about 10^{-5} s after the Big Bang, in which the various kinds of lepton were the main contributors to the density of the Universe. Pairs of leptons and antileptons were created in large numbers in the early Universe but, as the Universe cooled, most lepton species were annihilated. The lepton era immediately followed the hadron era, and is a subdivision of the radiation era. The end of the lepton era is usually taken to have occurred when most electron-positron pairs were annihilated, at a temperature of 5×10^9 K, about 1 second after the Big Bang.

Lepus A constellation of the southern sky, representing a hare. Its brightest star is Alpha Leporis.R Leporis is Hind's Crimson Star, a deep-red variable. M79 is a globular cluster; NGC 2017 is a small open cluster, actually a complex multiple star with five components visible in moderate-sized telescopes.

Leto family A small, well-defined family of asteroids in the main asteroid belt about 2.8 AU from the Sun. The three largest members of the family, (68) Leto, (236) Honoria, and (858) El Djezair, are of S class. Leto itself, diameter 128 km, was discovered in 1861 by the German astronomer Robert Luther.Leto's orbit has a semimajor axis of 2.783 AU, period 4.64 years, perihelion 2.27 AU, aphelion 3.30 AU, and inclination 8°.0.

Levee Zones in a lava flow where the lava between the zones is moving faster than the lava outside the zones.

Lexell, comet A lost periodic comet, discovered by C. J. Messier on 1770 June 14. It passed within 0.015 AU (2.2 million km) of Earth on July 1, appearing of 2nd magnitude with a coma over 2° wide. Calculations in 1779 by the Swedish astronomer Anders Johan Lexell after whom the comet is named, indicated that it was in an elliptical orbit of period 5.6 years. The comet's orbit has been considerably modified by close approaches to Jupiter. A passage only 0.0015 AU from Jupiter in 1779 increased its perihelion distance to 5.2 AU, and may even have placed it into a hyperbolic orbit, effectively ejecting it from the Solar System. It has not been seen since.

LHEA Laboratory for High Energy Astrophysics, located at NASA's Goddard Space Flight Center. The scientists, programmers and technicians working here study the astrophysics of objects which emit cosmic ray, x-ray and gamma-ray radiation.

Libra A constellation of the zodiac, representing a pair of scales. The Sun lies in Libra for the first three weeks of November. Alpha Librae is a wide double. Beta Librae is the constellation's brightest star, magnitude 2.6. The two star names come from the Arabic meaning 'southern claw' and 'northern claw', and arise because Libra once represented the claws of the scorpion, Scorpius, which it adjoins. Delta Librae is an Algol Star, an eclipsing binary that varies between magnitudes 4.9 and 5.9 in a period of 2 days 8 hours.

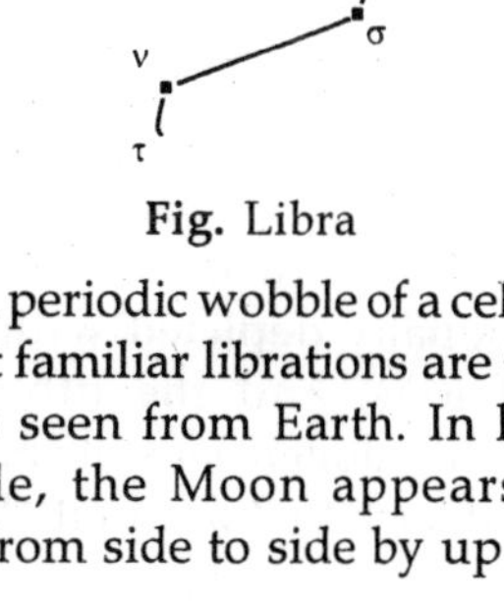

Fig. Libra

Libration A periodic wobble of a celestial body. The most familiar librations are those of the Moon as seen from Earth. In libration in longitude, the Moon appears to swing slightly from side to side by up to 7° 45' in each direction. This occurs because the Moon's speed along its elliptical orbit varies with its distance from Earth, while its axial rotation remains constant. In libration in latitude, the Moon appears to nod from north to south by up to 5° 09' in each direction. This is because the Moon's axis of rotation is not perpendicular to its orbital plane. As a result of these two librations, we can see up to 59% of the Moon's surface. A third libration, diurnal libration, occurs because we view the Moon from different sides of the Earth at moonrise and moonset, so we can again see slightly round the eastern and western limbs, although by less than 1°. Libration occurs when two motions are locked in resonance.For this reason, the planet Mercury also exhibits a libration, as do some planetary satellit.

Lick observatory The Lick Observatory is a University of California observatory located at Mount Hamilton, near San Jose, California. It has a 120 inch (3 m) reflecting telescope and a 36 inch (91 cm) refracting telescope.

Light Electromagnetic radiation that can be seen by the human eye. It lies between the ultraviolet and infrared regions of the electromagnetic spectrum. Different wavelengths of light appear as different colours. Visible radiation ranges from wavelengths of about 750 nm at the red end to around 380 nm at the violet end.

Light bridge Observed in white light, a bright tongue or streaks penetrating or crossing sunspot umbrae. The appearance of a light bridge is frequently a sign of impending region division or dissolution.

Light bucket Another name for a flux collector.

Light cone A means of illustrating the past and future of an event in space-time. Space-time diagrams in three dimensions are conventionally depicted with time as the vertical axis, and the other two axes representing dimensions in space. The axes are scaled such that one unit along the time axis is one second, and one unitalong either of the spatial axes is 300 000 km (because the speed of light is approximately 300 000 km/s). On such a diagram, the paths of rays of light from an event slant upwards at 45° along the sides of a cone whose vertex is at the event and whose axis is vertical. This is the future light cone of the event. Any and all events occurring after the first event which are to be affected in any way by that first event must lie within the cone. Events outside the cone would require fasterthan-light travel if they were to be affected by the initial event. Conversely, any earlier events which are to affect the event in question must be contained within a 45° cone going down from the event.

Light-curve A graph of the variations in brightness of an object, particularly a variable star or a rotating asteroid, plotted against time.

Light dispersion Breaking light into its various wavelengths is light dispersion. Transparent substances disperse light because red light (longer wavelengths) is refracted less than blue light (shorter wavelengths).

Light-gathering power The ability of a telescope to collect light; the larger a telescope's aperture, the greater its light-gathering power

Light pollution The intrusion of extraneous illumination, to the detriment of astronomical observations. Artificial sources of light are a significant problem for astronomers near urban areas. The pall of artificial light reflected upwards and sideways from towns and cities by suspended particles or water vapour in the atmosphere can be a nuisance, even at considerable distances from the source, fogging photographic emulsions and degrading the performance of telescopes. From some urban locations only the brightest stars are visible. Increasing awareness of the energy wastage, along with improved design of lighting, has led to some limitation of light pollution. Natural sources of light pollution also exist, including twilight and aurora (both particularly at high latitudes), moonlight, and airglow.

Light spectrum Light can be broken up into into its component colors (for example, by passing light through a prism) - this is a spectrum. Light from the Sun can be broken up into the colors of the rainbow (red, orange, yellow, green, blue, indigo, and violet). The different colors of light correspond to the different energies of the light waves: for example, blue light has more energy than red light. Different astronomical objects emit different electromagnetic spectra (spectra is the plural of spectrum), and these spectra are used to analyze the body.

Light, speed of 299792.5 km (186 291 miles) per second in a vacuum, the fastest speed in the Universe.

Light time The time taken for light to travel from an object to the observer. Light time must be taken into account when calculating the real times of occurrence of events such as eclipses, transits, and occultations of planetary satellites. The light time corresponding to one astronomical unit is 499 s.

Light waves Electromagnetic radiation at light frequencies (about 10^{15} cycles per second), or a wavelength between 400 and 700 billionths of a meter.

Light year A unit of length used in astronomy which equals the distance light travels in a year. At the rate of 300,000 kilometers per second (671 million miles per hour), 1 light-year is equivalent to 9.46053×10^{12} km, 5,880,000,000,000 miles or 63,240 AU.

Light, velocity of (symbol c) The velocity of light in a vacuum is 299 792.5 km/s. The velocity is less in other media. All electromagnetic radiation travels at this speed in a vacuum.

Limb The rim of the visible disk of a celestial body as seen from Earth. A body's leading edge as it moves across the sky is the preceding limb, while the trailing edge is the following limb. The edges of the Moon's visible face are known as the limb regions.

Limb brightening A phenomenon in which an object appears to become brighter towards the edges. Limb brightening can be observed at certain wavelengths in objects that are surrounded by hot but tenuous gas, such as the Sun's corona. It is particularly noticeable at ultraviolet wavelengths, and to a small extent at centimetre and millimetre radio wavelengths.

Limb darkening The decrease in brightness of an object towards its edges. Limb darkening is most noticeable at optical wavelengths in the Sun, where we see into deeper, hotter (and therefore brighter) layers at the centre of the Sun's disk than at the limb. Characteristic features in the light curves of certain eclipsing binaries demonstrate the existence of limb darkening in these stars. Limb darkening is also noticeable in the gaseous giant planets.

Limb flare A solar flare seen at the edge (limb) of the sun.

Limiting magnitude 1. The faintest apparent magnitude that can be detected with a given instrument under prevailing conditions. **2**. The faintest magnitude of object included in an astronomical catalogue.

Lindblad resonance A resonance that can affect stars in a galaxy's disk at certain distances from the galaxy's centre. It occurs when the natural frequency of the radial (i.e. in and out in the direction of the centre) component of motion of a star in its orbit is the same as the frequency of passage of the star through the maxima of the gravitational potential associated with the spiral pattern. If the star is moving around the centre faster than, and overtakes, the spiral pattern, then an inner Lindblad resonance occurs; if the pattern is moving faster than the star around the centre, then an outer Lindblad resonance occurs. At an inner resonance, energy is fed into the orbits of the stars from the spiral pattern, and vice versa at an outer resonance. The effect is named after B. Lindblad.

Lindblad, Bertil (1895-1965) Swedish astronomer. His early work on luminosity criteria made possible a photometric survey which revealed the distribution of stars in the Sun's part of the Galaxy. He went on to study stellar statistics and dynamics on an increasingly larger scale, and showed how J. C. Kapteyn's discovery of two star streams provided evidence that the entire Galaxy was rotating, with differential rotation (soon confirmed by J. H. Oort). Lindblad also showed that the Galaxy is a spiral, and went on to study the dynamics of other galaxies. In the 1940s he proposed the density wave theory to explain spiral structure. In his galactic work he was aided by his son Per Olof Lindblad.

Line blanketing The dimming of a star's spectrum by the presence of many hundreds or thousands of weak absorption lines too faint and close together to be resolved. In cool stars there can be so many different absorbing atoms or molecules with lines in the visible or ultraviolet part of the spectrum that the spectrum appears to be 'blanketed' with them. The energy absorbed must be re-radiated, but this is usually at lower energies (i.e. longer wavelengths); hence the red or infrared part of the spectrum will appear enhanced relative to a star with a non-blanketed spectrum.

Line broadening The broadening of absorption or emission lines in a spectrum by one of four mechanisms: Doppler broadening, due to the motion of the emitting gas; pressure broadening or Stark broadening, due to collisions with other atoms and molecules in the star's atmosphere; or Zeeman splitting, as a result of the presence of a strong magnetic field.

Line of nodes The straight line joining the ascending and descending nodes of an orbit.

Line profile The shape of an absorption or emission line in a spectrum across the narrow wavelength range it occupies, usually requiring high spectral resolution to be studied. The shape is different for the different line-broadening mechanisms, so the line profile can be used to infer the physical properties of the region where the line is formed.

Line ratio The ratio of the strengths of two particular absorption or emission lines. Because these lines correspond to different energy levels or arise from entirely different elements, they can be sensitive to the density and/or temperature of the gas. For example, in a hot gas, the ratio of the Ha to the Hß line strength is sensitive to the density of the gas.

Line receiver A radio receiver designed to detect radio waves in a narrow band around a known spectral line, such as those from interstellar molecules and the 21-cm hydrogen line.

Line spectrum A spectrum consisting of bright emission lines, or dark absorption lines superimposed on a continuous spectrum. Such a line spectrum will convey information about the physical nature of the emitting or absorbing gas.

Line wing The region of an absorption or emission line that is far removed from the line's peak. Extended line wings indicate the presence of high-velocity to the main region, formed in a high-density gas.

Linea A linear feature or elongated marking on a planetary surface; pl. lineae. The name, which means 'thread' or 'plumb-line', is not a geological term, but is used in the nomenclature of individual features, for example Minos Linea on Europa and Palatine Linea on Dione.

Liner The nucleus of a galaxy that shows a characteristic emission-line spectrum; it is an acronym for low-ionization nuclear emission region. A liner's spectrum is dominated by emission lines from low ionization states (e.g.

0 II, N II, S II), with only weak emission lines from higher ionization states (e.g. O III, Ne III, He II). The spectrum is evidence of unusual activity in the nucleus, probably not associated with normal stars, and may result from the heating of interstellar matter, either by radiation from a central source or by shock waves generated by supernova explosions. The linewidths are similar to those seen in Seyfert galaxies and, like Seyferts, the activity seems to occur more often in S0, Sa, and Sb galaxies than in other types.

Liner galaxy A low-ionization nuclear emission-line region galaxy belongs to a common class of otherwise normal galaxies that display low-ionization line emissions near their central regions

Linewidth The width of an absorption or emission line in a spectrum. It is usually quoted as full width at half maximum (FWHM), and gives an indication of the temperature or velocity of the emitting or absorbing region.

Lippershey, Hans Hans Lippershey (1570-1619) was a German-born Dutch lens maker who demonstrated the first refracting telescope in 1608, made from two lenses; he applied for a patent for this optical refracting telescope (using 2 lenses) in 1608, intending it for use as a military device.

Liquid Liquid is a phase of matter in which the molecules are close together and can move around slowly.

LISA (laser interferometer space antenna) A NASA mission which will detect gravitational waves. LISA will consist of three satellites that use laser interferometry to monitor their positions relative to each other. Gravitational waves passing by the satellites cause small changes in the distances between the satellites.

Lithium star An unusual giant star of spectral type G, K, or M that shows lithium in its spectrum. Nuclear reactions in or near the core of the evolving star produce beryllium, which is transported by convection to higher layers where it captures an electron to become lithium. The term is also sometimes used to refer to a T Tauri star in these cases the lithium is likely to have been present in the gas from which the star formed, and will soon be destroyed once the star reaches the main sequence.

Lithosiderite An alternative, now largely obsolete, name for a stony-iron meteorite.

Lithosphere The outer rigid shell of a planetary body, including the crust and part of the upper mantle. On the Earth it is distinguished from the asthenosphere beneath, which is a weaker, deformable part of a planetary body. The Earth's lithosphere, which includes the crust and the upper mantle, is about 150 km deep beneath the continents, and 80 km beneath the oceans.

Little dipper Popular name for the shape formed by the seven main stars of the constellation Ursa Minor, which resembles a smaller version of the Big Dipper, or Plough, formed by seven stars in Ursa Major.

Little dumbbell A planetary nebula in Perseus, also known as M76 or NGC 650-1. It is a smaller and fainter version of the Dumbbell Nebula in Vulpecula and, at 12th magnitude, is the faintest Messier object. It lies about 3500 L.Y away.

Lobate ridge A ridge on a planetary surface with projecting lobes on one or both sides. Many wrinkle ridges are lobate; this has been interpreted as evidence of lava extrusion from the ridges, but it now seems more likely that the lobes ate caused by buckling and folding of the surface during tectonic movement.

Lobate scarp A scarp on a planetary surface that takes the form of a series of lobes. Lobate scarps are a notable feature on Mercury, where they are interpreted as low-angle thrust faults, and may be evidence of global contraction.

Lobe 1. A bright, diffuse area of radio emission seen on one or both sides of the nucleus of a radio galaxy or other active galaxy and situated far outside the visible confines of the galaxy. The emission is mainly synchrotron radiation. The lobe is believed to consist of material ejected from the nucleus of the galaxy and transported into intergalactic space along a jet.

2. A region of high sensitivity appearing as a bulge in the antenna pattern of a radio telescope.

Local arm The Local Arm is the arm of the Milky Way Galaxy where our solar system is located. It is also called the Orion Arm.

Local bubble A region of low density (approximately 0.07 atom/cm^3) in the interstellar matter surrounding the Solar System. The bubble is approximately 100 pc across and encompasses the stars of the immediate solar neighbourhood. The Solar System seems to lie about 10-20 pc from the bubble's edge. The low gas density in the local bubble may be a result of the shock wave from an ancient supernova sweeping through the region.

Local group The Local Group is the cluster of galaxies to which we belong. It is a group of about 30 galaxies that is about 5 million light-years across. The largest of the galaxies are Andromeda galaxy, Triangulum, and our Milky Way. The Local Group also includes Fornax, the Large and Small Magellanic Clouds, M32, M33, M101, and 9 dwarf spheroidal galaxies. The center of the Local Group is roughly between the Milky Way and the Andromeda Galaxy, M31. The galaxies in the Local Group are moving in concert with each other, independent of the "Hubble flow" expansion.

Local hour angle The hour angle of a celestial object as measured with reference to the observer's local meridian.

Local mean time The mean solar time on the meridian of an observing site. It is the local hour angle of the mean sun plus 12 hours.

Local oscillator Practical radio astronomy receivers incorporate frequency conversion because it is technically more convenient to amplify, filter, delay and cross-correlate signals at a fixed *intermediate frequency (i.f.)* than at their original, changeable (and often higher) radio frequencies. The frequency conversion is done in one or more *mixers*, in which a non-linear element (usually a diode) forms signals at the sum and difference frequencies between the astronomical signal and a tunable *local oscillator (l.o.)* signal.

Local sidereal time The sidereal time determined with reference to the observer's own local meridian. It is the local hour angle of the vernal equinox. It differs from the Greenwich Sidereal Time by the observer's longitude, being 4 min greater than GST for every degree east of Greenwich, and 4 min less for every degree west.

Local standard of rest (LSR) The sun has a systematic motion relative to nearby stars, the mean depending on the spectral type of the stars used for comparison. The *standard solar motion* is defined to be the average velocity of spectral types A through G as found in general catalogs of radial velocity, regardless of luminosity class. This motion is 19.5 km/s toward 18 hrs right ascension and 30° declination for epoch 1900.0 (galactic co-ordinates l=56°, b=23°). *Basic solar motion* is the most probable velocity of stars in the solar neighborhood, so it is weighted more heavily by the radial velocities of stars of the most common spectral types (A, gK, dM) in the solar vicinity. In this system, the sun moves at 15.4 km/s toward l=51°, b=23°.

The conventional *local standard of rest* used for galactic studies is essentially based on the *standard* solar motion. It assumes the sun to move at the rounded velocity of 20.0 km/s toward 18 hrs right ascension and 30° declination for epoch 1900.0. This choice presumes that.

Local supercluster The supercluster of galaxies that contains the Local Group, which lies near its edge. It is also known as the Virgo Supercluster since it is centred on the Virgo Cluster of galaxies. The local supercluster has a flattened shape and is probably around 100 million L.Y across. It probably contains at least 10 000 galaxies.

Local thermodynamic equilibrium A balance achieved in a hot gas, in which the radiation emitted is determined by local values of the gas temperature and density. LTE is assumed to apply in stellar atmospheres where strict thermodynamic equilibrium does not hold because of radiation loss to space.

Local time The time at an observing site determined with reference to the observer's own local meridian. The local time could be the local mean time, the local sidereal time, or even the local apparent solar time. All involve hour angles measured with reference to the observer's meridian.

Lockyer Norman English scientist and solar physicist. He was the first to study sunspots spectroscopically, finding Doppler shifts caused by convection currents in the Sun's gases. In 1868, independently of P. J. C. Janssen, he observed the spectra of solar prominences and developed a spectrohelioscope. Janssen found a new line in the solar spectrum, which Lockyer attributed to a hitherto unknown element. He named it helium; it was discovered in the Earth's atmosphere in 1895. Lockyer also studied archaeoastronomical sites such as Stonehenge, and founded and edited the science journal Nature.

Lodranite A very rare class of stony-iron meteorite, named after the town of Lodran in Pakistan, where this type was first identified. Lodranite meteorites consist primarily of olivine and bronzite mixed with metallic nickel-iron. Lodranites show some similarity to the ureilites.

Long-period In astronomy, a comet with an orbital period greater than 200 years.

Long-period comet A comet with an orbital period greater than 200 years. Examples include Comets Donati, Tebbutt, and West. The Kreutz sungrazers are also long-period comets. The shortest-known period for a long-period comet is 226 years (C/1905 F1 Giacobini).

Long-baseline interferometry A technique in which radio telescopes hundreds of kilometres apart can be joined electronically to form an array. Signals received by the various antennas in the array are transmitted by microwave links to a control centre. LBI is ultimately limited by variable propagation conditions in the ionosphere and along the microwave beams. A well-known example is the Merlin array.

Longitude at the epoch (symbol L) The longitude that a planet has at a particular date and time. It is the angle measured along the ecliptic from the vernal equinox eastwards to the ascending node (i.e. to the longitude of the ascending node), and then eastwards along the orbital plane to the planet's position at that instant. In the diagram, the longitude at the epoch is the angle VSN plus the angle NSP.

Longitude of perigee For the Moon or an artificial Earth satellite, the angle measured from the direction of the vernal equinox eastwards along the celestial equator to the ascending node of the satellite's orbit, and then continued eastwards along the orbital plane to the satellite's perigee.

Longitude of perihelion The angle measured from the vernal equinox eastwards along the ecliptic to the ascending node of a planet's orbit, and then continued eastwards along the orbital plane to the perihelion. In other words, it is the longitude of the ascending node added to the argument of perihelion..

Longitude of the ascending node (symbol O) The angle measured along the ecliptic from the vernal equinox eastwards to the ascending node of a planet's orbit. It defines the point at which an object's orbit crosses the plane of the Earth's orbit.

Long-period variable A red-giant pulsating variable, with a well-defined period in excess of 100 days, otherwise known as a Mira star. The term was previously applied to any late-type variable with a long period, including the types currently known as RV Tauri stars and semiregular variables, but it is now generally restricted to the Mira type.

Lookback time The time taken for light from a distant object to reach the Earth. Light from nearby galaxies takes several million years to reach us, but the lookback time to very distant galaxies and quasars may be of the same order as the Hubble time. Galaxies which are so far away that their lookback time would be greater than the age of the Universe can never be observed.

Loop prominence A very bright active prominence in the form of a loop above the site of a very large flare, seen on the Sun's limb in Ha light. A series of loop prominences usually occurs, progressively larger loops forming above one another. The smaller, earlier ones eventually fade from sight after an hour or so. The maximum height attained is 100 000 km, and a complete loop prominence system may last more than a day. X-ray loops, observed simultaneously on occasion, are located on the outer edge of Ha loops, giving the impression that loop prominences somehow condense out of the hot coronal material.

Loop prominence system (LPS) A system of loop prominences associated with major flares.

Lorentz transformation A set of equations relating the time and position of an event as seen by one observer to the time and position of that same event seen by a second observer who is moving at constant velocity relative to the first. The equations are named after the Dutch theoretical physicist Hendrik Antoon Lorentz who initially derived them. They were later incorporated in special relativity.

Lorentz-fitzgerald contraction The contraction in length of a moving body in its direction of motion. The effect was proposed independently by the Dutch physicist Hendrik Antoon Lorentz in 1895 and the Irish physicist George Francis Fitzgerald in 1893 to account for the inability of the Michelson-Morley experiment to detect the Earth's motion through the ether. It was later incorporated into the special theory of relativity. The contraction becomes noticeable only at velocities near that of light; at the speed of light itself, the object's length would in theory contract to zero.

Lost city meteorite The second meteorite fall to be photographed by a camera network, thus enabling its trajectory and orbit to be determined and the meteorite to be recovered. A magnitude -15 fireball was photographed by the Smithsonian Prairie Network over Oklahoma on 1970 January 4. Four fragments, totalling 17 kg, of the H5 ordinary chondrite meteorite were subsequently found near the predicted impact site, about 4.5 km east of Lost City, Oklahoma. Its orbit had an aphelion in the asteroid belt.

Lovell telescope A 76-m fully steerable radio telescope at the Nuffield Radio Astronomy Laboratories in Cheshire, England. The telescope was completed in 1957, and was known as the Mark I until 1971 when it was substantially rebuilt with a new bowl and supporting structure, and became the Mark IA. In 1987 it was named the Lovell Telescope in honour of A. C. B. Lovell.

Lovell, Bernard English radio astronomer. His daytime radar observations of the Giacobinid meteor shower in 1946 proved the worth of the technique. From 1950 he studied radio emission from space, showing that its fluctuations were atmospheric effects akin to the scintillation of starlight.

He supervised the building of the Mark I radio telescope at Jodrell Bank (now known as the Lovell Telescope), and attracted publicity by using it to monitor the first Soviet space probes. Lovell also studied radio emission from flare stars.

Low earth orbit A satellite is in low Earth orbit (LEO) when it orbits near the top of the Earth's atmosphere, about 100-200 miles (160-320 km) up. These satellites orbit the Earth in about 90 minutes.

Low frequency (LF) That portion of the radio frequency spectrum from 30 to 300 kHz.

Low mass X-ray binary (LMXB) A binary star system consisting of a compact object and a low mass companion (i.e. with a mass less than the Sun's mass).

Low surface brightness galaxy A type of galaxy whose density of stars is so low that it is difficult to detect against the sky background. The proportion of low surface brightness galaxies relative to normal galaxies is unknown, and they may represent a significant component of the Universe. Many of these dim galaxies are dwarfs, found particularly in clusters of galaxies; some are as massive as large spirals, for example Malin-1.

Lowell observatory An observatory at an altitude of 2210 m on Mars Hill, Flagstaff, Arizona, founded in 1894 by P. Lowell. It contains a 24-inch (0.61-m) refractor installed by Lowell in 1896, and the 0.33-m astrograph with which Pluto was discovered. In 1961 a dark-sky outstation was opened on Anderson Mesa, 19 km southeast of Flagstaff, altitude 2200 m. This contains the 1.83-m Perkins reflector of Ohio State and Ohio Wesleyan Universities, moved here in 1961; the 1.1-m John S. Hall reflector opened in 1968; and a 0.8-m reflector. The Navy. Prototype Optical Interferometer a joint project of the Naval Research Laboratory, the US Naval Observatory, and Lowell Observatory, began operation on Anderson Mesa in 1995.

Lowell, Percival Percival Lowell (1855-1916) was an American astronomer and mathematician who founded the Lowell Observatory at Flagstaff, Arizona in 1894. Lowell studied Mars extensively, especially its surface markings, which he thought were canals. He also thought that the bright areas were deserts and the dark ones were areas containing vegetation (this was not true). Lowell published three books on Mars: Mars (1895), Mars and Its Canals (1906), and Mars As the Abode of Life (1908). Lowell also calculated that an unknown planet, dubbed Planet X, must orbit beyond Neptune. Percival Lowell calculated the rough location of Planet "X's" orbit, but died in 1916, before it was found. This planet was eventually found by the American astronomer Clyde W. Tombaugh in 1930 and named Pluto). Tombaugh did his observations at the Lowell Observatory.

Lower culmination The point on an observer's meridian at which a celestial body reaches its lowest altitude..

Low-luminosity star A loose term that can encompass red dwarfs, subdwarfs, white dwarfs, and brown dwarfs. The difficulty of detecting stars of low luminosity means that their total numbers are uncertain. However, they may constitute a significant fraction of the total mass of the Galaxy.

Low-mass star A loose term, sometimes including stars of mass a little greater than that of the Sun, and sometimes used only of stars of less than a few tenths of a solar mass, but still with sufficient mass to burn hydrogen in their cores (i.e. at least 0.08 solar masses). The first definition separates stars with radiative cores from higher-mass stars with convective cores; the latter restricts the term to red dwarfs.

Low-velocity star A star whose speed relative to stars in the solar neighbourhood is small, and which is therefore in a similar orbit around the galactic centre.

Lucy algorithm A technique for estimating corrections to data in statistical astronomy. It is used for removing the effects of diffraction in radio observations and seeing in ground-based optical images, and was used for sharpening the aberrated images from the Hubble Space Telescope before its new optics were installed. The technique is named after its inventor, the British astronomer Leon Brian Lucy.

Luminosity (symbol L) The amount of radiation a star emits, corrected for interstellar absorption. It is expressed in watts, or in terms of the luminosity of the Sun, approximately 3.9×10^{26} watts. Luminosity is related to the absolute bolometric magnitude of the star, Mbol, by the equation. For X-ray sources it is usual to define luminosity not bolometrically but in particular passbands.

Luminosity class A classification of stars according to their luminosity, which can vary widely for a given spectral type. Luminosity class indicates, for example, whether a star is a supergiant, a giant, or a dwarf. The luminosity class is assigned by examination of the star's spectrum, looking at luminosity-sensitive lines or line ratios.

Luminosity function The luminosity function for stars within 20pc of the Sun: the number of stars plotted against their absolute magnitude. There are very few stars on the plot brighter than M=0; stars this luminous are rare. The number of stars of low luminosity is considerably under-represented because they are difficult to detect.

Although it missed its target; Luna 2 was the first man-made object to hit another body in space; Luna 3 took the first pictures of the Moon's farside. Luna 9 made the first soft landing on the Moon and sent back the first pictures from its surface, while Luna 10 was the first probe to go into orbit around the Moon. Lunas 16, 20, and 24 brought lunar samples back to Earth; Lunas 17 and 21 delivered the Lunokhod rovers to the Moon's surface.

Luminosity-volume test A method of testing whether or not galaxies evolve with time. It involves calculating the volume of a sphere with the Earth at its centre and a given galaxy or quasar on its perimeter. The calculated volume is then divided by the volume of a sphere in which the source is so distant as to be just undetectable. The calculation is repeated for each member of a sample of galaxies or quasars. The average value of this ratio for the complete sample should be 0.5 if there is no evolution in the properties of the sources. Quasar samples give a value of about 0.7, which indicates that there are many more quasars at high redshift than would be expected without evolution.

Luna One of two successful series of Soviet unpiloted lunar missions. There were 24 missions to the Moon in the Luna series launched from 1959 to 1976. (The other series, Zond, had 5 lunar missions launched in 1965 to 1970.) The first image of the farside of the Moon was taken by the Luna 3 spacecraft in October, 1959. Three Luna landers (Luna 16, 20, and 24) collected a total of approximately 300 grams of lunar samples and returned them to Earth.

Lunar-A A Japanese space probe that will go into orbit around the Moon and fire three penetrators into its crust, two on the nearside and one on the farside, to measure moonquakes and the Moon's internal temperature. It is scheduled for launch in 1999.

Lunar eclipse The passage of the Moon through the Earth's shadow, resulting in a darkening which may range from barely detectable in a penumbral eclipse to very dark when the Moon enters the Earth's umbra. Lunar eclipses can occur only at full Moon when the Moon is close to the node of its orbit around the Earth. When the Moon is slightly to the north or south of the node, it does not pass centrally through the umbra, and may undergo only a partial eclipse. During umbral eclipses the Moon appears darkened to varying degrees determined principally by the clarity of the Earth's atmosphere. Refraction of sunlight through the atmosphere on to the eclipsed Moon produces a coppery or reddish coloration in some events, while darker eclipses may appear steelgrey. Coloured fringes may be seen around the edges of the umbra during the partial phases. A lunar eclipse is visible over the entire hemisphere of the Earth from which the Moon is above the horizon.

Lunar halo A lunar halo is a luminous ring that is sometimes seen surrounding the moon. Some parts of the halo are very bright, others are not very bright. Sometimes, only a part of the ring is visible. The halo is produced as moonlight is reflected and refracted through tiny, flat ice crystals in the atmosphere. Halos always are at an angle of 22° away from the moon, due to the hexagonal structure of the ice crystals. The diameter of the halo is about an eighth of the sky.

Lunar inequality A departure in the Moon's orbital motion from its behaviour according to Kepler's laws. It results not only from the perturbing presence of the Sun and the planets, but also from tidal forces and the non-spherical shapes of the Earth and Moon. Lunar theory takes all these factors into account to produce a mathematical description of the Moon's past, present, and future geocentric positions and velocities over thousands of years. Such a description contains not only terms describing the elliptical orbit of the Moon, but also hundreds of periodic terms— inequalities— due to those other factors. Among those factors are evection, variation and the annual equation.

Lunar meteorites Meteorites that have been identified through their chemistry and minerals as being from the Moon.

Lunar module The spacecraft in which Apollo astronauts landed on the Moon. It consisted of two halves: the descent stage, with four legs and an engine to control the landing on the Moon; and the ascent stage, in which two astronauts rode. After exploring the Moon's surface, the astronauts took off in the ascent stage to rejoin the third astronaut in the Command Module orbiting above.

Lunar month A lunar month (also called a synodic month) is the time between successive new or full moons. It lasts 29 days, 12 hours, and 44 minutes.

Lunar orbiter A series of five NASA space probes that photographed the Moon from orbit. Their main targets were potential landing sites for the Apollo missions. The first three of the series orbited around the Moon's equator; the last two were put into polar orbit from which they photographed the entire lunar surface. Tracking the Lunar Orbiters revealed the existence of mascons, areas of denser rock within the Moon.

Lunar parallax The angle subtended at the Moon by the Earth's equatorial radius.

Lunar phases As the moon circles the Earth, the shape of the moon appears to change; this is because different amounts of the illuminated part of the moon are facing us. The shape varies from a full moon (when the Earth is between the Sun and the moon) to a new moon (when the moon is between the sun and the Earth). The phases of the Moon include: the new moon, waxing crescent, first quarter, waxing gibbous, full moon, waning gibbous, last quarter, waning crescent, and new moon again.

Lunar prospector A NASA space probe that will go into polar orbit around the Moon, surveying the composition of its crust in search of frozen water at the poles and other resources. It will also monitor for volcanic emissions and map the Moon's gravitational field. Lunar Prospector is due for launch in 1997 September.

Lunar rover The lunar rover (also known as the "moon buggy" or the Lunar Roving Vehicle) was a Jeep-like vehicle that the astronauts used to drive on the moon during the Apollo 15, 16, and 17 missions. It weighed 450 pounds, (or 75 pounds in the Moon's gravity). It could carry 1000 Earth-pounds. On the flight from the Earth to the moon, the lunar rover was folded up in the lunar module. There are three abandoned lunar rovers on the Moon.

Lunar transient phenomenon A temporary event observed on the Moon's surface, taking the form of a local abnormal brightening, obscuration, or colour change; also known as a transient lunar phenomenon.Such events have been reported by many visual observers from the time of F. W. Herschel, and have been taken as evidence of volcanic activity or outgassing at the lunar surface. However, there has been no unequivocal photographic or photometric record of such an event, and some authorities doubt their reality.

Lunation The period of time taken for the Moon to go through a complete cycle of phases. It is the same as a synodic month.

Lundmark, Knut Emil Swedish astronomer. In the 1920s he argued that the lack of significant rotational motion in some 'spiral nebulae' indicated their extreme distance. Using redshift data for over 40 galaxies, he derived an early form of distance-velocity relation.He was one of the first to point out that many of the brighter galaxies were distributed around a great circle on the celestial sphere, later shown by G. H. de Vaucouleurs to reflect groupings in clusters and superclusters. In 1925 Lundmark introduced an 'upper class' of novae for the brightest novae observed in other galaxies.

Lunisolar precession The effect of the attractions of the Sun and Moon on the Earth's equatorial bulge, which causes the Earth's pole to sweep out a circle on the sky around the pole of the ecliptic. Simultaneously, the position of the equinoxes moves westwards along the ecliptic.

Lunitidal interval The time between the transit of the Moon across the local meridian and the time of the next high tide.

Lunokhod Two unmanned lunar rovers, launched by the former Soviet Union. The eight-wheeled vehicles carried TV cameras and instruments to analyse the Moon's soil, and were driven over the lunar surface by remote control from Earth. Lunokhod 1 was delivered to Mare Imbrium aboard Luna 17 in 1970 November. It had covered a total of 10½ km by the end of its mission 11 months later. Lunokhod 2, an improved version with twice the top speed, landed aboard Luna 21 in 1973 January in the partially ruined crater Le Monnier on the edge of Mare Serenitatis, and covered a total of 37 km in its five-month lifetime.

Lupus A constellation of the southern sky, representing a wolf. Its brightest star is Alpha Lupi, of magnitude 2.3. The constellation contains a variety of double stars, notably Epsilon, Eta, Kappa, Mu, Xi, and Pi Lupi; NGC 5822 is a 6th-magnitude open cluster, and NGC 5986 a 7th-magnitude globular cluster.

Luyten, Willem Jacob Dutch-American astronomer, born in Java, in modern Indonesia. In the early 1920s he began a search for stars with large proper motion, in the hope of detecting nearby white dwarfs, only three of which were known in 1920. The search eventually yielded the proper motions of 120000 stars, and added considerably to the number of known white dwarfs, which by 1960 had risen to 400, most found by Luyten. In 1947 he discovered a star in Cetus with high proper motion; this became the first flare star to be identified when photographs taken for measuring its parallax showed it to brighten and fade suddenly.

Lydia family A small family of asteroids at a mean distance of about 2.75 AU from the Sun. The family is unusual in that its members are of varying composition. The largest member is the T-class (308) Polyxo, diameter 138 km. The family is named after (110) Lydia, an M-type asteroid, diameter 76 km, discovered in 1870 by the French astronomer Louis Alphonse Nicolas Borrelly.. Lydia's orbit has a semimajor axis of 2.732 AU, period 4.52 years, perihelion 2.52 AU, aphelion 2.95 AU, and inclination 6°.0.

Lyman limit The short-wavelength end of the hydrogen Lyman series, at 91.2 nm. It corresponds to the energy required for an electron in the hydrogen ground state to jump completely out of the atom, leaving the atom ionized.

Lyman series A sequence of absorption or emission lines in the ultraviolet part of the spectrum, due to hydrogen. They range from Lyman-a at 121.6 nm towards shorter wavelengths, the spacing between the lines diminishing as they converge on the Lyman limit at 91.2 nm. The Lyman series is caused by electron jumps between the ground state and higher levels of the hydrogen atom. The term is also used to describe certain lines in the spectrum of singly ionized helium. The He II Lyman lines have almost exactly one-quarter the wavelength of their hydrogen equivalents: for example, He II Lyman-a is at 30.4 nm, and the corresponding Lyman limit is at 22.7 nm. It is named after the American physicist Theodore Lyman.

Lyman-a forest A close series of narrow absorption lines seen at shorter wavelengths than the strong Lyman-a emission line in the spectra of quasars. They are believed to be caused by absorption in cool hydrogen clouds, perhaps in the haloes of galaxies, between us and the quasar. All these absorption lines are Lyman-a lines, but redshifted by different amounts depending on the distance of the absorbing cloud. Since the clouds are between us and the quasar, their redshifts are less than that of the Lyman-a emission from the quasar itself.

Lynx A constellation of the northern sky, representing a lynx. Its brightest star, Alpha Lyncis, is of magnitude 3.2. The other stars are faint, but include several interesting doubles and triples.

Lyot, Bernard Ferdinand French solar astronomer and scientific instrument designer. In the 1920s he designed and built a highly sensitive polariscope for studying the Moon and planets in polarized light. Among his findings were that Venus has a high-altitude haze layer, and that Mars has a thin atmosphere and dusty surface. In 1930 he invented the coronagraph for observing the Sun's corona, and various filters for solar observation. In 1935 he made the first cine films of the corona.

Lyra A prominent constellation of the northern sky, representing a lyre. Its brightest star is Vega. Beta Lyrae is an eclipsing binary, the prototype of the Beta Lyrae stars; it varies between magnitudes 3.3 and 4.3 in a period of 12.9 days. It is also a double star, with a fainter component of magnitude 7.2. Delta Lyrae is a wide double, magnitudes 4.2 and 5.6, while Epsilon Lyrae is a famous quadruple star, the Double Double. Zeta Lyrae is another double star, magnitudes 4.4 and 5.7. RR Lyrae is the prototype of the RR Lyrae stars, an important class of pulsating variables. M57 is a well-known planetary nebula, the Ring Nebula. The Lyrid meteors radiate from the constellation every April.

Lyrid meteors A meteor shower active from April 19 to 25, with maximum on April 21 from a radiant at RA 18h 08m, dec. +32°, southwest of Vega. In most years the shower is not particularly prolific, with a peak ZHR of 10-15. Short-lived outbursts were seen in 1922 and 1982. Lyrids are fast (geocentric velocity 49 km/s), with a fair proportion of bright events and meteors leaving persistent trains. The Lyrids are associated with Comet Thatcher

Lysithea A satellite of Jupiter, eleventh from the planet; also known as Jupiter X. Its distance is 11720000 km, close to that of Elara, and its orbital period is 259.2 days. Its diameter is 36 km. Lysithea was discovered in 1938 by the American astronomer Seth Barnes Nicholson.

M

M 3000 The optimum high frequency radio wave with a 3000 km range, which reflects only once from the ionosphere (single hop transmission).

M magnitude The magnitude of a star as measured through the infrared M filter of the Johnson photometry system. It has an effective wavelength of 4750 nm and a bandwidth of 460 nm. MPAT Abbr. for Multiple Mirror Telescope. mock Moon Popular name for a parselene. mock Sun Popular name for a parhelion.

M region The name originally given to areas on the Sun that were assumed to cause geomagnetic disturbances not associated with observed active regions; 'M' stands for magnetically effective. Data from the Skylab space station in the early 1970s revealed that M regions are actually coronal holes at low solar latitudes.

M star A star of spectral type M, with a very cool surface (temperature below 3900 K), appearing reddish in colour and emitting most of its radiation in the infrared. M-type dwarfs, popularly known as red dwarfs, lie at the bottom end of the main sequence. They have masses under 0.5 solar masses, and luminosities up to 0.08 times the Sun's, too faint for even the closest, Proxima Centauri and Barnard's Star, to be seen with the naked eye. They have potential lifetimes longer than the present age of the Universe. Many are flare stars. M-type giants, however, are of 1.2-1.3 solar masses and luminosities of 300 or more times the Sun's. M-type supergiants, such as Betelgeuse and Antares, are of 13-25 solar masses and luminosities 40 000-500 000 times the Sun's. The low temperatures of M-type supergiants are a consequence of their inflated physical size, as large as the orbit of Jupiter; they also tend to vary in size and brightness. M-type spectra are completely dominated by broad molecular absorption bands, part

M0 A representation of the mass of an object in terms of Solar mass. The average mass of the Sun is about $2x10^{33}$ grams. Astronomers often express units for other objects in terms of solar units, since it makes the resulting numbers smaller and easier to deal with.

M31 M31 (commonly known as Andromeda Galaxy) is the closest major galaxy to us. It is a spiral galaxy (like our galaxy, the Milky Way). It can just be seen with the naked eye.

MACHO Abbr. for massive compact halo object, a hypothetical object presumed to exist in the halo of galaxies, including our own. MACHOs are invisible but they make a significant contribution to the overall mass of a galaxy; they are therefore a form of dark matter. MACHOs could be low-luminosity stars, Jupiter-like planets, or black holes. Observed microlensing of stars in the Large Magellanic Cloud could be due to MACHOs around our Galaxy.

Mach's principle The proposal that the inertial mass of a particle is determined by the gravitational effect of all the other matter in the Universe. Hence the concept of mass would be meaningless in an empty universe. The principle is named after the Austrian physicist Ernst Mach.A. Einstein was influenced by Mach's principle, but his theory of general relativity failed to incorporate it. Many theorists have sought to remedy this failing, but with only limited success.

Macroscopic A physical entity or process of large scale, the scale of ordinary human experience. Specifically, any phenomena in which the individual molecules and atoms are neither measured, nor explicitly considered in the description of the phenomena. See: microscopic.

Macrospicules Large spicule-like features that radiate from the limb at the Sun's polar regions. Macrospicules extend to much greater heights than ordinary spicules (40 000 km instead of 9000 km). They are visible in spectroheliograms of the ionized helium ultraviolet line at 30.4 nm.

Macroturbuience The broadening of a star's spectral lines when viewed from a distance, or at low resolution, so that the star's individual surface features are no longer distinguishable. The broadening is caused by Doppler shifts from motions of different parts of the star's surface.

Macula A dark spot on a planetary surface; pl. maculae. The name is not a geological term, but is used in the nomenclature of individual features, for example Tyre Macula on Europa.

Maffei galaxies Two galaxies just beyond the Local Group, discovered by the Italian astronomer Paolo Maffei in 1968. Both lie in Cassiopeia, near the border with Perseus, and are heavily obscured by dust in the plane of our Galaxy. Maffei 1 is an elliptical galaxy about 10 million L.Y away, and Maffei 2 is a barred spiral about 15 million L.Y away. Both belong to a group of at least a dozen members which includes the spiral galaxy IC 342, Dwingeloo 1 and two probable companions to Maffei 1: MB 1, a late-type spiral, and MB 2, a dwarf irregular galaxy.

Magellan A NASA space probe to Venus, launched from the Space Shuttle Atlantis in 1989 May. It went into a highly elliptical near-polar orbit around Venus in 1990 August and mapped the entire planet by radar with an average resolution better than 300 m. In 1993 its orbit was deliberately lowered and circularized by atmospheric drag, a process termed aerobraking, the first time such a manoeuvre had been carried out at another planet. In its new orbit, Magellan completed a map of the gravity field of Venus. The probe burned up in the planet's atmosphere in 1994 October.

Magellan telescopes A pair of 6.5-m reflectors at the Las Campanas Observatory, Chile, jointly owned and operated by the Carnegie Institution of Washington, the Universities of Arizona, Harvard, Michigan, and the Massachusetts Institute of Technology. The two telescopes are due for completion in 1998 and 2002.

Magellanic clouds The Magellanic Clouds are irregular-shaped galaxies, congregations of millions of stars. The irregular shape may be the result of a disturbance, perhaps a collision of two galaxies. The Large Magellanic Cloud (LMC) is near the constellation Dorado, and is 163,000 light-years away. The Small Magellanic Cloud (SMC) is near the constellation Tucana, and is under 200,000 light-years away. These two clouds are satellites of our Milky Way Galaxy.

Magellanic stream A thin trail of neutral hydrogen which stretches in a great circle around the sky for at least 110° from the Magellanic Clouds, passing near the south galactic pole. It may be debris pulled out in a close encounter between the Magellanic Clouds and our Galaxy about 200 million years ago.

Magflare A geomagnetic and/or cosmic storm has been associated with this flare.

Magic value blanking The process of marking an invalid value by placing a special value (typically an IEEE NaN) into it. Unlike a *Mask,* magic value blanking is irreversible.

Magma Molten rock within a planetary body. When magma erupts onto the surface it separates into lava and gas.

Magma ocean Term used to describe the layer of magma, hundreds of kilometers thick; thought to have covered the Moon 4.5 billion years ago.

Magnesium An element with atomic number 12; symbol: Mg. Magnesium is contained in the minerals olivine and pyroxene, and is very abundant in the mantles of the inner planets.

Magnetar A magnetar is a highly magnetic star. Magnetar have magnetic fields of about 10^{15} Gauss, about a thousand trillion times stronger than the Earth's. The strong magnetic field puts the star's surface under enormous stress, perhaps causing "starquakes" and resulting high energy bursts of radiation. These short-lived neutron stars were theorized to exist in 1992 by Robert C. Duncan and Christopher Thompson. Their existence was confirmed in 1998 by Chryssa Kouveliotou et al.

Magnetic bay A relatively smooth excursion of the H (horizontal) component of the geomagnetic field away from and returning to quiet levels.

Magnetic field A description of the strength of the magnetic force exerted by an object. Bar magnets have "di-polar" fields, as the force is exerted from the two ends of the bar. In simple terms, the earth, the sun, stars, pulsars all have dipolar magnetic fields.

Magnetic inversion line The line separating areas of opposite magnetic polarity on the Sun. Magnetic inversion lines are often marked by filaments, either quiescent filaments in quiet-Sun regions or active-region filaments.

Magnetic monopole A hypothetical point-like defect in the fabric of spacetime, produced in some models of the early Universe. They are so named because they would behave like the isolated north or south pole of a magnet. A magnetic monopole is the one-dimensional analogue of the cosmic string, domain wall, and cosmic texture. Monopoles, as well as the other types of defect, are a consequence of some forms of grand unified theory.No monopoles have yet been detected.

Magnetic pole Either of two limited regions in a magnet at which the magnet's field is most intense. The two regions have opposing polarities, which we label "north" and "south", after the two poles on the Earth.

Magnetic star A star with a strong magnetic field, as demonstrated by the Zeeman splitting of lines in its spectrum. Prominent examples are the peculiar A stars in which the strength of the field is also variable. More recently, the term 'magnetic star' has been applied to the AM Herculis stars, a class of cataclysmic variables which contains white dwarfs with extremely powerful magnetic fields (about 100 tesla).

Magnetic storm A magnetic storm is a temporary perturbation (disruption) of the Earth's magnetic field, caused by solar flares, which eject plasma from the Sun's chromosphere. This solar plasma travels at speeds of roughly 600 to 1,200 miles per second (1,000 to 2,000 km per second), and after about 21 hours, hit the Earth's atmosphere. Since they disturb the (the upper atmosphere), magnetic storms can disrupt radio transmissions. Magnetic storms usually last for a few days. During a magnetic storm, auroras increase over polar regions.

Magnetic variable An alternative term for an Alpha2 Canum Venaticorum star or SX Arietis star, types of variable that exhibit low-amplitude optical variations and a variable magnetic field. The term is also sometimes incorrectly applied to any variable in which magnetic processes are important, such as an AM Herculis star an RS Canum Venaticorum star, and certain types of RR Lyrae star.

Magnetite A black, strongly magnetic iron-oxide mineral; Fe_3O_4.

Magnetobremsstrahlung A class of radiation mechanisms in which a charged particle follows a spiral path in a magnetic field and thereby radiates energy in the form of electromagnetic waves. The name means 'magnetic braking radiation'. If the velocity of the particle is much less than the speed of light, the emission is known as cyclotron radiation. If the particle's speed is an appreciable fraction of the speed of light the emission is known as gyrosynchrotron radiation, and for speeds very close to that of light it is termed synchrotron radiation. In some usages, magnetobremsstrahlung is regarded as synonymous with synchrotron radiation.

Magnetofossils Small grains of magnetite occurring in sedimentary rocks and thought to have been produced by bacteria. (Some bacteria, called magnetobacteria, make magnetite, presumably as an aid to determining directions.)

Magnetogram A map of the magnetic field across the Sun's photosphere, measured by the Zeeman splitting of Fraunhofer lines. Daily magnetograms are produced by a number of observatories. Those from the National Solar Observatory at Kitt Peak, Arizona, plot magnetic fields down to 5×10^{-5} tesla in areas as small as 3600 km square.

Magnetograph An instrument that maps the strength, distribution, and direction of magnetic fields on the sun's disk.

Magnetohydrodynamics The study of the interactions between a plasma and a magnetic field. Plasma, being ionized, is a good electrical conductor. An electric current is induced in a plasma when it tries to cross lines of magnetic force, and the plasma then tends to follow the magnetic field lines. The magnetic field is said to be 'frozen in' to the plasma, or vice versa. Depending on the relative energies in the plasma and the magnetic field, the plasma may be forced to move only along the magnetic field lines, as in a quiescent solar prominence, or the magnetic field lines may be dragged along by the motion of the plasma, as in the Earth's magnetosphere and magnetotail. Waves propagating in a plasma frozen into a magnetic field are called Alfvén waves.

Magnetometer Sensitive scientific instrument used to measure the intensity at various points of Earth's (or another planetary body's) magnetic field.

Magnetopause The boundary between a planet's magnetosphere and the external magnetic field of the solar wind. The upwind region of the magnetopause usually lies below a bow shock, whose distance from the planet depends on the solar wind's intensity and field direction. The Earth's magnetopause normally lies about 64000 km (10 Earth radii) upwind, but may be driven closer by disturbed solar wind conditions following flares or coronal mass ejections.

Magnetosheath A region of a planet's magnetosphere close to the boundary marked by the magnetopause in which significant plasma circulation occurs.

Magnetospher The magnetosphere is the Earth's magnetic environment. The Earth is a huge dipole (2-pole) magnet. The Earth's magnetic field is probably cause by its molten iron-nickel core. This magnetic field is aligned with the north and south poles, and has reversed many times during geologic history. Charged particles become trapped on the field lines of the magnetosphere. William Gilbert hypothesized that the Earth was a giant magnet in 1600. Thomas Gold proposed the name "magnetosphere" in 1959. The Earth's magnetosphere extends far into space and is influenced by the solar wind (ions and electrons emitted from the sun). It extends into space from 60 to 37,280 miles (100 to 60,000 km) towards the Sun, and over 186,500 miles (300,000 km) away from the Sun, forming the Earth's magnetotail.

Magnetotail The part of a planet's magnetosphere that is elongated in the direction of the solar wind (i.e. away from the Sun). The Earth's magnetotail extends beyond the Moon's orbit, and that of Jupiter extends beyond Saturn's orbit. Reconnection between field lines in the Earth's magnetotail under disturbed geomagnetic conditions results in the acceleration of electrons into the high atmosphere, producing active and sometimes extensive displays of the aurora.

Magnification Two kinds of magnification are useful to describe optical systems and they must not be confused, since they aren't synonymous. Any optical system which produces a real image from a real object is described by its *linear magnification*. Any system which one looks through to view a virtual image is described by its *angular magnification*. These have different definitions, and are based on fundamentally different concepts.

Linear Magnification is the ratio of the size of the object to the size of the image.

Angular Magnification is the ratio of the angular size of the object as seen through the instrument to the angular size of the object as seen *with the 'naked eye'*. The 'naked eye' view is *without* use of the optical instrument, but under optimal viewing conditions.

Certain 'gotchas' lurk here. What are 'optimal' conditions? Usually this means the conditions in which the object's details can be seen most clearly.

Magnitude A measure of the brightness of a star. Ancient Greek astronomers defined the brightest stars as being of the first magnitude because they were the first to appear after sunset. The magnitude scale continued in steps of decreasing brightness down to sixth magnitude, for those stars which were visible only in total darkness. From its crude beginnings, the magnitude scale has been extended and is now on a strictly defined footing so that a difference of one magnitude corresponds to a difference in brightness of a factor of 2.512, and 5 magnitudes equals a brightness difference of exactly a hundredfold. Ancient magnitude estimates depended solely on the human eye, corresponding roughly to the modern V magnitude. The apparent magnitude of a star is its brightness as seen from Earth, whereas the absolute magnitude is a measure of its actual (i.e. intrinsic) brightness; the two differ because the intensity of light falls off with distance, and because of interstellar absorption.

Magnitude of an eclipse A measure of the extent of an eclipse. In a solar eclipse, the magnitude is the fraction of the Sun's diameter that is covered by the Moon. If the eclipse is total, magnitude is replaced by the ratio of the apparent diameters of the Moon and Sun, which is always 1.00 or more during totality. The magnitude of a lunar eclipse is the fraction of the Moon's diameter covered by the Earth's umbra; at a total eclipse it can be much larger than 1.00 (or 100%) because the Earth's shadow is much larger than the Moon. Eclipse magnitudes are expressed either as a decimal or as a percentage; the magnitude of a partial eclipse may be denoted as, for example, 0.59 or 59%.

Maiasaura The first dinosaur in space was *Maiasaura peeblesorum* (a duck-billed dinosaur). A piece of bone from a baby Maiasaura and a Maiasaura eggshell were taken into space by astronaut Loren Acton on an 8-day NASA mission (Spacelab 2) in 1985. The historic Maiasaura fossils now reside at the Museum of the Rockies in Bozeman, Montana, USA.

Main beam The direction in which a radio telescope is most sensitive; also called main lobe. Technically, it is the central lobe of the telescope's antenna pattern.

Main sequence A diagonal strip on the Hertzsprung-Russell diagram, from top left to lower right. A star is said to be on the main sequence at the stage of its life when it shines by converting hydrogen to helium at its centre. A star's position on the main sequence depends on its mass, with the most massive stars at the upper left and the least massive at the lower right. The Sun, an average star, lies about half-way along the main sequence. Whatever its mass, a star on the main sequence is termed a dwarf. Stars spend most of their lives on the main sequence, remaining at roughly constant temperature and luminosity, but the time they spend there depends on their mass. For very massive stars, the main-sequence stage lasts only about a million years, but for the least massive stars it is potentially longer than the age of the Universe.

Main sequence stars Main sequence stars are the central band of stars on the Hertzsprung-Russell Diagram. These stars' energy comes from nuclear fusion, as they convert Hydrogen to Helium. Most stars are Main Sequence Stars. For these stars, the hotter they are the brighter. The sun is a typical Main Sequence star.

Main sequence turnoff The main sequence turnoff is the place on the Hertzsprung-Russell Diagram where an aging (evolving) star begins to run out of hydrogen in the core and expand, thus moving off the main sequence.

Maksutov-Cassegrain telescope A Maksutov-Cassegrain telescope (MCT) is a wide-angle reflecting telescope with a curved correcting lens (called a Meniscus Corrector Shell.) that minimizes spherical aberration and a concave mirror that receives light and focuses an image. A second mirror reflects the light through a gap in the primary mirror, allowing the eyepiece or camera to be mounted at the back end of the tube. The Cassegrain telescope (named for the French sculptor Sieur Guillaume Cassegrain) was developed in 1672; the Meniscus Corrector Shell (a curved lens that corrected spherical aberration and was easier to produce than the Schmidt's correction lens) was added in 1941 by Dmitry Maksutov from Russia and A. Bouwers of Holland.

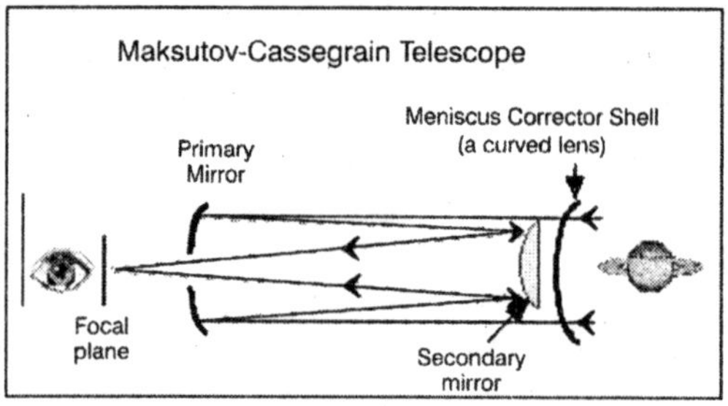

Fig. Maksutov-Cassegrain telescope

Main-belt asteroid An asteroid that orbits within the main asteroid belt. main lobe Another name for main beam.

Major axis The longest diameter of an ellipse, passing through the two foci also known as the line of apsides.

Major cycle In the Clark CLEAN algorithm for deconvolution, a number of minor cycles, or inner iterations, followed by the computation (by the FFT algorithm) of the full residual image, comprise a major cycle.

Majorflare This flare is the basis for the forecast of geomagstorm, cosmic storm and/or protons in the earth's vicinity.

Maksutov telescope A type of telescope that uses mirrors and lenses. Maksutov's are in fact Cassegrain telescopes: In the case of a Maksutov the light enters the front of the telescope via a curved corrector plate or lens (curved towards the interior of the telescope) which directs the light onto the concave primary mirror (which has a hole at its centre) which lies at the rear of the telescope. The primary mirror then reflects the light back toward the corrector plate which has a small convex secondary mirror coated onto the centre of the corrector plate. This secondary mirror then directs the light through the hole in the centre of the primary mirror and on into the eyepiece which is attached to the rear of the telescope. Like all Cassegrain telescopes the Maksutov has the advantage to being able to have a longer focal ratio than a Newtonian reflector of equal optical size but at the expence of a more complicated optical train.

Malin-1 A huge, low surface brightness galaxy, about 1 billion L.Y away in the constellation Coma Berenices. It has a diameter of nearly 600 000 L.Y and a total mass of over 2×10^{12} solar masses, but the gas-rich disk is so faint that it was discovered only in 1986 on specially processed photographic plates by the British astronomer David Frederick Malin

Malmquist bias A statistical selection effect that arises in surveys of galaxies or other objects to which an apparent magnitude limit is applied. At large distances from the observer, only galaxies that are intrinsically extremely bright can be seen. Nearer the observer, galaxies with average or below-average luminosity can also be seen. The statistical properties of the sample will therefore depend on the distance from the observer in a complicated way. It is named after the Swedish astronomer Karl Gunnar Malmquist.

Manganese star A chemically peculiar star with an unusually high ratio of manganese to iron, and a temperature corresponding to a late B spectral type. They are main-sequence stars, similar to Ap stars, but without evidence for a strong magnetic field.

Mantle A mostly solid layer of Earth lying beneath the crust and above the core; consisting mostly of iron, magnesium, silicon, and oxygen. On other planets and moons, the mantle is the layer between the crust and core. Mantles contain information about a planet's total composition, a key parameter in understanding planet formation and how the planets vary in composition with distance from the Sun. A second definition of mantle is a smooth blanket of surface material (generally fine-grained, sometimes layered) that was deposited by some process such as wind or water and that now obscures the underlying land.

Map MAP (Microwave Anisotropy Probe) is a NASA satellite that will be launched in April, 2001, and will orbit for 27 months. MAP was designed to help answer the cosmological questions about the early universe: "What are the values of the cosmological parameters of the Big Bang theory?", "When did the first structures of galaxies form?", and "How did structures of galaxies form in the universe?". MAP will measure minute temperature fluctuations (anisotropy) in the cosmic microwave background radiation (the radiant heat left over from the Big Bang) over the entire sky (these were originally detected by the COBE satellite). MAP will orbit at the unstable Sun-Earth Lagrange point L2; the Sun, Earth, and MAP will always be in a straight line.

Mare A large dcark or low area on a planetary surface; pl. maria. The name, which means 'sea', is not a geological term; it was originally used in the 17th century, when the dark plains of the Moon were thought to be water, and was subsequently applied to the dark patches on Mars in the 19th century. The lunar maria are dark, smooth lowland plains of lava that erupted between the end of the late heavy bombardment 3.9 billion years ago and about 2 billion years ago. Samples of the maria have been obtained by the Apollo missions and unmanned Luna landers. The maria consist of basaltic lavas, chemically and mineralogically different from the highlands, being enriched in iron and titanium, and with abundant pyroxene. The lunar mare lavas differ from those on Earth in being very fluid, with a viscosity similar to that of motor oil, so that they flow for great distances.

Mare basalt Rocks making up the dark, smooth, mare areas of the Moon.

Maria family A Hirayama family of asteroids at a mean distance of 2.55 AU from the Sun, differing from other main-belt asteroids in their high orbital inclination, about 15°. Its members are mainly S class. The largest family members are (170) Maria, (472) Roma, (660) Crescentia, (695) Bella, and (714) Ulula, all with diameters of 40-55 km. Maria itself, an S-class asteroid of diameter 40 km, was discovered in 1877 by the Frenchman (Henri) Joseph (Anastase) Perrotin.Its orbit has a semimajor axis of 2.552 AU, period 4.08 years, perihelion 2.39 AU, aphelion 2.72 AU, and inclination 14°.4.

Mariner A series of US planetary probes. Mariner 2, which flew past Venus, was the first space probe to reach another planet. Mariner 4 provided our first close-up views of Mars, and Mariner 9 was the first probe to go into orbit around another planet. Mariner 10, the last of the series, was the first probe to visit two planets, Venus and Mercury.

Marius, Simon German astronomer who gave Jupiter's "Galilean" moons their names. He and Galileo both claimed to have discovered them in 1610 and likely did so independently. Marius was also the first to observe the Andromeda Nebula with a telescope and one of the first to observe sunspots.

Markab The star Alpha Pegasi, magnitude 2.5. It is a giant or subgiant of type A0, lying 74 L.Y away.

Markarian galaxy A type of galaxy with unusually strong emission at nearultraviolet wavelengths; most also show strong emission lines in the optical region. Galaxies in which the source of the ultraviolet radiation lies in the nucleus, which is often the case, can be subdivided into Seyfert galaxies, N galaxies, and quasars. Alternatively, the ultraviolet radiation may originate from a more extended region, as in dwarf galaxies undergoing intense bursts of star formation. The galaxies are named after the Armenian astronomer Benjamin Eghishe Markarian who first detected them in the 1960s.

Mars The fourth planet from the Sun. It appears distinctly reddish to the naked eye. Its mean opposition magnitude is -2.0, although at perihelic oppositions it can reach -2.8, brighter than all planets except Venus and Jupiter. It is slightly ellipsoidal in shape (equatorial diameter 6794 km, polar diameter 6750 km). Mars has a thin atmosphere composed of about 95% carbon dioxide, 2.7% nitrogen, 1.6% argon, 0.1% oxygen, 0.1% carbon monoxide, and small variable traces of water vapour. The atmospheric pressure at the surface is about 6 mbar. Surface temperatures vary between about 0 and -125 °C, averaging about -50 °C. White clouds of condensed water vapour or carbon dioxide are relatively common, particularly near the terminator and in polar latitudes. There are two permanent water-ice caps at the poles which never melt. In winter these are augmented by caps of thin frozen carbon dioxide up to 60 across.

Mars exploration rovers U. S. missions to Mars with twin rovers equipped with tools to study a diverse collection of rocks and soils that may hold clues to past water activity on Mars. MER-A, named "Spirit," launched on June 10, 2003 for a seven-month journey to a landing site in Gusev Crater and MER-B, named "Opportunity," launched on July 7, 2003 toward a landing site in Terra Meridiani.

Mars face This photograph of the Cydonia Mense region of Mars was taken by NASA's Mars Global Surveyor in 1998. It is a coincidental alignment of rocks and other geologic formations that happens to look like a human face from this angle.

Mars global surveyor U. S. mission launched in 1996 to orbit Mars. Main instruments include cameras (MOC), a laser altimeter (MOLA), thermal emission spectrometer TES, and magnetometer. During the primary mission (through January 2001), the spacecraft circled the planet once every 118 minutes at an average altitude of 378 kilometers (235 miles). The extended mission phase began February 1, 2001 and is expected to continue to at least September 2004. After its science mission, the spacecraft will function as a communications satellite to relay data back to Earth from surface landers launched as part of future Mars missions.

Mars orbiter camera (MOC) Camera system onboard the Mars Global Surveyor spacecraft. MOC consists of three cameras: A narrow angle system that provides grayscale high resolution views of the planet's surface (typically, 1.5 to 12 meters/ pixel), and red and blue wide angle cameras.

Mars orbiter laser altimeter (MOLA) Instrument onboard the Mars Global Surveyor spacecraft that has been obtaining topographic data since March 1999. MOLA transmits focused infrared laser pulses toward Mars at a rate of ten times per second and measures the round-trip reflection time to determine the range of the MGS spacecraft to the Martian surface. Scientists use the range measurements to construct precise topographic profiles and maps of Mars in unsurpassed detail.

Mars pathfinder The second in NASA's Discovery program of low-cost spacecraft with highly focused science goals, Mars Pathfinder landed successfully on Mars on July 4, 1997. Images, science results, and engineering data were released on the World Wide Web throughout, and beyond, the primary mission dates (July 4 - August 3, 1997.) The Jet Propulsion Laboratory, Pasadena, CA, developed and managed the Mars Pathfinder mission for NASA's Office of Space Science, Washington, D.C. JPL is a division of the California Institute of Technology (Caltech).

Mars physical data diameter (equatorial) Oblateness Inclination of equator to orbit Axial rotation period (sidereal) 6794 km 0.0065 25°0.19 24.623 hours Mean density Mass Volume Mean albedo Escape (Earth = 1) (Earth = 1) (geometric) velocity 3.94 g/cm^3 0.11 0.15 0.15 5.02 km/s

Orbital data Mean distance from Sun Eccentricity Inclination Orbital 106 km AU of orbit of orbit to period ecliptic (sidereal)

227.941 1.524 0.093.1U+00B0.9 686.980 days

Mars's surface is a volcanic basalt with a high iron content. Oxidation of this iron gives Mars its distinctive rust-red colour. Dark and bright markings can be seen through telescopes, but these do not always correspond to topographic features or terrain types; dark patches appear to be due to areas of dark surface dust. These may slowly change over the years as the dust is transported by winds. The most prominent dark marking, Syrtis Major, is an unremarkable east-facing slope with a gradient of less than 1°. There are many areas of sand dunes; the largest surround the polar caps, and constitute the largest dune field of the Solar System.

Mars probes A series of probes to the planet Mars, launched by the former Soviet Union. Technical problems meant that most of them were complete or partial failures, although Mars 5 did send some photographs from orbit around the planet.

Mars surveyor A series of NASA space probes to Mars to be launched every 26 months, for successive oppositions. The first were Mars Global Surveyor and Mars Pathfinder in 1996. Further spacecraft will be launched over the following decade. The Mars Surveyor 98 Orbiter will be sent in 1998 December to monitor weather changes on Mars, while the Mars Surveyor 98 Lander, planned for launch in 1999 January, will touch down near the planet's south polar cap.

Marshall space flight center A NASA establishment at Huntsville, Alabama, founded in 1960 by the transfer to NASA of part of the missile work of the US Army's Redstone Arsenal. MSFC developed the Saturn series of launch rockets and the Skylab space station under the leadership of Wernher Magnus Maximilian von Braun and subsequently played a leading role in the development of the Space Shuttle and its rocket engines. MSFC houses the operations control centre for flights of the Shuttle-borne Spacelab and is involved in developing the International Space Station.

Mascon An area on the Moon either composed of, or underlain by, denser material, as evidenced by an increased gravitational pull; the word is a contraction of mass concentration. Mascons lie beneath many of the major mare basins.

Maser "Microwave-amplified stimulated emission of radiation". An amplifier of radio waves (similar to a laser, which amplifies visible light). This may be a natural feature, such as water molecules in space, or can be created by using special properties of certain crystals, such as ruby, at temperatures near absolute zero and in strong magnetic fields. Water molecules in space can form masers that help astronomers study radio emission from objects that would normally be too faint to detect.

Maser amplifier An amplifier used in radio astronomy which employs the principle of maser amplification. Maser amplifiers have extremely low noise and high sensitivity.

Maser source A radio source in which the spectral lines of an atom, ion, or molecule are greatly amplified by maser action to produce an intense source of radio emission. The first maser to be identified in, space was the OH maser at 1665 MHz, in 1965. Maser emission is now known from many other molecules, including water (H_2O), silicon monoxide (SiO), methanol (CH_3OH), ammonia (NH_3), formaldehyde (H_2CO), hydrogen cyanide (HCN), and silicon sulphide (SiS), and from hydrogen ions. Sometimes many spectral lines from the same molecule can be observed as masers. For example, over 30 different maser lines of SiO have been observed in the circumstellar envelope of the star VY Canis Majoris. Masers occur in star-forming regions (interstellar masers), in the circumstellar envelopes of red giant stars (circumsterar masers), in comets, in some planetary atmospheres, and in some active galactic nuclei, where they may be so bright that they are classed as megamasers.

Mask A collection of Boolean values which are used to mark whether a value is valid or invalid. Can be used to implement blanking.

Maskelyne, Nevil English astronomer. On a voyage to St Helena in 1761 he successfully tested tables of the Moon's position compiled by the German astronomer Johan Tobias Mayer for finding longitude at sea. When in 1765 he became the fifth Astronomer Royal, Maskelyne included tables of lunar positions for navigation in The Nautical Almanac, which he founded in 1766. He observed the 1769 transit of Venus and from it calculated the Sun's distance to an accuracy of 1%. In 1774 he measured the deflection of a plumbline near the Scottish mountain Schiehallion, from which he calculated the Earth's average density to be 4.7 g/cm^3, the first good approximation to the true value (5.52 g/cm^3).

Mass A measure of the amount of matter in a body. Mass gives rise to the inertia of an object, i.e. its resistance to change in its motion or state of rest; this is known as the inertial mass. It also produces a gravitational force. Although mass is formally defined in terms of its inertia, it is usually measured by the effects of its gravitation. The weight of a body is the force by which it is gravitationally attracted to the Earth. Although the terms 'mass' and 'weight' are often used interchangeably on Earth, in space a body may be weightless but still retain its inertia, i.e. it still requires a force to change its motion. According to the special theory of relativity, the mass of a body increases as its velocity approaches that of light. At the speed of light itself the mass would be infinite.

Mass defect The difference between the mass of an atomic nucleus and the masses of its individual protons and neutrons when in an unbound state. The mass defect arises because energy is given up when protons and neutrons bind together to form a nucleus. The energy equivalent of the mass defect is known as the binding energy.

Mass discrepancy The difference between the amount of directly observed luminous matter in an object and the mass inferred from the object's speed of rotation or its motion. Many objects, from galaxies up to superclusters, have a significant mass discrepancy, and this is usually taken to be evidence for dark matter.

Mass extinction Mass extinction is the process in which huge numbers of species die out suddenly. The dinosaurs (and many other species) went extinct during the K-T extinction, which was probably caused by an asteroid colliding with the Earth.

Mass function A mathematical relationship between the masses of the components in a spectroscopic binary and the orbital inclination. It is the only information about the masses that can be derived when the spectrum of just one star is visible. If that star is of mass m 1 (in solar units), then the mass function is given by (m 2 sini)3/(m 1 + m 2)3, where m 2 is the mass of the unseen star and i is the inclination of the orbit.

Mass loss The loss of material from a star, which may occur at different evolutionary stages and by various processes. A protostar may be the source of a bipolar flow. A T Tauri, giant, or supergiant star may lose mass through a vigorous stellar wind. Ejected material creates a disk around a Be star, or a shell around a Gamma Cassiopeiae star (a shell star); at a late stage of stellar evolution it may give rise to a planetary nebula. Violent ejection occurs in a nova or supernova explosion. Mass loss may also occur through the outer Lagrangian point in a contact binary or overcontact binary system.

Mass-luminosity relation The relationship between the luminosity of a star on the main sequence and its mass. Factors involved include whether radiation or convection is the dominant energy transport process, and, if the former, the type of opacity involved. For stars of less than 0.4 solar masses, which are fully convective, the luminosity varies as the square of the mass. For stars of mass similar to or slightly less than the Sun, the luminosity is approximately proportional to the mass to the fifth power (e.g. a star with twice the mass will have 32 times the luminosity). For more massive stars, the luminosity varies approximately as the mass cubed (e.g. a star with twice the mass will have about 8 times the luminosity).

Mass ratio An expression of the relative masses of the components of a binary star when both spectra are visible. It is derived from the radial velocity curves for both components and may be written as M 2 / M 1 = a 1 sini/a 2 sin i, where M 1 and M 2 are the masses of the components, and a 1 and a 2 are the amplitudes of the velocity curves. If the orbital inclination, i, can be determined the actual masses of the components, and the dimensions of the stars and their orbits, can be obtained.

Mass spectrometer Instrument that produces and measures, usually by electrical means, a mass spectrum. It separates ions according to the ratio of their mass to charge, allowing scientists to determine the abundances of each isotope.

Mass spectrum The pattern of the relative abundances of ions of different atomic or molecular mass (mass-to-charge ratio) within a sample. It frequently refers to the measured relative abundances of isotopes of a given element.

Mass transfer The transfer of gas between stars in a binary system. The component that is losing material may have evolved to fill its Roche lobe it may be the source of a strong stellar wind. The component accreting material does so either directly from the stellar wind or a stream of gas, or indirectly through an accretion disk. The transferred gas is generally hydrogen-rich because it originates in the outer layers of the donor star.

Mass-to-light ratio The ratio of the mass of an object to its total luminosity. The mass and luminosity are usually measured in terms of solar mass and solar luminosity; hence the mass-to-light ratio of the Sun is 1. Most extragalactic objects have mass-to-light ratios greater than 1, indicating that not all their mass is in the form of visible stars. The high values inferred for galaxies (up to 30) and clusters of galaxies (up to 300) indicates the existence of considerable amounts of dark matter.

Matter Physical objects are made of matter. Matter can exist in four phases (solid, liquid, gas, and plasma) and a few other extreme phases, like critical fluids and degenerate gases.

Matter era In the Big Bang theory, the era that started when the gravitational effect of matter began to dominate the effect of radiation pressure. Although radiation is massless, it has a gravitational effect which increases with the intensity of the radiation. Moreover, at high energies, matter itself behaves like electromagnetic radiation because it is moving at a speed close to that of light. In the very early Universe, the expansion rate was dominated by the gravitational effect of radiation pressure but, as the Universe cooled, this effect became less important than the gravitational effect of matter. Matter is thought to have become predominant at a temperature of around 10^4 K, roughly 30 000 years after the Big Bang. This marked the start of the matter era.

Mauna kea observatories A complex of observatories and telescopes at an altitude of approximately 4200 m on the summit of Mauna Kea island, Hawaii. The site, founded in 1964, is owned and managed by the University of Hawaii's Institute for Astronomy. It is the world's highest observatory site and is generally regarded by astronomers as the best location for ground-based optical, infrared, and millimetre/submillimetre observations. More major telescopes are located on Mauna Kea than at any other site. The University of Hawaii has its own 0.6-m and 2.2-m reflectors on Mauna Kea, opened in 1968 and 1970 respectively. Optical and infrared telescopes from other organizations are the NASA Infrared Telescope Facility; the Canada-France-Hawaii Telescope; the United Kingdom Infrared Telescope; the Gemini North telescope the Japanese Subaru Telescope; and the W. M. Keck Observatory. Radio and millimetre-wave instruments on Mauna Kea include the Caltech Submillimeter Observatory; the James Clerk Maxwell Telescope.

Maunder minimum The period from 1645 to 1715 when scarcely any sunspots or aurorae were seen. E. W. Maunder concluded that there had been a real decline in solar activity then. Further evidence for the Maunder minimum is provided by an increased carbon-14 content in tree rings during that period, since the cosmic rays that produce carbon-14 reach the Earth in greater numbers when the Sun's activity is low. There was also a lengthy cold spell on the Earth from 1550 to 1700, known as the Little Ice Age, which roughly corresponds to a period including the Maunder minimum and the earlier Spörer minimum. This cold period could be explained by a decrease in solar output of about 1%.

Maunder, Walter English solar astronomer. From the Royal Observatory at Greenwich, he observed the Sun and recorded sunspots on every possible day for forty years. He was assisted by his second wife, Annie Scott Dill Maunder, née Russell.He established the Sun's differential rotation from the motions of sunspots at different latitudes, and identified the relationship between solar activity and disturbances in the Earth's magnetic field. The Maunders also studied the history of astronomy. Walter's research revealed what is now called the Maunder minimum, in papers published in 1894 and 1922.

Maxwell gap A division in Saturn's rings, in the outer part of the C Ring (the crêpe ring). It is about 87 500 km from Saturn's centre, and is 270 km wide. The gap was discovered in 1980 by Voyager 1 and was named after J. C. Maxwell.

Maxwell montes A range of mountains on Venus at around +65° lat., 3° long. Maxwell Montes are nearly 800 km wide and lie near the centre of Ishtar Terra. They contain the highest peak on Venus, 11 km above the mean surface level. On the eastern slopes of Maxwell Montes lies the volcanic caldera Cleopatra, 100 km in diameter. Maxwell Montes showed up on the first. Earth-based radar maps of Venus, and were named after J. C. Maxwell.

Maxwell, James clerk Scottish physicist. In the late 1850s he explained that Saturn's rings must consist of many small objects following independent concentric orbits, and that a solid or fluid ring would be unstable. Of his many contributions to theoretical physics, the most important is perhaps the concept of electromagnetic radiation. In 1865 he published what are now called Maxwell's equations, which unite the phenomena of electricity and magnetism into a single theory.

Mayall telescope The largest telescope at Kitt Peak National Observatory, Arizona, aperture 4 m, opened in 1973. It is named after a former director of Kitt Peak, Nicholas Ulrich Mayall.

Mcauliffe, Christa Sharon Christa Corrigan McAuliffe (1948-1986) was an American schoolteacher who was chosen to be first teacher in space. She was killed, along with her six fellow astronauts (Francis R. Scobee, Michael J. Smith, Judith A. Resnik, Ellison S. Onizuka, Ronald E. McNair, and Gregory B. Jarvis), when the NASA's Space Shuttle Challenger Mission 51-L exploded only 73 seconds after its launch on the morning of January 28, 1986. McAuliffe was born on September 2, 1948, in Boston, Massachusetts. She taught at Concord High School in New Hampshire before being chosen for the Space Shuttle mission (she was chosen from over 11,000 applicants). McAuliffe was married and had two children.

McDonald observatory The observatory of the University of Texas, on the adjacent peaks of Mount Locke (altitude 2070 m) and Mount Fowlkes near Fort Davis in western Texas. It was founded in 1932 and is named after William Johnson McDonald a banker who gave money for its construction. Its largest instruments are the 11-m Hobby-Eberly Telescope on Mount Fowlkes; a 2.7-m reflector opened in 1969 on Mount Locke; and the 2.1-m Otto Struve reflector, opened in 1939 on Mount Locke.

M-class asteroid A class of asteroid, fairly common in the main asteroid belt, having featureless reflectance spectra, flat to slightly reddish over the wavelength range 0.3-1.1 ¼m. M-class asteroids are distinguished from the spectrally identical classes E and P by their moderate albedos (0.10-0.18). The 'M' is for metallic, since they are believed to have metallic compositions. Members of this class include (16) Psyche, diameter 248 km, and (21) Lutetia, diameter 108 km.

Mcmath-Pierce solar telescope The world's largest solar telescope, part of the National Solar Observatory, opened in 1962 on Kitt Peak, Arizona. The main telescope is housed in a diagonal shaft, 152 m long, aligned parallel with the Earth's axis. A heliostat on a tower 33.5 m high reflects sunlight down this shaft to a 1.5-m mirror, some 50 m below ground level. Rays are bounced from this mirror part-way up the shaft again to a flat mirror, which diverts the light vertically. downwards to form a solar image 85 cm wide in an observation room below ground. It is named after the American solar physicists Robert Raynolds McMath and A(ustin) Keith Pierce.

Me star A star of spectral type M which shows emission lines of hydrogenMe stars are often Mira stars (giants), but such emission lines are frequently found in M dwarfs.

Mean anomaly (symbol M) The angle between the periapsis of an orbit and the position of an imaginary body that orbits in the same period as the real one but at a constant angular speed. The angular speed assigned to the imaginary body is the average angular velocity (the mean motion) of the real orbiting body. The angle of mean anomaly is measured from the body being orbited, in the direction of orbital motion.

Mean daily motion (symbol n) The average movement of a body along its orbit in one day, usually expressed in degrees. The mean daily motion of a planet about the Sun can be accurately calculated from theory, but it can also be found by direct observation. For the Moon, ancient eclipse records extending back 2000 years or more have given us a very accurate value of its mean daily motion.

Mean density of matter The density of material that would be obtained if all the matter contained in galaxies were smoothed out across the Universe. Although stars and planets have densities greater than the density of water (about 1 g/cm^3), the cosmological mean density is extremely low (less than 10^{-29} g/cm^3, or 10^{-5} $atoms/cm^3$) because the Universe consists mostly of virtually empty space between galaxies. The mean density of matter determines whether the Universe will continue to expand..

Mean equator The great circle perpendicular to the direction of the mean pole.

Mean equinox The direction to the equinox at a particular epoch, with the effect of nutation subtracted. The mean equinox therefore moves smoothly across the sky due to precession alone, without short-term oscillations due to nutation.

Mean motion (symbol n) The average angular velocity of a body in an elliptical orbit, often measured in degrees per day. If the orbit is perturbed, allowance must be made for changes in the period of revolution. For an unperturbed orbit, the mean motion is the same as the angular velocity that the body's radius vector would have if the orbit were circular and of radius equal to the semimajor axis of the ellipse.

Mean place The coordinates of a celestial object referred to the mean equator and equinox at some standard epoch, such as the beginning of a year. In calculating a star's mean place, the effects of atmospheric refraction, parallax, aberration, and the gravitational deflection of light must be removed.

Mean pole The direction on the sky towards which the Earth's axis points at a particular epoch, with the oscillations due to nutation removed. It moves smoothly across the sky, affected only by lunisolar precession.

Mean sidereal time The time since the mean equinox crossed the meridian, i.e. the hour angle of the mean equinox. The equinox is not fixed with respect to the stars, but changes its position due to precession and nutation. It is convenient to define a mean equinox with the effect of nutation removed. The mean sidereal time defined by this equinox will, unlike apparent sidereal time, proceed uniformly.

Mean solar day The average length of the apparent solar day; also, the interval between successive transits of the mean sun. The apparent solar day is not constant in length due to the changing value of the equation of time. The number of mean solar days in the year is exactly one fewer than the number of sidereal days.

Mean solar time The time shown on a clock; technically, it is the time since the mean sun crossed the meridian with 12 hours added to make the day begin at midnight rather than noon. As the variations of apparent solar time due to the equation of time have been averaged out, mean solar time proceeds uniformly except for small irregularities in the gradual slowing of the Earth's rotation. Mean solar time is closely related to mean sidereal time, and one can be calculated if the other is known. The mean solar time on the Greenwich meridian is Universal Time.

Mean sun A fictitious body whose position on the celestial sphere defines the mean solar time. It is defined as moving around the equator in one year at a regular rate. The hour angle of the mean sun plus 12 hours gives mean solar time. The mean sun is necessary for uniform timekeeping since the motion of the true Sun along the ecliptic varies due to the eccentricity of the Earth's orbit around the Sun. Moreover, the ecliptic is inclined to the celestial equator, so that even a uniform motion on the ecliptic would not produce a uniform time-scale. For these two reasons the true Sun is not a suitable reference body for timekeeping.

Mean time A smoothly progressing form of time, with the short-term irregularities of apparent time averaged out. In mean sidereal time the effect of nutation is removed; in mean solar time a correction is made for the irregular movement of the Sun due to the Earth's elliptical orbit and the inclination of the ecliptic to the celestial equator. Greenwich mean time, which is the mean solar time at Greenwich, is the same thing as Universal Time.

Medium frequency (MF) That portion of the radio frequency spectrum from 0.3 to 3 MHz.

Medium-band photometry A little-used synonym for intermediate-band photometry.

Megahertz (MHz) A frequency of one million cycles per second.

Megamaser An extremely powerful maser source in the nucleus of an active galaxy. Megamasers can be more than a million times as powerful as maser sources within our own Galaxy. The most powerful are the OH and H_2O megamasers, which can generate more than 10^{30} watts of radio power (enough to power 10 000 Suns).

Megaparsec A megaparsec (Mpc) is a unit of distance that is equal to one million parsecs, 3.26×10^6 light-years or 3.085678×10^{19} kilometers. The Local Group of galaxies is roughly a megaparsec in diameter.

Megaregolith A layer of fractured or fragmented material that underlies the surface regolith of a rocky body, such as a planetary satellite or asteroid. A megaregolith may be formed by a major collision that causes extensive fracturing. On the Moon, the megaregolith is the thicker layer of ejecta from basin-forming impacts that underlies the surface regolith. It varies in thickness with distance from the lunar basins from a few hundred metres to a kilometre or more.

Mega-ton A unit of energy used to describe nuclear warheads. The same amount energy as 1 million tons of TNT.

1 mega-ton = 4×10^{22} ergs = 4×10^{15} joules.

Megrez The star Delta Ursae Majoris, magnitude 3.3. It is an AZ dwarf, 53 L.Y away.

Member function In object-oriented programming, a function that belongs to a class and is able to manipulate class data. Only member functions are allowed to directly manipulate class data. This is the mechanism by which encapsulation is enforced. The set of public member functions of a class make up its interface.

Meniscus lens A lens with one concave and one convex surface, so that the glass forms part of a shell. A positive meniscus lens is thicker at the centre than the edge and is a converging lens. A negative meniscus lens is thinner in the centre than at the edge and is a diverging lens. A meniscus mirror is used in lightweight telescopes, and is usually supported only by its centre as it has a comparatively thin edge.

Meniscus Schmidt telescope A reflecting telescope with a spheroidal primary mirror, a meniscus lens whose surfaces are concentric with the primary mirror, and an additional corrector plate in front of the meniscus lens. This provides a wide field of view and small focal ratio.

Menkalinan The star Beta Aurigae. It is an eclipsing binary co. nsisting of a pair of A2 subgiants that varies between magnitudes 1.9 and 2.0 with a period of 3.96 days. It lies 55 L.Y away. Menkar The star Alpha Ceti, magnitude 2.5. It is an Ml.5 giant, 200. L.Y away.

Menkent The star Theta Centauri, magnitude 2.1. It is a K0 giant, 50 L.Y away.

Mensa A small, faint constellation near the south celestial pole, representing Table Mountain at the Cape of Good Hope, South Africa. Its brightest star, Alpha Mensae, is of magnitude 5.1. It contains part of the Large Magellanic Cloud.

Menzel, Donald Howard American physicist and astronomer. In the 1920s he developed theoretical astrophysics with H. N. Russell. In 1927 he obtained accurate measurements of the surface temperatures of Mercury and Mars. He studied the composition of the Sun's atmosphere, establishing the preponderance of hydrogen from a study of the flash spectrum in 1929. With F. L. Whipple he proposed in 1954 that the surface of Venus is covered with water.

Merak The star Beta Ursae Majoris, magnitude 2.4. It is an A0 subgiant or dwarf and lies 62 L.Y away. With Dubhe, it forms the Pointers that indicate the direction of the Pole Star, Polaris.

Mercury has no permanent atmosphere Although some hydrogen and helium from the solar wind is temporarily captured. The surface temperatures average about 170 C, but Mercury has the most extreme temperature range of any planet in the Solar System, becoming extremely hot during the day, nearly 430 °C at the subsolar point at perihelion, and rapidly dropping below -183 °C during the long night. Dark and bright surface markings can be glimpsed through a telescope, but they have much lower contrast than the markings on Mars or the Moon. The Mariner 10 probe photographed half the planet in detail, revealing a lunar-like landscape heavily scarred with impact craters, many with bright rays. There are some slight differences in cratering style: on Mercury, secondary craters fall closer to the main crater than they do on the Moon because of the higher gravity, and inner rings are seen in smaller craters than on the Moon. The largest impact structure on Mercury, the Caloris Basin, is 1300 km across, similar in size to t

Mercury project The first American space programme to launch humans into orbit around the Earth. Astronauts flew in the single-seater Mercury capsule. The first two manned flights, in May and July 1961, were sub-orbital, that is, the spacecraft was launched to the edge of space and fell back again without going into orbit. The first orbital Mercury flight, by John Herschel Glenn was on 1962 February 20. The series ended after the fourth orbital flight in 1963 May.

Mercury-manganese star A form of manganese star that has a spectral line at wavelength 398.4 nm, identified with ionized mercury; also known as a manganese-mercury star.

Mercury's high density suggests that it is composed of about 70% iron, probably concentrated in a central core, and 30% rock. The iron-rich core probably has a diameter 75% of that of the planet, proportionally the largest of any planetary body known. The planet has a slight but significant magnetic field with a strength of 3.5×10^{-7} tesla at the surface, the only terrestrial planet apart from the Earth to have one. It has no natural satellites.

Meridian An imaginary line on the sky that runs due north and south and passes through your zenith.

Meridian angle An alternative name, only occasionally used, for hour angle.

Meridian astronomy The measurement of positions of celestial objects based on observation of the times of their transit across the meridian and of their zenith distance at those times. The intention is to obtain star positions which are self-consistent over large areas of sky, but accuracy is limited by the thermal and mechanical instability of the instrument, and uncertainty in the calculation of the effect of atmospheric refraction.

Meridian circle A telescope with an accurately calibrated circle which is constrained to rotate in the plane of the meridian about an east-west horizontal axis. The altitude at which the telescope is aimed at the time of an observation is obtained by measuring the circle divisions with a series of fixed microscopes, usually four or six.

Meridian passage Another name for a meridian transit.

Meridian, celestial A great circle on the celestial sphere that joins the north and south points on an observer's horizon, passing through the zenith and the celestial poles. When a celestial object crosses an observer's meridian it is said to be at culmination or in transit.

Meridian, terrestrial A plane through the Earth's poles, defining a particular longitude on Earth.

Meson A class of elementary particle. Mesons are a subdivision of the hadrons, and are composed of two quarks. Mesons are involved in the processes that hold atomic nuclei together, and are also a major component of secondary cosmic-ray showers. They are unstable, with half-lives in the range 10^{-8} to 10^{-16} s, and decay into stable particles. They can be positively or negatively charged, or electrically neutral. Mesons have masses intermediate between those of the electron and proton, and include the kaon, pion, and psi particles. The mu-meson was an old name for a muon; it is not a meson at all, but a lepton.

Mesopause The upper boundary of the mesosphere, 85 km above Earth's surface. Temperatures close to the mesopause are low, typically around -110 °C. Temperatures as low as -162 °C have been recorded close to the mesopause by rocket-borne instruments flown into noctilucent clouds, whose condensation close to this atmospheric level during the summer months is apparently favoured by such conditions.

Mesosiderite A class of stony-iron meteorite, apparently related to the howardites. Mesosiderites resemble lunar surface breccias, and were apparently formed on the surface of their parent body by impact processes. They consist of fragments of eucrite and diogenite material, mixed with a large proportion of metallic nickel-iron alloy (typically 40-60% by mass). Their textures indicate that they were subjected to shock melting and metamorphism at temperatures up to 1000 °C.

Mesosphere The middle layer of Earth's atmosphere, lying above the stratosphere and below the thermosphere, at altitudes of 50-85 km. Temperatures in the mesosphere fall with increasing altitude, reaching a minimum at the mesopause. Measurements from sounding rockets indicate the mesosphere to be extremely dry, but sufficient traces of water vapour are present during the summer months for noctilucent clouds to form.

Messier catalogue A list of celestial objects that might be mistaken for comets, compiled by the French comet-hunter C. J. Messier. His list first appeared in 1771, containing 45 objects; a revised list in 1780 added another 23 objects; and his final list, with 103 objects, appeared in 1781. Many of the objects were actually discovered by others, notably his compatriot Pierre François André Méchain.The objects catalogued by Messier include star clusters, nebulae, and galaxies. Astronomers still refer to these so-called Messier objects by their Messier, or M, numbers. Later observers have extended the catalogue beyond the 103 objects listed by Messier.

Messier, Charles While hunting for comets in the skies above France, 18th century astronomer Charles Messier made a list of the positions of about 100 fuzzy, diffuse looking objects which appeared at fixed positions in the sky. Although these objects looked like comets, Messier knew that since they did not move with respect to the background stars they could not be the undiscovered comets he was searching for. These objects are now well known to modern astronomers to be among the brightest and most striking gaseous nebulae, star clusters, and galaxies. Objects on Messier's list are still referred to by their "Messier number". For example the Andromeda Galaxy, the 31st object on the list, is known as M31.

Metagalaxy An archaic term for the entire Universe, deriving from the time before it was realized that the spiral nebulae were in fact separate galaxies of stars beyond the Milky Way.

Metal Used by astrophysicists to refer to all elements except hydrogen and helium, as in: "The universe is composed of hydrogen, helium and traces of metals;".

Metallic hydrogen A form of hydrogen in which the atoms are highly compressed by intense pressure, as in the interiors of the massive gaseous planets Jupiter and Saturn. Under such conditions, hydrogen behaves like a liquid metal and hence can conduct electricity and generate a magnetic field.

Metal-poor star A star with a far smaller proportion of heavy elements than the Sun, perhaps as little as 1% or less. They are Population II stars, and are usually found in the halo of our Galaxy or in globular clusters; they are very old stars, formed before the Galaxy was chemically enriched by the first generations of supernova explosions.

Metal-rich stars Stars with a high proportion of heavy elements such as calcium, iron, and titanium. They are members of Population I, and are found in the disk and spiral arms of our Galaxy.

Metamorphic Rocks that have recrystallized in a solid state as a result of changes in temperature, pressure, and chemical environment.

Metamorphic rock Metamorphic rocks are compacted by pressure and heat from deep inside the earth.

Metastable state An excited energy level in an atom, ion, or molecule, from which transitions to lower energy levels are said to be forbidden. A forbidden transition is one of such low probability that it is not observed under laboratory conditions. However, in the low-density conditions of interstellar space and nebulae, electrons cannot escape from the metastable state by upward excitation to a higher level by collisions. The metastable state can then become highly overpopulated, and emission lines from the forbidden transitions can be the strongest lines in the spectrum of a nebula.

Meteor A brief streak of light in the Earth's upper atmosphere between altitudes of 85 and 115 km, produced by the high-speed entry of a small fragment of interplanetary debris.An estimated 100 minion meteors are visible to the naked eye over the whole Earth in an average 24-hour period. Meteoroids enter the atmosphere at velocities of 11-72 km/s. A typical naked-eye meteor of magnitude +2 is produced by a meteoroid about 8 mm in diameter. Over 0.1-0.2 s, its kinetic energy is converted principally to heat and ionization; only a small proportion is converted to visible light. The surface of a meteoroid is rapidly vaporized by the process of ablation. Material eroded from the meteoroid's surface goes on to collide further with atmospheric particles, producing excitation and ionization along a column perhaps 20-30 km long. The excess energy imparted to the atmospheric particles is re-emitted in a fraction of a second as visible light.

Meteor crater One of the best-preserved and the most famous terrestrial meteorite impact crater, 55 km east of Flagstaff, Arizona; also known as Barringer Crater. The crater is 1.2 km in diameter, 175 m deep, and its rim rises an average of 45 m above the surrounding plains. It was formed about 50 000 years ago when a nickel-iron meteorite, about 50 m across and weighing several hundred thousand tonnes, hit the desert at about 16 km/s. The energy of the impact, equivalent to about 20 megatonnes of TNT, excavated over 175 million tonnes of limestone and sandstone to produce the crater. The meteorite vaporized or melted almost completely on impact, any remaining material being scattered around the crater. Over 10000 nickel-iron octahedrite meteorite fragments have been recovered, up to 7 km from the crater. The crater's impact origin was recognized in the 1930s, mainly through the efforts of the American mining engineer Daniel Moreau Barringer.

Meteor storm An extremely rare and typically short-lived event, usually associated with a regular meteor shower, during which rates exceed 1000 meteors/h. Such events can occur when the Earth runs through a meteor swarm, or a denser filament within a meteor stream, normally located relatively close to the parent body. For instance, the Leonid meteor storm of 1966 closely followed the passage of the parent comet, Tempel-Tuttle, through the descending node of the stream orbit, and resulted from recently ejected debris not yet dispersed into the main body of the stream.

Meteor stream A trail of debris laid down by a larger parent body, comprising small particles which pursue a common orbit around the Sun. Most meteor streams originate from periodic comets, whose nuclei release dust when near perihelion. Over time, the meteoroids spread out along the comet's orbit. When a meteor stream's orbit approaches within about 0.1 AU of the Earth, a meteor shower may occur. Filaments of debris released by the parent body over several successive returns give rise to variable activity from year to year, particularly in younger showers such as the Perseids; older streams such as the Taurids show a smoother, wider distribution of material, and less variable activity.

Meteor swarm A denser concentration of meteoroid particles within a meteor stream. Particles comprising a meteor swarm will usually have been released from the parent body comparatively recently, and have not yet become dispersed by gravitational and other perturbations. Examples of streams in which meteor swarms are found include the Giacobinids and Leonids. Passage of the Earth through these more highly populated regions in the streams gives rise to periodic meteor storms.

Meteorite A natural object from space that hits the surface of the Earth or other planetary body. Impacts by large meteorites are believed to have created most of the craters on the planets and their satellites. An estimated 40-50 tonnes of cosmic debris enters the Earth's atmosphere every day, but only about a tonne reaches the ground. The fate of a body entering the Earth's atmosphere depends mainly on its mass and velocity. The smallest objects (micrometeorites) are decelerated and drift slowly down to the surface. Objects with masses between about 10^{-6} g and 1 kg burn up to produce meteors. Bodies with masses between about 1 kg and 1000 tonnes are substantially slowed by atmospheric drag, but penetrate the atmosphere. The atmosphere has no significant slowing effect on bodies of over 1000 tonnes. The average entry velocity of incoming bodies which fall as meteorites is about 20 km/s.

Meteorites from mars Many meteorites from Mars have landed on Earth, one with a suspected microbial fossil.

Meteoritic bombardment Intensive and prolonged impacts of a surface by meteorites or other impactors.

Meteoroid A small particle from a comet or asteroid in orbit around the Sun. While the distinction between a large meteoroid and a small asteroid may be somewhat blurred, the term meteoroid most commonly refers to a particle which gives rise to a meteor on entering Earth's atmosphere. Typical meteoroids in a meteor stream, such as that which produces the Perseid meteor shower, are a few millimetres in diameter, and have low densities (0.2-0.3 g/cm^3). Meteoroids in the Geminid stream, whose parent body is the asteroid Phaethon, have a higher density, 2 g/cm^3.

Methane A compound composed of carbon and hydrogen: CH_4.

Methanol maser A maser source in which the methanol molecule (CH 3 OH) is excited to maser action. Methanol has many maser lines, of which the most important are at frequencies of 6.7, 12.2, and 44.1 GHz. Methanol masers have so far been found only in star-forming regions, where they are second only to H_2O (water) masers in brightness.

Metis Metis is the closest of Jupiter's 16 moons. Metis is 25 miles (40 km) in diameter and orbits 79,500 miles (128,000 km) from Jupiter, within Jupiter's main ring. Metis and the next moon, Adrastea, are probably the source of the dust in this ring. Metis has a mass of 9 x 10^{16}kg. It orbits Jupiter in 0.294780 (Earth) days; this is faster than Jupiter rotates on its axis. Metis was discovered by Stephen Synnott (Voyager 2) in 1980.

Metonic cycle A period of 19 years after which the Moon's phases repeat on approximately the same calendar dates. The cycle was discovered by the ancient Greek astronomer Meton in the 5th century BC and is used in constructing lunisolar calendars. It occurs because there are 235 synodic months (lunations) in 19 tropical years.

Meudon observatory A branch of the Paris Observatory, 4 km south of Paris, founded in 1876 by P. J. C. Janssen. It is the site of the 0.83-m Grande Lunette refractor, opened in 1895.

Mev Mega (million) electronvolt. A unit of energy used to describe the total energy carried by a particle or photon.

Miaplacidus The star Beta Carinae, magnitude 1.7, an A1 giant 64. L.Y away.

Mice, the A pair of interacting 14th-magnitude spiral galaxies in Coma Berenices; also known as NGC 4676A and B, or IC 819 and 820. The gravitational interaction has drawn off long tails of stars and gas from each galaxy, hence the name.

Michelson interferometer An apparatus used to measure very precise lengths, such as the wavelength of light, and to analyse details of spectral lines. It consists of a half-silvered flat mirror which divides a beam of light into two. These beams are returned along the same paths by additional mirrors, then recombined. The interference of light creates dark and light bands or fringes whose positions depend on the lengths of the two light paths. It is named after its inventor, A. A. Michelson.

Michelson, Albert Abraham German-American physicist, born in modern Poland. In 1878 he constructed a rotating-mirror apparatus for measuring the velocity of light, on which he collaborated with S. Newcomb. The interferometer he built for the Michelson-Morley experiment of the 1880s led him to construct highly accurate spectroscopes and diffraction gratings for astronomy, and also the Michelson interferometer. In the 1920s he measured the angular diameter of Betelgeuse with a stellar interferometer he built and installed on the 100-inch (2.5-m) telescope at Mount Wilson Observatory.

Michelson-Morley experiment An experiment performed in 1887 by the American physicists A. A. Michelson and Edward Williams Morley to establish the presence or absence of an ether, the medium through which light was supposed to travel. The experiment attempted to detect the motion of the Earth relative to the ether by measuring the speed of light in two directions at right angles to each other. No such motion was found. An explanation was provided by the Lorentz-Fitzgerald contraction, and subsequently the special theory of relativity showed that the concept of an ether was unnecessary.

Microcalorimeter A type of X-ray detector which measures the heat input due to an incoming X-ray.

Microchannel plate A type of detector composed of layers of reactive material divided into narrow channels, incoming photons react with the material to produce an electric signal

Microchannel plate detector A device for producing images of high-energy photons. It consists of one or more arrays of tiny hollow glass tubes, typically 12.5-25 μm in diameter and 1-2 mm long, each operating as an individual photomultiplier. Around 10 million tubes are stacked side by side to form a thin plate structure. These microchannel plates can be used singly or stacked in pairs or triplets to detect ultraviolet, extreme ultraviolet, and X-ray photons directly, when coated with a suitable material. They are also used as optical image intensifiers in combination with a phosphor screen.

Microdensitometer An instrument for measuring the photographic density at different points on a photographic emulsion or paper. For light passing through the emulsion, the transmitted density is measured; for prints the reflected density is measured. Modern computer-controlled measuring machines such as APM and SuperCOSMOS are sophisticated versions of the microdensitometer, designed to measure accurately the centres of star images and to distinguish stars from galaxies.

Microgravity Microgravity is a state in which gravity is reduced to virtually negligible levels. For example, when an object is in free fall, it experiences microgravity.

Microlensing A small-scale gravitational lens effect. In microlensing, the gravitational field of the lensing object is not strong enough to form distinct images of the background source; instead, it causes an apparent brightening of the source. Stars are expected to vary in brightness in a characteristic manner if low-mass stars or planets pass in front of them, and this effect has been detected for stars in the Large Magellanic Cloud and in the central bulge of our Galaxy.

Micrometeorite A cosmic dust particle that has survived its passage through the Earth's atmosphere without burning up as a meteor. Particles smaller than 0.1 mm in diameter, with masses under 10-6 g, are slowed by the upper atmosphere without melting, and drift down to the Earth's surface. About one micrometeorite of diameter 0.1 mm falls on each square metre of the Earth's surface per year, while for diameters of 0.01 mm the rate rises to about one per square metre per day. Micrometeorites can be collected in the stratosphere, providing samples of material from comets and asteroids for analysis. On the Moon and asteroids, the lack of an atmosphere means that micrometeorites impact the surfaces of these bodies at high speed.

Micrometer An instrument used in astronomy for measuring angular distances and relative positions on the sky. Micrometers are used particularly for measuring the orbital motions of double stars, but can also be used to measure the locations of features on planets. There are many types of micrometer, but all introduce standard reference points into the field of view of an eyepiece. These references may be illuminated reticles or wires, although the wires may be spider thread and are referred to as webs. The webs are located in the same plane as the field stop, so that they are always in focus. The reticle may be fixed, as in the crosswire micrometer, in which case the daily movement of the sky is used for reference; or there may be a movable portion, as in the filar micrometer, with scales which should be calibrated against the sky's daily motion. Other types of micrometer provide movable artificial stars or double images of the stars in the field of view.

Micron One thousandth of a millimetre. The symbol used is μ (mu).

Microquasar Microquasars are stellar mass black holes, that display characteristics of the supermassive black holes found at the centers of some galaxies. For instance, they have radio jets - something not every black hole has.

Microscopic A physical entity or process of small scale, too small to directly experience with our senses. Specifically, any phenomena on the molecular and atomic scale, or smaller.

Microscopium An insignificant constellation of the southern sky, representing a microscope. Its brightest stars are Gamma and Epsilon Microscopii, both of magnitude 4.7.

Microturbulence Turbulence of an emitting gas in which the turbulent eddies are very small and unresolvable, giving rise to the broadening of a spectral line in a star's spectrum by Doppler shifts. If a Fraunhofer line is broader than expected from the motions of emitting atoms and other known causes, the excess broadening is sometimes attributed to unseen microturbulence, regardless of whether it actually exists.

Microvariable A variable star in which the amplitude of the fluctuations amounts to a few thousandths of a magnitude. The variations may arise from a number of causes (e.g. rotation in conjunction with non-uniform surface brightness). The Sun may be considered a microvariable.

Microwave anisotropy probe A NASA space probe to study small fluctuations in the cosmic microwave background left over from the Big Bang, in succession to the Cosmic Background Explorer. It is due for launch in 2000 and will be positioned at the L2 Lagrangian point of the Earth's orbit, where it can observe the Universe without interference from the Sun, Earth, or Moon.

Microwave background radiation The radiation from the glowing of the hot early universe, now so greatly red-shifted that it appears not as light but as microwaves (radio waves with a wavelength of a few centimeters).

Microwave Electromagnetic radiation which has a longer wavelength (between 1 mm and 30 cm) than visible light. Microwaves can be used to study the Universe, communicate with satellites in Earth orbit, and cook popcorn.

Microwave burst A radiowave signal associated with optical and/or X-ray flares.

Microwave radiation Microwave radiation is a type of electromagnetic radiation that has a wavelength between 1 mm and 30 cm.

Middle latitudes With specific reference to zones of geomagnetic activity, "middle latitudes" refers to 20 deg. to 50 deg. geomagnetic.

Mie scattering Scattering of light by particles whose size is similar to the wavelength of light. It occurs in interstellar space and in the Earth's atmosphere. For wavelengths much shorter than a given particle size, Mie scattering is a complex function of wavelength. Scattering drops to a minimum at a wavelength half the particle size, rises to a maximum when it is the same as the particle size, and then decays to zero towards longer wavelengths. Mie scattering can thus make objects appear either redder or bluer according to the size of the scattering particle. In the Earth's atmosphere there is usually a wide range of particle sizes, so that these colour effects are blurred into a nondescript grey. Very occasionally, atmospheric effects will combine so that the dust is mostly 900 nm in size and the minimum in scattering is at 450 nm. This is the wavelength of blue light that causes the phenomenon of the blue Moon (1). Mie scattering is named after the German physicist Gustav Mie.

Milankovitch theory The Milankovitch Theory attempts to explain major temperatures changes on Earth, especially ice ages, by slight variations in the amount of sunlight reaching the Earth caused by the eccentricity of the Earth's orbit around the Sun. The Earth's orbital eccentricity changes the Earth's average distance from the sun and therefore slightly changes the amount of sunlight reaching the Earth. Milankovitch looked over the past 600,000 years and correlated summer temperature mimnima with four major ice ages in this time. Eccentricity cycles last over 100,000 years. This theory was proposed by Milutin Milankovitch in 1938. Recently, scientists have found that changes in the axal tilt of the Earth's orbit more closely match glacial cycles for the past million years.

Milky way galaxy The Milky Way Galaxy is a spiral galaxy; our sun and solar system are a small part of it. Most of the stars that we can see are in the Milky Way Galaxy. The main plane of the Milky Way looks like a faint band of white in the night sky. The Milky Way is about 100,000 light-years in diameter and 1,000 light-years thick. There are about 2 x 10 11 stars in the Milky Way. This spiral galaxy formed about 14 billion years ago. It takes the sun roughly 250 million years to orbit once around the Milky Way. The Earth is about 26,000 light-years from the center of the Milky Way Galaxy. The major arms of the Milky Way galaxy are the Perseus Arm, Sagittarius Arm, Centaurus Arm, and Cygnus Arm; our Solar System is in a minor arm called the Orion Spur.

Milky way, the (1) Another name for The Galaxy.

(2) The broad band of faint light that stretches across the sky (when the time is favourable) as seen from a dark site. It is the result of countless millions of stars of The Galaxy seen edge on from the viewpoint of Earth. The Milky Way is brightest in the Northern hemisphere in the constellation of Cygnus, and in the Southern hemisphere in the constellation of Crux.

Millibar 1/1000 of a bar.

Millikan Robert Andrews Millikan (1868-1953) was an American physicist who measured the charge of the electron (in his oil-drop experiment, 1911). Millikan named "cosmic rays" in 1928, while studing the radiation from space. He won the Nobel Prize for Physics in 1923 for his work on the electron's charge and his work on the photoelectric effect.

Millimetre waves Electromagnetic radiation with wavelengths in the approximate range 1-10 mm. Wavelengths shorter than 1 mm are known as submillimetre waves.

Millimetre-wave astronomy The observation of electromagnetic waves from the Universe at wavelengths of about 1-10 mm, a part of the spectrum rich in lines emitted by interstellar molecules. Millimetre-wave astronomy is one of the newest fields in astronomy and is used to study the chemistry of interstellar matter, and the complex processes in star-forming regions and dark clouds. The cosmic background radiation is brightest in this part of the spectrum. Major millimetre-wave facilities include those of the Institut de Radio Astronomie Millimétrique and Nobeyama Radio Observatory.

Millisecond pulsar A pulsar that flashes every few thousandths of a second. The first to be discovered, PSR 1937+21, has a period of 1.56 ms, which is still the shortest known period and close to the theoretical minimum for a spinning neutron star. More than fifty pulsars with periods of less than 20 ms have been discovered, many of them in globular clusters. Millisecond pulsars are extremely stable rotators and keep time better than atomic clocks.

Mills cross A design of radio interferometer consisting of numerous antennas arranged in two arms in the form of a cross oriented north-south and east-west. It can be configured to produce either a pencil beam or a fan beam. The original Mills Cross, in Sydney, had 500 half-wave dipoles placed along two arms 457 m long. The Mills cross is named after its designer, the Australian radio astronomer Bernard Yarnton Mills.

Milne, Edward Arthur English mathematical astrophysicist and cosmologist. His work on radiative equilibrium and stellar atmospheres in the 1920s yielded a temperature scale for the sequence of spectral types. With the English mathematician Ralph Howard Fowler he developed equations which allowed C. H. Payne-Gaposchkin to show that stars are composed largely of hydrogen. In later work on stellar structure he was one of the first to associate nova outbursts with stellar collapse and the emission of a gas shell. In 1932 he turned to relativity and cosmology, showing that there are valid, non-Einsteinian ways of modelling the expanding universe and space-time.

Milne-Eddington approximation A simple model of the outer layers of a star, used in studying the formation of absorption lines in stellar spectra. It assumes that absorption occurs throughout the outer region of the star, rather than in just the cool outermost layers, and that the ratio of line absorption to continuum absorption is constant. It successfully predicts the strength of lines of ionized metals such as calcium and iron in solar-type and similar stars. The model is named after E. A. Milne and A. S. Eddington.

Mimas Mimas is one of larger of the 18 moons of Saturn. It has a diameter of 235 miles (390 km) and it orbits at about 116,000 miles from Saturn. It has a very cratered surface, including one huge crater, Herschel, that is about one quarter of the moon's diameter (80 miles = 130 km wide). Herschel's central mountain is 4 miles (6 km) tall, taller than Mount Everest. Mimas has very low temperatures (-200° C = -328°F). Its orbital period is about 22.5 Earth days. Mimas was discovered by William Herschel in 1789.

Mineral A mineral is a naturally-occurring solid of definite chemical composition whose atoms usually form a regular pattern.

Mini black hole One of many tiny black holes with the mass of a mountain but the size of an atom that are believed to have been created in the Big Bang

Minkowski, Hermann German mathematician, born in modern Lithuania. He was the first to see that the principle of relativity set out by the Dutch physicist Hendrik Antoon Lorentz and A. Einstein implied that space and time were not separate entities, but components of a fourdimensional space-time. His mathematical model of space-time provided the foundation for all subsequent developments of relativity, and allowed Einstein to move from spęcial to general relativity.

Minkowski, Rudolph leo bernhard German-American astrophysicist, born in Strasbourg.In 1941 he classified supernovae into Types I and II according to their spectra. In 1954 he and W. Baade reported the first optical counterpart to an extragalactic radio source, Cygnus A. Minkowski went on to specialize in the optical identification of radio sources with high redshift. He also carried out a search for planetary nebulae, doubling the number known.

Minor axis The shortest diameter of an ellipse, at right angles to the major axis.

Minor cycle In the Clark CLEAN deconvolution algorithm, an inner iteration, in which the peak residual over a subregion (the CLEAN window) of the full residual image is found and is used to obtain the next successive iterate.

Minor planet A minor planet is another name for an asteroid.

Minor planet center The organization responsible for collecting and disseminating positional measurements and orbital data for asteroids.It was established by the International Astronomical Union at Cincinnati Observatory in 1947, and moved to the Smithsonian Astrophysical Observatory in 1978. The Minor Planet Center allocates preliminary designations to new discoveries and gives permanent numbers when satisfactory orbits have been determined. The information is published through the Minor Planet Circulars. The work of the Minor Planet Center complements that undertaken at the Institute for Theoretical Astronomy in St Petersburg, Russia, which publishes annual volumes of ephemerides of numbered asteroids.

Mintaka Mintaka (Delta Orionis) is the uppermost and faintest of the three major stars that form belt of Orion, a constellation in the Northern Hemisphere (the other two stars in the belt ate Alnitak and Alnilam). Mintaka (which means "belt") is located quite near the celestial equator. It is the seventh brightest star in Orion; it is a variable star of magnitude 2.22-2.5; it is also a double star. It is about approximately 915 light-years from Earth. Its spectral type is O9.5II.

Minute of arc (symbol ') A small unit of angular measure, equivalent to one-sixtieth of a degree.

MIPS Millions of instructions per second.

MIR Mir (which means "peace" in Russian) was a Russian space station, the first semi-permanent human habitat in orbit around the Earth. The first part of Mir (the Mir module) was launched on February 20, 1986; it orbited the Earth for over 15 years. Mir is composed of many of different modules that have been attached to it through the years; the modules perform different functions. Cosmonauts and astronauts from many countries have lived and performed scientific experiments on Mir. Mir fell back to Earth on March 23, 2001.

Mira The star Omicron Ceti, the prototype Mira star or long-period variable. It was first

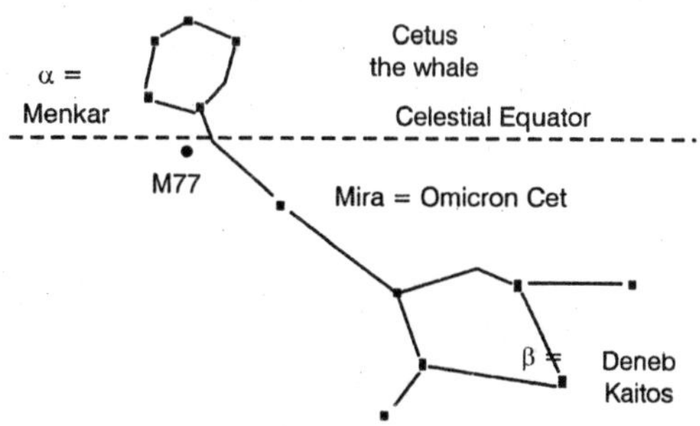

Fig. Mira

observed by D. Fabricius in 1596, and was originally believed to be a nova. It was found to be periodic by the Dutch astronomer Jan Fokkens in 1638, and was named Mira by J. Hevelius. Mira has a period of 331.96 days, a range of 2.0-10.1 mag., and is a red giant of spectral type M5e-M9e. Its diameter is about 300 times that of the Sun. It forms a binary with VZ Ceti, a white dwarf that is accreting material from the giant's stellar wind. Mira is about 250 L.Y away.

Mira star A type of red giant or supergiant pulsating variable; abbr. M. Mira stars have highly characteristic late-type spectra with molecular bands, and there is extensive mass loss. Their light-curves may be approximately sinusoidal, but most show a steep rise to maximum and slower decline; amplitudes range from 2.5 to as much as 11 magnitudes, with periods from 80 to 1000 days. Because of their relatively high luminosities, large amplitudes, and well-defined features, Mira stars are readily detectable. More of them are known than any other type of variable star.

Mirach The star Beta Andromedae, magnitude 2.1. It is an M0 giant 170 L.Y away.

Miranda Miranda is one of the 18 moons of Uranus. Its surface is marked by huge grooves, and appears to have shattered many times. It has a bright, V-shaped groove on its surface, the Inverness Corona, which has been nicknamed the chevron. Miranda's diameter is about 290 miles (470 km) and its mass is 6.3×10^{19} kg. It orbits at a distance of about 80,000 miles (129,780 km) from Uranus, and orbits once every 33.8 hours. It was discovered by Gerard Kuiper in 1948.

Mirfak The star Alpha Persei, also known as Algenib, magnitude 1.8. It is an F5 supergiant 630 L.Y away, and is surrounded by a scattering of stars known as the Alpha Persei Cluster or Melotte 20.

Mirror An optical element that reflects light A mirror may be flat which simply laterally inverts an image on reflection, or a concave or convex mirror, which forms an image. Mirrors in an optical path are always coated by aluminizing their front surface, unlike domestic mirrors which are coated on their backs. Astronomical images usually have to be very sharp, so the surface shape of the mirror must be precise. In general, the shape of an astronomical mirror should alter the reflected wavefront of light by less than a quarter of a wavelength of light across its entire surface.

Mirror blank A disk of glass or other suitable material which is to be turned into a mirror. It may be cast as a disk or cut from a sheet. The ratio of diameter to thickness is usually at least 6: 1 so that the mirror will not deform. Low-expansion glass such as Pyrex is advisable, to prevent distortion of the finished mirror's precise curve as a result of temperature changes.

Mirzam The star Beta Canis Majoris. It is a B1 giant of magnitude 2.0, and a variable of Beta Cephei type, changing by less than 0.1 mag. in a period of 6 hours. It lies 750. L.Y away.

Missing mass The additional, unseen matter whose gravitational effect is needed to explain the rotation speeds of galaxies, and also to keep clusters of galaxies together. Such mass is thought to reside in massive dark haloes around galaxies. In addition, if gravity is to prevent the Universe from expanding indefinitely, there must be considerably more matter in intergalactic space than is contained in visible stars and galaxies.

Mitchell, Maria Maria Mitchell (August 1, 1818 -June 28, 1889) was the first woman Professor of Astronomy of the United States. In 1865, Maria Mitchell became a professor of astronomy at Vassar College in Poughkeepsie, New York (she had previously been a librarian). She discovered a comet (in 1847), studied the planets Jupiter and Saturn, and photographed many stars. Despite her accomplishments, when she visited the Vatican Observatory in Italy, she was only allowed to enter the observatory during the day. Maria Mitchell was the first woman accepted into the American Academy of Arts and Sciences (1848), the Association for the Advancement of Science (1850), and the American Philosophical Society (1869). Mitchell was one of the founders of the American Association for the Advancement of Women (1873).

Mittenzwey eyepiece A modified form of Huygenian eyepiece consisting of convex meniscus lenses. It is named after its inventor, the German optician Moritz Mittenzwey.

Mixing ratio A relationship which indicates how effectively energy is transported by convection in stars. It is defined as the length of eddies in the convective current divided by the distance over which the gas pressure varies significantly. Convection in stars is not fully understood, so calculations of the sizes of convective stars are uncertain by about 15%.

Mizar The star Zeta Ursae Majoris, an A1 dwarf of magnitude 2.3, distance 74 L.Y It forms a naked-eye double with Alcor, but the two are not a binary pair. However, a closer companion of magnitude 4.0 is connected. In 1889 E. C. Pickering found that Mizar is a spectroscopic binary. Mizar was therefore the first telescopic binary and the first spectroscopic binary to be discovered. The 4th-magnitude companion is also a spectroscopic binary.

MK classification Abbr. for Morgan-Keenan classification.

Mock sun An effect caused by ice crystals in Earth's atmosphere which refract sunlight and cause the appearance of two diffuse patches of light 22° either side of the Sun. These patches are termed *Parhelia* or *Sundogs* and often appear on the rim of a halo surrounding the Sun. The patches often have coloured fringes, red on one side, blue on the other.

Modified Julian date A more convenient form of the Julian date, in which the zero point is 1858 November 17.0. Hence Modified Julian Date = Julian Date - 2 400 000.5 days. In the modified Julian date, the day starts at midnight. Orbital data for artificial Earth satellites are often expressed in Modified Julian Date Numbers.

Mohorovicic discontinuity The Mohorovicic discontinuity separates the crust and the upper mantle of the Earth.

Mohorovicic, Andrija Andrija Mohorovicic (1857 - 1936) was a Yugoslavian geophysicist. After examining seismic waves from the 1909 Kulpa Valley earthquake, Mohorovicic theorized that a boundary between the Earth's crust and the upper mantle existed (about 50 km beneath the surface) in which the speed of earthquake waves became very rapid. This region is now called the Mohorovicic discontinuity. A crater on the far side of the moon was also named for Mohorovicic.

Moldavite Moldavite is a rare, glassy, translucent, dark green gemstone. Moldavite is a silica-based tektite, a mineral formed when a meteorite (a rock from space) struck the Earth's surface and melted and fused the surrounding rock. Moldavite is only found in Bohemia (the Czech Republic) in the Ries Crater in the Moldau River valley (which it was named for). Moldavite was discovered in the late 1800's; the meteorite from which it formed hit the Earth about 14.7 million years ago. Moldavite has a Mohs hardness of 5.5-6.6. Inclusions of gas bubbles and iron/nickel spherules are common. This natural glass has been used for jewelry, religious articles, and decorative objects since prehistoric times.

Mole The term mole is short for the name gram-molar-weight; it is not a shortened form of the word molecule. (However, the word molecule does also derive from the word molar.) See: Avogadro's constant.

Molecular cloud A cloud of interstellar gas and dust composed mainly of molecules. The smallest known molecular clouds contain less than 1 solar mass of gas, whereas giant molecular clouds have masses of up to 107 Solar masses. Most of the molecular gas is very cold, with typical temperatures of around 20 K. Molecular clouds can be detected optically as dark nebulae. Infrared telescopes detect the small amount of heat emitted by the dust. Radio telescopes detect numerous spectral lines from the interstellar molecules in such clouds.

Molecular hydrogen (symbol H_2) The normal form of hydrogen gas found on Earth, with two hydrogen atoms bound together by two shared electrons. Hydrogen molecules cannot survive in most regions of space, because they are easily by ultraviolet light from stars, but they do occur in cold, dense molecular clouds where they are shielded by dust particles. Molecular hydrogen is the main constituent of molecular clouds, but it is difficult to observe directly as it is a symmetrical molecule with no strong radio or millimetre-wave spectral lines.

Molecular line A narrow spectral emission or absorption feature produced by an interstellar molecule. Most molecules of astrophysical interest emit and absorb in the infrared, millimetre, and radio parts of the spectrum.

Molecular mass The *molecular mass* of something is the mass of one mole of it (in cgs units), or one kilomole of it (in MKS units). The units of molecular mass are gram and kilogram, respectively. The cgs and MKS values of molecular mass are numerically equal. The molecular mass is not the mass of one molecule. Some books still call this the *molecular weight*.

One dictionary definition of *molar* is 'Pertaining to a body of matter as a whole: contrasted with *molecular* and *atomic*.' The mole is a measure appropriate for a *macro*scopic amount of material, as contrasted with a *micro*scopic amount (a few atoms or molecules).

Molecule The smallest part of a compound, which is a substance formed by the chemical combination of one or more types of atom. For example, water consists of two atoms of hydrogen and one atom of oxygen, hence it has the formula H_2O.

Molonglo radio observatory A radio observatory at Hoskinstown, New South Wales, 30 km east of Canberra, Australia. It was founded in 1962 and is owned and operated by the University of Sydney. It contains the Molonglo Observatory Synthesis Telescope,opened in 1981. This is an aperture synthesis telescope that consists of two parabolic reflectors, each 778 m long and 11.6 m wide, oriented east-west, originally one arm of a Mills cross that operated 1965-78.

Moment of inertia (symbol I) For a system of several bodies, the mass of each body multiplied by the square of its distance from the centre of the system, the products all being added together. For a single solid body, the moment of inertia is the sum of the mass of each particle in that body multiplied by the square of its distance from the body's axis of rotation. The shape of the Moon, for example, is an ellipsoid with three different axes, the longest of them pointing towards the Earth. The moments of inertia of the Moon about these three axes are slightly different, which can produce measurable effects in the orbit of an artificial lunar satellite. Moment of inertia is a measure of a body's ability to resist a change in its state of rest or angular velocity.

Momentum Momentum is an object's mass times its velocity.

Monocentric eyepiece An eyepiece design consisting of three thick glass elements cemented together, with no air gaps, the outer two being of flint glass and the inner one of crown glass. The design, like other solid eyepieces such as the Tolles, is free from ghost reflections of bright objects in the field of view and gives bright images with good contrast. However, the field of view is rather small, about 30°, which makes such eyepieces suitable only for planetary and double-star observations. An adaptation is known as the Hastings triplet.

Monoceros A constellation on the celestial equator, representing a unicorn. It contains the massive binary Plaskett's Star. Its brightest star is Alpha Monocerotis, magnitude 3.9. Beta Monocerotis is a superb triple star, magnitudes 4.6, 5.4, and 5.6. M50 and NGC 2232 are open clusters. NGC 2244 is a star cluster within the Rosette Nebula. NGC 2261 is Hubble's Variable Nebula. NGC 2264 is a star cluster associated with the Cone Nebula, including the highly luminous star S Monocerotis, magnitude 4.7 but slightly variable.

Monochromatic light Light of one colour only, having a single wavelength. A common source of monochromatic light is an emission lamp, such as a mercury or low-pressure sodium streetlight, filtered to remove unwanted emission lines.

Monochromatic magnitude The brightness of a star at a single wavelength. Such magnitudes are important in assessing the variations in magnitude to be expected when a given star is observed through different filters. The monochromatic magnitude is a mathematical ideal that cannot be measured in practice, but can be approximated by narrow-band photometry, provided there are no strong spectral features in the band.

Monochromator A device for selecting a very narrow range of wavelengths of light. It can be a spectrograph in which incoming light is dispersed by a diffraction grating, and a small part of the resulting spectrum selected by passing it through a narrow exit slit. Alternatively, it can be an interference filter that allows only a narrow band of wavelengths to pass.

Monochrome Literally "one color". Used in computing to characterize a video display unit that uses only black and *one* contrasting color (white/green/amber), as distinct from a *color* monitor. Each pixel on a monochrome display corresponds to a single bit in video memory, and is therefore either "on" or "off".

The term is sometimes also applied to a gray scale display using more than one bit per pixel but no hue.

Monogem ring A region of soft X-ray emission about 25° wide extending from Gemini into Monoceros; also known as the Gemini-Monoceros X-ray enhancement. It is thought to be a supernova remnant about 1000 L.Y away with a true diameter of about 150 L.Y

Mons A mountain on a planetary surface; pl. montes. The name, which was first used for features on the Moon in the 17th century, is not a geological term, but is used in the nomenclature of individual mountains irrespective of their origin. For example, the Montes Apenninus on the Moon are part of the walls of a large impact basin, while Olympus Mons on Mars is a giant volcano.

Month A period of time based on the orbit of the Moon around the Earth. Several types of month are defined. The synodic month is equivalent to one lunation, the time taken for the Moon to go through one cycle of phases; its mean length is 29.530 59 days. The anomalistic month is the time between successive passages of the Moon through perigee; it lasts 27.554 64 days. The sidereal month is the time taken for the Moon to return to the same position against the star background; it lasts 27.321 66 days. The tropical month is the time between successive passages of the Moon through the vernal equinox; it lasts 27.321 58 days is the time between successive passages of the Moon through its ascending node; it lasts 27.2122 days. The calendar month is an artificial unit consisting of a whole number of days.

Month, anomalistic The interval between two successive perigee passages of the Moon, equal to 27.55 days.

Month, nodical or draconic The interval between successive passages of the Moon through one of its nodes, equal to 27.21 days.

Month, sidereal The revolution period of the Moon relative to the stars, equal to 27.32 days.

Month, synodic The interval between two successive New Moon's (a lunation), equal to 29.53 days.

Month, tropical The time taken for the Moon to return to the same celestial longitude (7 seconds shorter than the sidereal month).

Moon The only natural satellite of the Earth. The magnitude of the full Moon is -12.7, but its surface is actually dark, with a mean geometric albedo of only 0.12, lower than for all the planets except Mercury. It is the fifth-largest satellite in the Solar System (diameter 3476 km), over a quarter the diameter of the Earth and about 1/81 the Earth's mass. Being so similar in size, the Earth and Moon are often considered a double planet. The Moon's sidereal period of axial rotation, 27.322 days, is the same as its orbital period, so that it keeps the same face towards the Earth. Its equator is inclined by 1°.53 to the plane of the ecliptic. Surface temperatures vary from extremes of 123 °C during the day down to -233 °C at night; typical values are 107 °C (day) and -153 °C (night).

The Moon shows two distinctly different types of terrain with very different densities of impact craters: the brighter highlands and the darker lowland mare areas.

Moon, blue The second Full Moon that occurs during a calendar month. Since the Moon takes 29.53 days to complete a cycle of phases a Blue Moon cannot occur in February. The last Blue Moon was on 30th November 2001, there are none during 2002 and 2003. The next Blue Moon occurs on 31st July 2004. Blue Moon's occur roughly every two and a half years so are not very common. Hence the term "Once in a Blue Moon".

Moon buggy The Moon buggy (also known as the lunar rover or the Lunar Roving Vehicle) was a Jeep-like vehicle that the astronauts used to drive on the moon during the Apollo 15, 16, and 17 missions. It weighed 450 pounds, (or 75 pounds in the Moon's gravity). It could carry 1000 Earth-pounds. On the flight from the Earth to the moon, the lunar rover was folded up in the lunar module. There are three abandoned lunar rovers on the Moon.

Moon landing NASA's Apollo missions sent people to the moon for the first time. Apollo 11's LEM (Lunar Excursion Module) was the first to land on the moon, on July 20, 1969, with Neil Armstrong and Buzz Aldrin (Michael Collins was in the orbiter). Apollo missions 12 through 17 continued lunar exploration.

Moon, harvest The Full Moon that occurs nearest the time of the autumnal equinox. So-called because this will occur around the time of harvest.

Moonquake Sudden motion or trembling of the Moon caused by the abrupt release of slowly accumulated elastic energy in rocks.

Moonrise The moon rises and sets every day, appearing on the horizon just like the sun. The time depends on the phase of the moon. It rises about 30 to 70 minutes later each day than the previous day, so the moon is out during daytime as often is it's out at night. At the time of the new moon, the moon rises at about the same time the sun rises, and it sets at about the same time the sun sets. As the days go by (as it waxes to become a crescent moon, a half moon, and a gibbous moon, on the way to a full moon), the moon rises during daytime (after the sun rises), rising later each day, and it sets at nighttime, setting later and later each night. At the full moon, the times of moonrise and moonset have advanced so that the moon rises about the same time the sun sets, and the moon sets at about the same time the sun rises. As the moon wanes (becoming a half moon and a crescent moon, on the way to a new moon), the moon rises during the night, after sunset, rising later each night.

Moore, Patrick Patrick Moore (1923-) is an English astronomer who has written over 60 books on astronomy and made regular BBC television appearances popularizing astronomy. He has done work on lunar mapping.

Moreton wave A shock wave in the Sun's chromosphere moving outwards from a large solar flare at a velocity of about 1000 km/s. A Moreton wave is visible in Ha light as an arc, sometimes moving for several hundred thousand kilometres, and may cause the winking of dark filaments in its path. Metre-wave radio bursts invariably accompany Moreton waves. They are named after the American solar astronomer Gail Ernest Moreton.

Morgan, William Wilson American astronomer. With Philip Childs Keenan he developed the Morgan-Keenan classification of stellar spectra, set out in An Atlas of Stellar Spectra.Morgan and Keenan devised a series of standard spectra against which a star's spectrum may be identified by visual comparison to yield its mass and luminosity (and hence distance). In 1951, he inferred the spiral structure of the Galaxy from the distribution of O- and Btype stars near the galactic equator. In 1953, he helped originate the system of UBV photometry by which a star's colour index may be determined.He also developed classification systems for galaxies and clusters of galaxies.

Morgan-Keenan classification A system of classifying stellar spectra developed at Yerkes Observatory in 1943 by W. W. Morgan, Philip Childs Keenan and Edith Marie Kellman also known as the MK classification or the Yerkes system. The Morgan-Keenan system retained the sequence of stellar spectral types O, B, A, F, G, K, M introduced in the Harvard classification, but with a more precise observational definition of each type. To these spectral types were added a range of luminosity classes which indicate whether the star is a supergiant, giant, dwarf, or some intermediate class. Unusual stars which do not fit this system, such as the carbon stars stars, white dwarfs and Wolf-Rayet stars, have their own individual classification schemes.

Morgan's classification A classification scheme for galaxies. It is based primarily on the spectrum of the galaxy's central regions, which correlates with the degree of concentration of light towards the centre. A notation is also given for the simple structural form of the galaxy. Lower-case letters af, f, fg, g, gk, and k indicate the spectral type of star that most closely matches the spectrum of the galaxy's nuclear region. Structural forms are E (elliptical), S (spiral), B (barred spiral), I (irregular), D (dustless elliptical with extended envelope of stars), L (low surface brightness), and N (small brilliant nucleus on faint background). An index from 1 (face-on) to 7 (edge-on) shows the inclination of the plane of the galaxy to the line of sight. For example, M33 is classified as fS3. The scheme was devised by W. W. Morgan.

Morning star Not actually a star, but the planet Venus shining brightly in the morning sky.

Mottles Features that form the chromospheric network on the Sun, visible in Ha and calcium K-line spectrograms. They cover the entire Sun, quiet and active regions alike. Coarse mottles are bright, up to about 20 000 km across, and often elongated. They merge to form plages. Fine mottles are narrow (700 km wide, up to 7000 km long), dark or bright, and are probably spicules seen against the disk. Fine mottles form clusters with their bases rooted in the coarse mottles. All coincide with regions of strong magnetic field on the photosphere.

Mount Graham international observatory An observatory at an altitude of 3260 m on Mount Graham in the Pinaleño mountains near Safford, Arizona, 125 km northeast of Tucson, founded in 1988. It is owned and operated by the University of Arizona's Steward Observatory, and hosts telescopes from various other institutions. Its largest instrument is the Large Binocular Telescope. It is also the site of the Vatican Advanced Technology Telescope and the Submillimeter Telescope Observatory.

Mount Stromlo and siding spring observatories Two observatories owned and operated by the Australian National University, Canberra. Mount Stromlo Observatory, founded in 1924 as a solar observatory, lies 11 km west of Canberra at an altitude of 770 m. Its main telescope is a 1.88-m reflector, opened in 1955. Other telescopes are a 1.27-m reflector opened in 1954, and a 0.76-m reflector opened in 1927. Siding Spring Observatory was founded in 1962 as a field station of Mount Stromlo.

Mount Wilson observatory An observatory at an altitude of 1740 m on Mount Wilson in the San Gabriel Mountains about 30 km northwest of Los Angeles, founded by G. E. Hale in 1904 for solar observation. Mount Wilson Observatory is owned by the Carnegie Institute of Washington and has been operated since 1991 by the Mount Wilson Institute. Its main instrument is the 100-inch (2.5-m) Hooker Telescope, opened in 1917. There is also a 60-inch (1.5-m) reflector.Two tower telescopes for solar observation, 18 m and 46 m high, were opened in 1907 and 1909; they have been operated since 1986 by the University of Southern California and the University of California at Los Angeles, respectively. Mount Wilson is also the site of an infrared interferometer operated by the University of California at Berkeley, and of the CHARA Array.

Mounting The structure that supports a telescope and allows it to be pointed at a chosen part of the sky. A mounting should hold the telescope firmly and allow it to be moved so that it can follow objects across the sky as the Earth rotates. This can be done either with a motor drive or with controls known as slow motions. Telescope mountings fall into two main categories: altazimuth and equatorial.

The part that allows the telescope to move is often referred to as the head, which may be mounted on either a pillar or a tripod.

Movie In radio astronomical image processing, a contiguous series of video displays shown in rapid succession to visualize a data cube as a sequence of images ordered along one axis of the cube.

Moving cluster A physically related group of stars which share a common motion in space, such as the Hyades. Because of perspective, the proper motions appear to be directed towards a single point, called the convergent point. If the linear velocity of the cluster is known from observations of radial velocity, then the distance of each star can be estimated from the total proper motion; this technique is known as moving cluster parallax.

Mullard radio astronomy observatory The radio astronomy observatory of the University of Cambridge, at Lord's Bridge near Cambridge, England. The technique of aperture synthesis was developed at MRAO. Many radio sources are referred to by their designations in the third, fourth, fifth, etc. Cambridge sky surveys (3C, 4C, 5C, etc.). The Observatory's main instrument is the Ryle Telescope, an aperture-synthesis array. The One-Mile Telescope, opened in 1964, was an earlier aperture-synthesis instrument, consisting of three 18-m dishes in an east-west line, but it is no longer used. Other instruments include the Cambridge Low Frequency Synthesis Telescope a survey instrument consisting of 60 Yagi antennas arranged along the line of the Ryle Telescope; and the Cosmic Anisotropy Telescope for studying the cosmic microwave background. An array of 4096 dipole antennas covering 3.6 hectares is used to study scintillation of radio sources; an earlier version of this instrument, half the size, detected the first pulsars

Muller, Johann Johann Müller, also known as Johann Regiomontanus (1436-1476) was a German astronomer and mathematician. He studied trigonometry, translating Ptolemy's *Almagest*, from the original Greek. Ironically, his translation helped overthrow the Ptolemaic view of the universe (in which the Earth was thought to be at the center of the universe). He also did work on plane and spherical trigonometry. Muller also observed the motion of the moon, planets, and comets. A 108 km diameter lunar crater, called Regiomontanus (Latitude: -28.3 degrees, Longitude: 1.0 degrees), was named for Muller.

Multicultural astronomy The variety of ways in which cultures of the past and present have observed, recorded, interperted, and made use of astronomy to structure their lives and/or satisfy their curiosity about the universe

Multiple mirror telescope A telescope consisting of six 1.8-m mirrors arranged on a common mounting, giving a collecting area equivalent to one 4.5-m mirror. It is jointly owned and operated by the Steward Observatory and the Smithsonian Astrophysical Observatory. The MMT was opened in 1979 on Mount Hopkins, Arizona, at an altitude of 2600 m, in the grounds of the Fred Lawrence Whipple Observatory. The six mirrors were due to be replaced by a single 6.5-m mirror in 1997.

Multiple star A system of three or more stars bound by their mutual gravitational attraction. It is estimated that about one-third of all known binaries are actually triple. The proportions decrease with greater multiplicity. Systems with six components are known, but they are rare (less than 1% of all multiples). In triple systems there is often one relatively close pair, with the third component in a wider orbit. Quadruple systems, however, occur approximately equally in 2: 2 (2 pairs) and 1 : 1 : 2 hierarchies.

Multiple star system A gravitationally bound system in which two or more stars orbit a common center of mass.

Multiplet A group of closely spaced spectral lines formed by the same element or ion, caused when electrons with different directions of spin jump from a common lower energy level.

Muon An elementary particle having the same charge and spin as an electron, but a mass 207 times greater. It is a lepton which decays into electrons and neutrinos with a half-life of two microseconds. Muons are present in cosmic-ray showers which are detected on Earth. Muons are produced by the decay of pions. Muons were once known as mu-mesons, but they are not mesons.

Mural circle An instrument used before the invention of the telescope to measure the angle above the horizon of astronomical objects as they crossed the meridian. It consisted of a large arc engraved with an angular scale, mounted vertically on a north-south wall. If the arc covered 90° (a quarter-circle) it was known as a mural quadrant. A sighting device allowed the altitude above the horizon of a chosen star to be measured.

Musca A small constellation of the southern sky, representing a fly. Its brightest star is Alpha Muscae, magnitude 2.7. Theta Muscae is a double star, magnitudes 5.7 and 7.3, the companion being the secondbrightest Wolf-Rayet star.

Muses-CN MUSES-CN is a Japanese nanorover, a small, book-sized rover designed to be sent to explore the Near-Earth asteroid 4660 Nereus. MUSES stands for the Mu Space Engineering Spacecraft; C indicates that it is third in a series, N is for NASA. It was launched from Japan in Jan. 2002; it is the smallest rover ever to fly on a space mission. It will analyse rock samples, send color video to Earth, collect rock samples, and will eventually return to Earth.

Muttnik Muttnik is a term that many people used for the dogs who were sent into space on Sputnik missions. Between 1958 and 1961, the USSR (now Russia) sent 13 dogs into space in preparation for future missions. Laika (meaning "barker" in Russian) was the first dog in space; she was launched in Sputnik 2 on November 3, 1957.

N

N galaxy A type of galaxy with a bright, star-like nucleus that has a strong emission-line spectrum, surrounded by a faint nebulous envelope. They have active galactic nuclei, and may represent lower-luminosity versions of quasars. The designation N galaxy comes from the Morgan classification.

N magnitude The magnitude of a star as measured through the infrared N filter of the Johnson photometry system. It has an effective wavelength of 10.4 μm and a bandwidth of 5.3 μm.

Nadir The point on the celestial sphere directly beneath an observer. It is the direction in which a plumb-line points. The nadir is 180° from the zenith.

Nagler eyepiece An ultra-wide-field eyepiece designed in 1980 by the American optician Albert Hirsch Nagler.Characteristically it is a complex sevenelement design (the Nagler-2 has eight elements) that produces an 82 apparent field of view while correcting for spherical and chromatic aberration, coma, astigmatism, and field curvature. Its eye relief is very generous, and the performance at the edge of field is excellent. The design is optimized to work with focal ratios as fast as f/4.

Naiad The innermost satellite of Neptune, orbiting outside the planet's Galle Ring every 0.294 days at a distance of 48 230 km; also known as Neptune III. Its diameter is about 58 km. Naiad was discovered in 1989 by the Voyager 2 spacecraft.

Naked eye The eye used without any optical aid other than normal spectacles or contact lenses. The naked-eye limiting magnitude is that of the faintest star visible to the eye alone.

Naked singularity A singularity that is not hidden by an event horizon. A singularity is a point of infinite density that theories predict should exist at the centre of a black hole. If the black hole is rotating then the area of the surrounding event horizon can be reduced to zero, leaving the singularity exposed to view. Some theorists have invoked the concept of cosmic censorship, which requires that all singularities be hidden behind event horizons at all times. If this is true, some unknown process must be operating in very rapidly rotating black holes to retain the event horizon.

Nakhlite A rare type of achondrite meteorite, named after the meteorite that fell at Nakhla, Egypt, in 1911, the first known fall of this type. The nakhlites consist of approximately 80% by weight of the mineral augite, and about 14% by weight of iron-rich olivine. Their textures suggest they formed within cooling magmas. Nakhlites belong to the class of SNC meteorites. In common with the Chassigny meteorite, the nakhlites have a formation age of 1.3 billion years and an exposure age of about 12 million years.

NaI detector An instrument for detecting high-energy X-rays and gamma rays. Crystalline sodium iodide (NaI) converts an X-ray or gamma-ray photon into a pulse of visible light, the brightness of which is proportional to the energy of the incoming photon. The optical pulse is detected by a photomultiplier tube. To obtain images, a two-dimensional array of photomultipliers is used.

Nançay radio astronomy observatory The radio astronomy station of Paris Observatory, founded in 1953, 200 km south of Paris. Its main instrument is the Large Decimetric Radio Telescope, opened in 1965, which consists of a fixed antenna 300 m long and 35 m high, shaped like a portion of a sphere, into which radio waves are reflected by a tiltable flat antenna 200 m long and 40 m wide. A radioheliograph, opened in 1982, consists of a T-shaped array with arms 3.2 and 1.25 km long. A Decametric Array was constructed in 1975-8 for detecting radio bursts from Jupiter and the Sun.

Nanometer A nanometer is unit of measure equal to 10 angstroms, which is one billionth (1×10^{-9}) of a meter.

Nanorover A nanorover (also called small science vehicle or ssv) is a small rover designed to be sent to explore small celestial bodies, like asteroids. The book-sized Japanese nanorover MUSES-CN (MUSES stands for the Mu Space Engineering Spacecraft) explored asteroid 4660 Nereus in Jan. 2002 and arrivied on Nereus on September 9, 2003; it is the smallest rover ever to fly on a space mission. It analyse rock samples, send color video to Earth, collect rock samples, and will eventually return to Earth.

Naos The star Zeta Puppis. It is an O5 supergiant, one of the hottest stars known (surface temperature 40 000 K), of magnitude 2.3. Its estimated distance is 2000 L.Y, and it is as luminous as 45 000 Suns.

Narrow-band photometry Photometry using filters with a bandwidth of 3-10 nm. For wavelength ranges narrower than 3 nm a spectrophotometer must be used, since it is difficult to make filters with such a narrow bandwidth. Narrowband systems are mostly used when it is necessary to isolate an individual spectroscopic line or molecular band.

NASA National Aeronautics and Space Administration.

Nasmyth focus A focal position for a reflecting telescope, to which light is reflected along the hollow declination axis or altitude axis of its mounting. At the Nasmyth focus, the image position remains stationary and heavy equipment can be used without affecting the balance of the telescope. The configuration was devised by the Scottish engineer James Nasmyth (1808-90).

National aeronautics and space administration A US government agency founded in 1958 for civil aeronautical research and space exploration, superseding the National Advisory Committee for Aeronautics.NASA headquarters are at Washington, DC. NASA operates several installations: Ames Research Center; Dryden Flight Research Facility at Edwards, California, used for flight testing and as a landing site for the Space Shuttle; Goddard Space Flight Center; Jet Propulsion Laboratory; Johnson Space Center; Kennedy Space Center; Langley Research Center at Hampton, Virginia, which carries out research in aeronautics and space technology; Lewis Research Center at Cleveland, Ohio, concerned with aircraft and rocket propulsion; Marshall Space Flight Center; Michoud Assembly Facility near New Orleans, Louisiana, which makes Space Shuttle external fuel tanks; the

Space Telescope Science Institute; Stennis Space Center, near Bay St Louis, Mississippi, for testing rocket engines; and Wallops Flight Facility.

National optical astronomy observatories A group of optical observatories operated by the Association of Universities for Research in Astronomy on behalf of the US National Science Foundation. NOAO was founded in 1984 and currently has four divisions: Cerro Tololo Inter-American Observatory, Chile; Kitt Peak National Observatory, Arizona; National Solar Observatory, with observing facilities at Kitt Peak and Sacramento Peak; and the US Gemini Telescopes office. NOAO's headquarters are in Tucson, Arizona.

National radio astronomy observatory The collective title for the government-owned radio astronomy facilities at various sites in the USA. NRAO is administered by Associated Universities, Inc., a consortium of nine universities, on behalf of the US National Science Foundation. NRAO's oldest site is at Green Bank, West Virginia, founded in 1937. Its main instruments there are a 43-m equatorially mounted dish, opened in 1965, the 100 x 110-m Green Bank Telescope, due for completion in 1998, and a 13.7-m dish for space VLBI, first used in 1997 with the VSOP satellite. NRAO has a 12-m millimetre-wave dish at Kitt Peak, Arizona, opened in 1967 and upgraded in 1983. NRAO also operates the Very Large Array and the Very Long Baseline Array. NRAO's headquarters are in Charlottesville, Virginia.

National solar observatory A division of the US National Optical Astronomy Observatories, founded in 1984. NSO operates solar telescopes at two main sites: Kitt Peak, Arizona, and Sacramento Peak Observatory, New Mexico. It also manages the Global Oscillation Network Group.The NSO instruments at Kitt Peak are the McMath-Pierce Solar Telescope, and the Solar Vacuum Telescope, 23 m high and with a 0.7-m main mirror, opened in 1973.

National space science data center A department of NASA's Goddard Space Flight Center, founded in 1966, that is the repository for data from US space missions, including images from planetary probes. Data from large-scale ground-based sky surveys are also held there. The associated Astronomical Data Center at Goddard maintains an archive of catalogues produced from ground-based observations.

Nautical almanac The A yearly publication containing tables of the Sun, Moon, planets, and stars, plus other data for marine navigation, prepared and issued jointly by the US Naval Observatory and the Royal Greenwich Observatory. Founded by N. Maskelyne, it was first published in Britain in 1766 under the title The Nautical Almanac and Astronomical Ephemeris. In 1914 it was divided into two versions, one for astronomers and an abridged version for navigators; the astronomical version is now known as The Astronomical Almanac. In 1960 The Nautical Almanac merged with The American Nautical Almanac, published separately in the USA since 1914.Related publications are The Air Almanac for air navigators and The Star Almanac for land surveyors.

Nautical twilight The period before sunrise and after sunset when the centre of the Sun's disk is between 6° and 12° below the horizon. It begins when the brightest stars become visible, and is the time during which the sea horizon remains visible for observing altitudes of celestial objects.

Navigation satellites Satellites that help aircrafts and ships to identify their exact locations.

N-body problem The mathematical problem of finding the velocities and positions of any number (n) of objects moving under their mutual gravitational attraction for any time in the past or future; also known as the many-body problem. It applies, for example, to the members of the Solar System and the members of a star cluster. As with the three-body problem, no general mathematical solution holding for all time has ever been found, but certain general results are known, true for any number of bodies. The centre of mass of the system of particles travels with constant velocity; the total energy of the system is constant; and the total angular momentum of the system does not change. With modern electronic computers, the bodies' velocities and positions can be calculated to any desired accuracy for finite lengths of time into the past or future but, for reasons such as round-off error and the chaotic nature of the problem, the accuracy deteriorates as the time increases.

Nea NEA stands for Near-Earth Asteroid. NEA's are asteroids whose orbits bring them within 1.3 AU (121 million miles/195 million kilometers) of the Sun are called Near-Earth Asteroids (NEA) or Earth-Approaching asteroids. These asteroids probably came from the main asteroid belt, but were jolted from the belt by collisions or interactions with other planets' gravitational fields. 433 Eros, pictured above, is a NEA.

Neap tide The tide raised in the Earth's oceans when the Moon is at either first or third quarter phase, and is pulling on the Earth at right angles to the Sun. The amplitude of neap tides (i.e. the difference in water level between high and low tide) is the smallest in the monthly tidal cycle.

NEAR (near earth asteroid rendezvous) U. S. Discovery-class mission launched on February 17, 1996. Primary scientific objectives were to return data on the bulk properties, composition, mineralogy, morphology, internal mass distribution, and magnetic field of 433 Eros as the spacecraft orbited the asteroid from February 14, 2000 to February 12, 2001. On February 12, 2001, the NEAR-Shoemaker spacecraft touched down on 433 Eros. Earlier in the mission, on June 27, 1997, NEAR-Shoemaker flew within 1200 km of the C-class asteroid 253 Mathilde, and it flew by the Earth on January 23, 1998.

Near infrared That part of the electromagnetic spectrum covering the shorter infrared wavelengths, but with no well-defined limit at the longer-wavelength end. Broadly, the near infrared extends from the red end of the visible spectrum (approximately 0.7 μm, or 700 nm) to about 35 μm. In practice, the term is often used to describe the infrared windows in the atmosphere between 1 and 5 μm. The 0.7-1.0 μm region is sometimes known as the photographic infrared.

Near project NEAR (Near-Earth Asteroid Rendezvous) is a spacecraft designed to fly by and examine near-earth asteroids. It was launched on February 17, 1996; its goal was to rendezvous with asteroid 433 Eros (pictured above). In 2001, NASA renamed the spacecraft NEAR Shoemaker to honor the planetary science pioneer Dr. Eugene M. Shoemaker, who died in 1997. It orbited Eros from Feb. 14, 2000, until Feb. 28, 2001, when it landed on Eros. After orbiting the asteroid Eros (433) for 3 years, NEAR Shoemaker landed on Eros on Feb 12, 2001 (this was the first spacecraft to land on an asteroid). After transmitting radio data, the last contact with NEAR Shoemaker was on Feb. 28, 2001 (it is powered by solar power).

Near ultraviolet The part of the ultraviolet spectrum adjacent to the visible band. It covers the wavelength range from approximately 200 to 350 μm.

Near-earth asteroid Any asteroid belonging to the Apollo, Amor, or Aten groups. Such asteroids have perihelion distances of less than 1.3 AU. The term near-Earth objects is also used, allowing for the fact that some may be extinct short-period comets. Members of the Amor group cross the orbit of Mars but do not quite reach the Earth's orbit, while the Apollo and Aten groups do cross the Earth's orbit. The three groups are not completely separate, since planetary perturbations may cause an Amor asteroid to become an Apollo, and vice versa. Atens are the least common NEAS, while Apollos and Amors are found in roughly equal numbers. Only three NEAs are larger than 10 km in diameter: (1036) Ganymed, (433) Eros, and (4954) Eric, all members of the Amor group. Near-Earth asteroids have finite lifetimes (typically 10 million years); most will be destroyed by collision with one of the inner planets, and the remainder ejected from the Solar System.

Near-earth asteroid rendezvous A NASA space probe launched in 1996 February to the asteroid Eros. On the way it passed and photographed (253) Mathilde, a C-class asteroid, in 1997 June, and will undergo an Earth gravity-assist in 1998 January. It was the first of NASA's Discovery class of low-cost space missions.

Near-infrared Light from the part of the infrared band of the electromagnetic spectrum closest to the visible range.

Nearside The side of the Moon that always faces Earth.

Nebula A cloud of gas and dust in space. The term was originally applied to any object with a fuzzy telescopic appearance, but with the advent of larger instruments many 'nebulae' were found to consist of faint stars. In 1864 W. Huggins discovered that true nebulae could be distinguished from those composed of stars on the basis of their spectra. Nowadays the term 'nebula' means a gaseous nebula. The term extragalactic nebula, originally used for galaxies, is now obsolete. There are three broad types of gaseous nebula: emission nebulae, which shine by their own light; reflection nebulae, which reflect light from nearby bright sources such as stars; and dark nebulae which appear dark against a brighter background. This broad classification scheme has been carried over to other wavelengths, giving rise to terms such as infrared reflection nebula. Emission nebulae include the diffuse nebulae or H II regions around young stars, planetary nebulae around old stars, and supernova remnants.

Nebula, dark A dense cloud of gas and dust in space that appears as a dark silhouette against a brighter background. In many cases dark nebulae are part of a larger nebulae complex which provides the light background against which the dark nebulae appears. At infra-red wavelengths young stars not visible at visual wavelengths are often found to be forming. Dark nebulae sometimes appears as a dark feature against a starry background but without any other associated nebulae.

Nebula, emission A gas cloud which is close to a star or stars. The radiation from the stars excites the gas so that it emits light of its own. Most of this gas is hydrogen which emits the red/orange light which is typical of this type of nebulae.

Nebula filter A filter that transmits only selected visual spectral lines from emission nebulae. A typical filter transmits a band 25 nm wide, including the O III (495.9 and 500.7 nm) and Hß (485.6 nm) lines. Narrow-band filters transmit bands just 10 nm wide, isolating either O III or Hß.

Nebula, reflection A cloud of gas and dust which lies close to a star or stars which reflects light from those stars. This type of nebulae is often part of a larger nebulae complex (although it is sometimes seen without the other nebulae types above) and the reflection is typically blue in colour due to the dust within the nebulae.

Nebular line An emission line that is found in the spectra of diffuse nebulae, notably the green lines due to doubly ionized oxygen, [O III], at 495.9 and 500.7 nm. Nebular lines are also forbidden lines, since they do not occur under conditions on Earth.

Nebular variable Any of the various types of eruptive variable intimately associated with nebulosity; also known as an Orion variable. Most are young stars of the FU Orionis or T Tauri types, together with the probably related UVN subtype of UV Ceti flare star.

Nebulosity A general term for any fuzzy area in the sky, usually a cloud of gas and dust. The use of the term may imply that the gas cloud is part of a larger nebula, is associated with a more readily resolvable star cluster, or that its form does not allow a more exact classification as a particular type of nebula.

Negative eyepiece An eyepiece whose focal point lies within it and which cannot, therefore, be used as a magnifying glass. The Huygenian eyepiece is an example.

Negative lens Another name for a diverging lens.

Nemesis Nemesis is a hypothetical companion dark star to our Sun. Once every 30 million years, this dark star would pass through the Oort cloud, triggering comets that perhaps cause periodic mass extinctions on Earth.

Neneutron star A neutron star is a very small, super-dense star which is composed mostly of tightly-packed neutrons. This hard-to-see body has a thin atmosphere of superhot hydrogen plasma and a crust. It has a diameter of about 5-10 miles (5-16 km) and a density of roughly 10 15 gm/cm^3. Neutron stars are formed from supernova explosions.

Neptune's Atmosphere Neptune's Atmosphere has dark belts or bands with bright zones between, similar to those on Jupiter and Saturn but fewer in number and less prominent. Dark and bright spots also occur, the most prominent being the Great Dark Spot discovered in 1989 by Voyager 2, similar to Jupiter's Great Red Spot but not as long-lived. There are also bright wispy clouds of methane ice resembling cirrus clouds on Earth, 50-100 km higher than Neptune's main cloud tops. One of them in the southern hemisphere, photographed by Voyager 2, had a much higher velocity than other clouds and was dubbed the Scooter.

Nereid Neptune's outermost moon is Nereid. It is the smaller of Neptune's 2 larger moons. It has a VERY elliptical orbit, going as close as 867,000 miles from Neptune and as far as 6 million miles from it; it may be a captured asteroid. It takes almost one Earth year for Nereid to orbit Neptune in this extreme orbit. Nereid's diameter is about 200 miles (320 km). It was discovered by G. Kuiper in 1949.

Nereus 4660 Nereus is a Near-Earth asteroid that is about 0.6 to 1.25 miles (1-2 km) in diameter. It has an orbital period of 1.82 yrs and a semimajor axis of 1.490 AU. Nereus may be an extinct comet.

Neumann lines A rectangular pattern of fine striations appearing when a slice of hexahedrite iron meteorite is cut, polished, and etched with dilute nitric acid; also called Neumann bands. The pattern is aligned along the faces of a cube. Neumann lines are produced by compressional shock waves. Although most obvious in hexahedrites, Neumann lines are present in other iron meteorites as well. They are named after their discoverer, the German mineralogist Franz Ernst Neumann.

Neutral atom An atom with an equal number of protons and electrons, hence no net electric charge

Neutral hydrogen Hydrogen gas made up of electrically neutral atoms.

Neutral line The line that separates longitudinal magnetic fields of opposite polarity.

Neutral point A point at which a particle would experience no net gravitational force. A neutral point exists between two static massive bodies. Its position depends on the bodies' separation and their masses, being closer to the more massive body. If the bodies are orbiting each other, as in reality, the problem is more complicated, and there exist five points at which the total acceleration due to gravitational and centrifugal forces is zero.

Neutrino An elementary particle with zero charge and a very small rest mass. Neutrinos travel at a very high velocity which, if the rest mass is zero, equals the velocity of light. Three types of neutrino are known: the muon neutrino, the electron neutrino, and the tau neutrino. Neutrinos have only a weak interaction with matter and consequently neutrinos produced in nuclear reactions at the centres of stars can escape without colliding with the overlying material. Some of the missing mass of the Universe may be accounted for if neutrinos have non-zero rest mass. Neutrinos are leptons.

Neutrino astronomy The observation of neutrinos emitted by celestial objects. Neutrinos pass through large quantities of matter without significant absorption. For example, a neutrino produced by the nuclear processes in the Sun's core has only one chance in 10^{10} of being absorbed in escaping from the Sun. However, observations have detected only about one-third the number of neutrinos from the Sun predicted by theory.Bursts of neutrinos are predicted to be produced in supernova explosions, and such a burst was detected from Supernova 1987A. Neutrinos can be detected in several ways. One uses the neutrino's interaction with the chlorine isotope37 Cl, which produces radioactive argon,37 Ar. Detectors have also been built which utilize the conversion of gallium to germanium by a neutrino.Neutrino telescopes detect the direction from which the neutrino comes, as well as its existence. These rely on the neutrino colliding with an electron inside a large tank of water.

Neutron degeneracy The state of degeneracy attained when the density of matter is so high that neutrons cannot be packed any more closely together. The effect is analogous with that of electron degeneracy, but the density required for neutron degeneracy (about 10^{17} kg/m^3) is much higher, and is encountered only in neutron stars.

Neutron star An extremely small, superdense object believed to be formed when a massive star undergoes a Type II supernova explosion. During the explosion, the core of the massive star collapses under its own gravity until, at a density of about 10^{17} kg/m^3, electrons and protons are so closely packed that they combine to form neutrons. The resultant object, consisting only of neutrons, is supported against further gravitational collapse by the degeneracy pressure of the neutrons, provided its mass is not greater than about 2 solar masses.If the object were more massive it would collapse further into a black hole. A typical neutron star, with a mass a little greater than the Sun's, would have a diameter of only about 30 km, and a density such that the mass of the entire human race would occupy the volume of a sugar cube. The greater the mass of a neutron star, the smaller its diameter. Neutron stars are believed to have an interior of superfluid neutrons (i.e. neutrons behaving like a fluid of zero viscosity)

New general catalog The New General Catalog (NGC) is a list of over 13,000 deep-sky celestial objects. It was developed in 1888. For example, the Great Nebula in Orion is NCG 1976 (and M42).

New moon The phase in which the moon is in the same direction as the sun in Earth's sky, so it is unilluminated and invisible

Newcomb, Simon Canadian-born American mathematical astronomer. At the US Naval Observatory's Nautical Almanac office, of which he became head in 1877, he initiated an extensive project to refine the orbits of the Moon and planets. The project involved the use of historical data, from which he discovered the nonNewtonian component of the advance of the perihelion of Mercury's orbit that was later accounted for by the general theory of relativity. Newcomb improved.

Newstar Netherlands east- west synthesis array reduction A software package for the reduction of radio interferometric data. NEWSTAR is designed and optimized for WSRT data, but can also handle data from other East-West arrays.

Newton, Isaac Isaac Newton (1642-1727) was an English mathematician and physicist who invented calculus (simultaneously, but independently of Leibniz), formulated the laws of gravitation, investigated the nature of light (he discovered that sunlight is made of light of different colors), and the laws of motion: 1. An object in uniform motion tends to remain in that state of motion unless an external force is applied to it (the Law of Inertia).

2. A force causes a change in the velocity (acceleration) of an object (F=ma). **3.** For every action there is an equal and opposite reaction. Newton also improved the design of the refracting telescope (using an objective mirror, instead of a lens), and it is now called a Newtonian telescope.

Newtonian focus A focal position for a reflecting telescope at the side of its tube, at right angles to the incoming light. Light reflected from the main mirror is diverted to the Newtonian focus by a flat diagonal mirror, or a prism, higher up inside the tube. In a standard Newtonian telescope, the focal ratio and focal length are the same for the Newtonian focus as for the prime focus.

Newtonian reflector A reflecting telescope in which the light is collected by a concave primary mirror at the rear of the telescope and then reflected back to a flat secondary mirror angled at 45 degrees mounted near the front of the telescope. The light is then reflected through the side of the telescope tube and on into an eyepiece. It is the simplest type of reflecting telescope.

Newtonian telescope A reflecting telescope in which light is collected by a paraboloidal primary mirror and then reflected by a secondary mirror or prism within the tube to a focal point at the side of the tube, at right angles to the direction of view of the telescope. The image is inverted but not laterally reversed. The focal ratio should be larger than $f/3$, otherwise coma becomes excessive. The design was invented in 1668 by I. Newton.

Newton's law of gravitation The law put forward by I. Newton in 1687 which states that any two bodies attract each other with a force that depends on the product of their masses divided by the square of the distance between them. Put mathematically, the gravitational force of attraction, F, between two masses m_1 and m_2, separated by distance r, is given by $F = Gm_1m_2/r^2$, where G is a constant known as the gravitational constant..

Newton's laws of motion Three laws published in 1687 by I. Newton concerning the motion of bodies: 1. A body continues in a state of uniform rest or motion unless acted upon by an external force.

2. The acceleration produced when a force acts is directly proportional to the force and takes place in the direction in which the force acts.

3. To every action there is an equal and opposite reaction.

NFRA(Netherlands foundation for research in astronomy) A national facility supported by the Netherlands Organization for Scientific Research (NWO). It operates the WSRT and Dwingeloo radio telescopes. It participates in operating the JCMT and UKIRT facilities in Hawaii, and several optical telescopes on La Palma

NGC The **NGC (New General Catalog)** is a list of over 13,000 deep-sky celestial objects. It was developed in 1888. For example, the Great Nebula in Orion is NCG 1976 (and M42). NGC4414 (pictured above) is a spiral galaxy 60 million light-years away.

Nicholson, Seth Barnes American astronomer; discovered Lysithea, Ananke, Carme and

Sinope; also did important work on sunspots.

Nicol prism A device for producing plane-polarized light. It consists of a rhombus-shaped prism of calcite crystal, cut diagonally and recemented with Canada balsam, a resin with similar optical properties to glass. Calcite naturally splits light into two separate beams, polarized at right angles to each other. The layer of Canada balsam within the prism cuts out one beam through total internal reflection. The advantage of a Nicol prism over Polaroid material as a polarizer is that all wavelengths of light are polarized and transmitted equally. It was devised by the Scottish geologist and physicist William Nicol.

Night sky brightness The very low background level of sky illumination present even in the absence of artificial or other light pollution. It results from atmospheric airglow emissions and, to a lesser extent, from starlight and the zodiacal light.

Night vision An alternative name for dark adaptation.

Nightglow Another name for airglow.

NIMS Near-Infrared Mapping Spectrometer An instrument carried to Jupiter on the Galileo planetary spacecraft, imaging in regions of the energy spectrum that humans cannot see.

Nitrogen An element with atomic number 7; symbol: N. It is common in Earth's atmosphere and along with hydrogen, carbon, and oxygen is essential for life.

Nix olympica The name by which Olympus Mons was known before 1973.

No boundary condition The idea that the universe is finite but has no boundary (in imaginary time).

Noachian epoch The geologic history of Mars has been divided into three broad time periods, or Epochs. From oldest to youngest, these are Noachian, Hesperian, and Amazonian (named after places on Mars). These Epochs are defined by the number of impact craters on the ground surface; older surfaces have more craters. The Noachian Epoch extends from the birth of Mars to between 3.8 and 3.5 billion years ago. The actual timing of the Epochs is not known because we have different models of the rates of meteorite falls on Mars.

NOAO Abbr. for National Optical Astronomy Observatories.

Nobeyama Radio Observatory The radio observatory of the National Astronomical Observatory of Japan, 120 km west of Tokyo at an altitude of 1350 m, founded in 1978. Its main instruments are a 45-m millimetre-wave dish opened in 1982, and the Nobeyama Millimetre Array, an aperture-synthesis array of six 10-m antennas movable along two intersecting 500-m baselines. The millimetre array was opened in 1986 with five dishes; the sixth dish was added in 1993. Also at the same site is the Nobeyama Radioheliograph, which consists of 84 dishes of 0.8-m aperture in a T-shaped array, opened in 1992.

Noble gas One of the inert gases helium, neon, argon, krypton, xenon, and radon. Each has a stable configuration of electrons in its atoms that renders it unreactive. Apart from argon, only traces of the noble gases are present in the Earth's atmosphere.

Noctilucent clouds Literally, 'night-shining' blue-silver clouds seen in late or early twilight during the summer from high temperate latitudes. Noctilucent clouds result from water vapour condensing on nuclei at altitudes of around 82-85 km, close to the mesopause. The clouds are extremely tenuous, and are seen only when the Sun lies between 6° and 16° below the horizon, appearing bright by contrast with the twilight sky. Noctilucent clouds lie five times higher than the loftiest clouds in the troposphere, and are distinct from cirrus clouds to which they bear a superficial resemblance. In particular, noctilucent clouds may be identified by their delicate, feathery, interwoven herring-bone structure.

Nocturnal A historical hand-held device for telling the time at night using the orientation of northern circumpolar constellations. It consisted of a dial marked with the hours, with indicators to be aligned with specific stars, notably the Pointers of Ursa Major. When oriented with the stars and with Polaris in the centre, the time was shown against the appropriate date.

Nodal line The straight line joining the nodes of an orbit.

Node Either of the two points at which an orbit intersects a reference plane such as the plane of the ecliptic or of the celestial equator. The line joining these points is known as the line of nodes or nodal line. At the ascending node an orbiting body moves from south to north of the reference plane; at the descending node it moves from north to south..

Nodical month An alternative name for the draconic month.

Noise The random fluctuations that are always associated with a measurement that is repeated many times over. Noise appears in astronomical images as fluctuations in the image background. These fluctuations do not represent any real sources of light in the sky, but rather are caused by the imperfections of the telescope. If the noise is too high, it may obscure the dimmest objects within the field of view.

Non-baryonic matter A hypothetical form of matter not containing baryons-that is, without protons or neutrons. An example would be the positron-electron 'atoms' that may constitute much of the Universe in the very distant future if protons decay. Non-baryonic matter has been suggested as a possible component of the missing mass of the Universe. In this case it could take the form of neutrinos, if they have non-zero rest mass, or of hypothetical particles called WIMPS.

Non-gravitational force An effect that accelerates or decelerates a comet's motion, changing its orbital period. Such forces are caused by jets of gas emerging from active regions on the surface of the nucleus, giving a rocket-like effect. Nongravitational forces are most marked when the nucleus is highly active, close to perihelion, leading to uncertainties in subsequent returns for some periodic comets. They are responsible for systematic changes in the perihelion time of Comet Encke, and the apparently delayed 1992 return of Comet Swift-Tuttle.

Non-radial pulsation A form of pulsation in which waves run in all directions across the star's surface, as opposed to spherically symmetrical expansion and contraction (radial pulsation). There are normally multiple periods, resulting in a complex pattern of nodes and antinodes over the surface. Non-radial pulsations are particularly prominent in white-dwarf ZZ Ceti stars.

Non-thermal emission Electromagnetic radiation produced by synchrotron radiation, maser line emissions from atoms and molecules, or the mechanisms not related to temperature.

Non-thermal radiation Radiation due to a cause other than the temperature of the emitting body. It has a different spectrum from that of black-body radiation. Examples of non-thermal radiation are synchrotron radiation, maser radiation, and artificially generated radio and TV signals.

Nordic optical telescope A 2.5-m reflector opened in 1989 at the n Roque de los Muchachos Observatory on La Palma in the Canary Islands, which uses a thin mirror figured to high accuracy. It is owned and operated by a consortium of research councils of Denmark, Finland, Norway, and Sweden.

Norite Igneous rock found in the lunar highlands composed of plagioclase and pyroxene.

Norma A constellation of the southern sky, representing a surveyor's level. Its brightest star, Gamma2 Normae, is of magnitude 4.0 and forms an optical double with the unrelated Gamma1 Normae, magnitude 5.0. NGC 6087 is a 5th-magnitude open star cluster.

Normal astrograph A photographic refractor with aperture 0.33 m and focal length 3.438 m, giving a plate scale of 0.006 to 1 mm. The usable field is about 2° × 2°. This type of telescope was designed in Paris by the brothers Paul Pierre Henry and Prosper Matthieu Henry in about 1880. Many such telescopes were constructed and were used for the Astrographic Catalogue and Carte du Ciel.

North America nebula A diffuse nebula in Cygnus, also known as NGC 7000, shaped like the continent of North America. Its dimensions are about 2° × 1½° and it lies 1500 L.Y away, similar to the distance of the bright star Deneb. However, its illuminating star is thought not to be Deneb but a 6th-magnitude hot, blue star within the nebula, HR 8023. Next to the North America Nebula lies the Pelican Nebula, actually part of the same enormous cloud which stretches for 100 L.Y

North celestial pole The point in the sky to which Earth's Geographical North Pole points

North galactic spur An arc of radio emission about 80° long, projecting from the Milky Way towards the north galactic pole; also known as the North Polar Spur. It is believed to be part of the expanding shell of an ancient supernova remnant a few hundred light years from the Sun.

North polar distance The angular distance of an object from the north celestial pole, measured along the hour circle passing through the object.

North polar sequence A series of stars within 2° of the north celestial pole that were used to provide standard magnitudes and an arbitrary zero point for the magnitude scale. The original sequence included 96 stars with photographic magnitudes down to 20.1; it was extended by a supplementary list of 56 other stars. A North Polar Sequence of Photovisual Magnitudes down to 17th magnitude was subsequently introduced. The sequences have been superseded by the use of UBV photometry to establish the zero point, used in conjunction with lists of accurately measured standard stars.

North star The north star is a star that is located almost due north and is useful for navigation. Polaris is currently the pole star of the Northern Hemisphere.

Northern coalsack Another name for part of the dark nebula known as the Great Rift.

Northern cross A popular name for the constellation Cygnus.

Northern lights Northern lights is another name for auroras that appear in the Northern Hemisphere. Auroras are beautiful undulating sheets of light in the near-polar sky. They are caused by gases that become excited after being hit by solar particles. Most auroras are 100 to 250 km above the ground.

Nova A type of cataclysmic variable that exhibits a sudden, unpredictable outburst, with a typical amplitude of 11-12 mag.; abbr. N. Novae are interacting binaries, and usually consist of a main-sequence or slightly evolved secondary and a white-dwarf primary. Mass transfer occurs from the secondary to an accretion disk and thence to the white dwarf. The accumulation of hydrogen-rich gas on the white dwarf's surface eventually leads to a thermonuclear runaway producing the outburst and ejecting much of the outer envelope from the system.

Novae are divided into three subtypes. Based on their speed of decline from maximum fast declining by 3 magnitudes in less than 100 days; slow 3 magnitudes in 150 days or more; very slow persisting at maximum for years. Very slow novae probably have a giant or supergiant secondary, and are sometimes referred to as symbiotic novae or RR Telescopii stars.

Nova-like variable A heterogeneous type of poorly studied cataclysmic binaries; abbr. NL. There are either small-amplitude outbursts or minor fluctuations, which together with the spectral details resemble those of a post-nova at minimum. Closer study usually reveals features that enable a nova-like object to be more firmly classified as a post-nova, a dwarf nova, a UX Ursae Majoris star, or some other related group.

NRAO (national radio astronomy observatory) The U.S. national radio astronomy facility, managed by Associated Universities Inc., a university consortium, under a co-operative agreement with the National Science Foundation. The NRAO operates the GBT, VLA, VLBA and several smaller radio telescopes.

NSFNet (national science foundation network) A high speed hierarchical "network of networks" in the U.S., funded by the National Science Foundation. At the highest level, it is a *backbone* network comprising 21 nodes connected to a 45Mb/s facility spanning the continental United States. Attached to that are mid-level networks and attached to the mid-levels are campus and local networks. NSFNET also connects out of the U.S. to Canada, Mexico, Europe, and the Pacific Rim. The NSFNET is part of the Internet.

Nubecula major, nubecula minor Other names for the Magellanic Clouds.

Nuclear bulge The nuclear bulge is the central, spherical part of a spiral galaxy. It is surrounded by a disk-shaped mass of stars with spiral arms.

Nuclear fusion Nuclear fusion is an atomic reaction in which many nuclei (the centers of atoms) combine together to make a larger one (which is a different element). The result of this process is the release of a lot of energy (the resultant nucleus is smaller in mass than the sum of the ones that made it; the difference in mass is converted into energy by the equation $E=mc^2$). Suns are powered by nuclear fusion, mostly converting hydrogen into helium. In this process, 4 hydrogen nuclei combine to form a single helium nucleus; about 0.3% of the original mass is converted into energy.

Nuclear reaction A reaction involving one or more nuclei of atoms in which there is a change to the nuclei. Nuclear reactions can release enormous amounts of energy; examples are nucleosynthesis, fission, fusion, and radioactive decay. However, other nuclear reactions require an input of energy to build up elements heavier than iron, such as the r-process and s-process.

Nuclear time-scale The time taken for a star to exhaust its supply of a particular nuclear fuel. The nuclear time-scale may be very short, as with the oxygen-burning phase of some massive stars, which can be less than a year, or very long, as with the hydrogen-burning phase of low-mass stars, which is potentially longer than the age of the Universe. The nuclear time-scale for hydrogen burning in the Sun is about 7 billion years, but becomes shorter with increasing stellar mass because the larger amount of fuel in massive stars is more than offset by the faster rate of burning.

Nucleon A proton or a neutron present in the nucleus of an atom.

Nucleosynthesis The process of creating elements by nuclear reactions. Helium was produced by nucleosynthesis in the few minutes following the Big Bang. Helium and heavier elements are built up by nucleosynthesis inside stars. First, hydrogen is converted to helium by the proton-proton reaction or the carbon-nitrogen cycle. When the hydrogen-to-helium phase ends, the triplealpha process takes over. Successively heavier elements up to iron are then synthesized, each in turn using the product of the previous reaction. If a star becomes a supernova, the heaviest possible nuclei are formed and are ejected into interstellar space. New generations of stars forming from the enriched medium have a higher heavy element content than old stars.

Nucleus, atomic The central core of an atom that contains most of its mass. The nuclei of atoms are positively charged and consist of one or more protons and neutrons. The number of protons in a nucleus defines the atomic number of the element. The simplest nucleus is that of the lightest isotope of hydrogen, consisting of one proton only. All other nuclei also contain one or more neutrons, which contribute to the mass but not the charge of the nucleus. The most massive nucleus that occurs in nature is 238U, containing 92 protons and 146 neutrons.

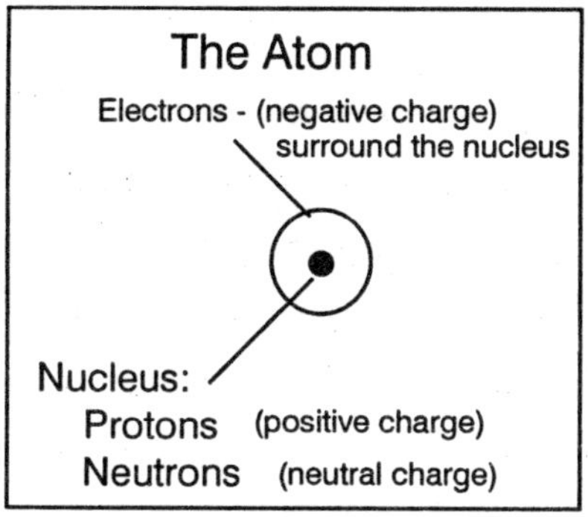

Fig. Nucleus

Nucleus, cometary The small solid body, composed of frozen water and gas plus embedded dusty material, at the centre of a comet's head. It is the source of activity in a comet, and contains essentially all its mass. The nucleus of Halley's Comet, so far the only directly observed example (by the Giotto space probe in 1986), is irregularly shaped, 16 × 8 km, with a dark crust; most nuclei are probably smaller. When very distant from the Sun, a comet's nucleus is inert. Within 3-4 AU, however, solar heating leads to gas sublimation, and production first of a coma, then a tail Near perihelion, surface temperatures may reach 25 °C. Cometary nuclei have low densities (typically 0.2 g/ cm^3), and are prone to fragmentation. Nuclei which have made several perihelion passages are thought to be covered by a dark crust, as Halley's Comet. Gas emerges through fissures in the crust, carrying off dust which pursues its own orbit as a meteor stream.

Nucleus, galactic The central region of a galaxy characterized by high densities of stars. The nucleus may also contain a supermassive black hole and may be the source of considerable high-energy, nonstellar luminosity.

Nuffield radio astronomy laboratories The radio astronomy observatory of the University of Manchester at Jodrell Bank in Cheshire, England, founded by A. C. B. Lovell in 1945. Its major instruments include the 76-m Lovell Telescope the 25 x 38-m elliptical Mark II telescope, built in 1964, and the MERLIN long-baseline interferometer.

Null geodesic A path or geodesic joining two events in space-time along which the separation is zero. The separation in space-time can be zero even if the two events actually occur at different points in space. This apparent contradiction can arise because in the formula for determining the separation of events in space-time, the time interval between the events has the opposite sign to that for their distance apart in space. Light in a vacuum travels along a null geodesic.

Nunki The star Sigma Sagittarii, magnitude 2.0. It is a B3 subgiant, 170 L.Y away.

Nutation A periodic oscillation of the Earth's pole about its mean position on the celestial sphere, due to the attractions of the Sun and Moon on the Earth's equatorial bulge. It causes a small periodic variation in the positions of stars, and is superimposed on the much larger effect of precession. The main oscillation has a period of 18.6 years, equal to the revolution period of the Moon's nodes around the ecliptic. In addition there are many smaller oscillations, some with periods as small as a few days. Nutation is conventionally resolved into nutation in longitude and nutation in obliquity, which are respectively parallel and perpendicular to the ecliptic.

Nutation in right ascension An alternative name for equation of the equinoxes.

Nyquist sampling rate The slowest rate of sampling which, according to the Shannon sampling theorem, would allow a band-limited function $f(t)$ to be recovered via the *Shannon series*. If the smallest symmetric interval which contains the support of the Fourier transform of f is the interval $[-a,a]$, then the Nyquist sampling rate for f is $2a$; i.e., the interval between samples (the *sampling period*) must be less than the *reciprocal bandwidth* $1/2a$. The terms *oversampling* and *undersampling* refer to sampling at rates faster or slower than the Nyquist rate. The difference between f and the Shannon series formed from too coarsely spaced samples is called aliasing.

Nysa family One of two subfamilies of asteroids at a mean distance of 2.42 AU from the Sun, the other being the Hertha family. They are close neighbours with a small but distinct difference in orbital inclination. Both consist of a single large asteroid (namely (44) Nysa, diameter 68 km, orbital inclination 3°.7, and (135) Hertha, diameter 80 km, inclination 2°.3) plus many small objects 20 km or less in diameter. Nysa is of E class and Hertha M class, while the minor members of the Nysa subfamily are of the rare F class. Some family members are very close to the Kirkwood gap at the 3 : 1 resonance with Jupiter at 2.50 AU, and could be thrown into the gap to be subsequently ejected into Earth-crossing orbits. This family may therefore be an important source of meteorites. Nysa itself was discovered in 1857 by the German astronomer Hermann Mayer Salomon Goldschmidt.Its orbit has a semimajor axis of 2.423 AU, period 3.77 years, perihelion 2.06 AU, aphelion 2.79 AU, and inclination 3°.7.

O

O magnitude The magnitude of a star measured with a filter with wavelength 11.5 μm and a bandwidth of 2 μm. This infrared magnitude is now little used because it has a wavelength close to the N filter of the Johnson photometry system, which admits more light.

O star A star of spectral type O, the brightest, hottest and most massive of all normal, hydrogen-burning stars on the main sequence, appearing blue in colour and emitting most of its energy in the ultraviolet. Their temperatures range from 30 000 K to above 50 000 K, and they have luminosities from 100 000 to more than a million times the Sun's. With masses of 20-100 solar masses or more, O stars burn their nuclear fuel at a prodigious rate and have lifetimes of only 3-6 million years. Because of their high temperatures, hydrogen lines in their spectra are weak, the dominant lines being those of singly ionized helium (the Pickering series). O stars are very rare, with only a handful of O3 and O4 stars known (no stars are classified as type O0-O2). The brightest naked-eye O stars are Delta and Zeta Orionis, both O9.5 supergiants. O stars are often found with B stars in OB associations

OAO 3 Orbiting Astronomical Observatory 3 - NASA ultraviolet/X-ray mission, also known as Copernicus.

OB association A group of dozens of young, massive, and hot stars of spectral types O and B found in the spiral arms of galaxies. OB associations can be tens or hundreds of light years across. The stars are moving apart at several kilometres per second, so associations disperse among the background stars within about 30 million years. Famous examples are the Orion Association and the Sco-Cen Association.

Oberon The outermost satellite of Uranus, distance 583 520 km, orbital period 13.463 days; also known as Uranus IV. It is the second-largest satellite of Uranus, 1524 km in diameter, and it keeps one face turned permanently towards the planet. Oberon was discovered in 1787 by F. W. Herschel, soon after the discovery of Uranus itself. Most of the surface photographed by the Voyager 2 probe is covered with impact craters.

Object glass The main lens of a refracting telescope, which brings light to a focus.

Object In object-oriented programming, a unique instance of a data structure (abstract data type) encapsulated with a set of routines, called methods which operate on the data. An object is defined according to the template provided by its class. Each object has its own values for the variables belonging to its class and can respond to the messages (methods) defined by its class.

Objective The main lens or mirror of a telescope, which brings light to a focus. The term is usually taken to mean the main lens of a refractor.

Objective grating A coarse grating or series of parallel rods placed over the front lens of a telescope to create fringes alongside star images. These fringes are short spectra, but if the spacing of the grating is very coarse (about 10 mm) they will effectively be so short that they are subsidiary star images at a known distance from a star. The grating can be rotated, enabling the position angle and separation of double stars to be measured. A fine grating may be used to create spectra, in the same manner as an objective prism.

Objective prism A wedge of glass placed over the aperture of a telescope, which disperses the light from each star in the field of view into a short spectrum. Lowresolution spectra of all the stars in the field of view can therefore be obtained in one exposure. In many cases, an objective prism is really a grism.

Oblate An oblate sphere is one that is flattened at its poles. Saturn is the most oblate planet in our Solar System; the difference in its equatorial and polar diameters is almost 10%.

Oblateness The measure of how much a rotating object deviates from being a perfect sphere. A perfect sphere would have an oblateness of 0.0 (0%). The Earth for example is oblate to a factor of 0.0034 (0.34%), while Saturn, the most oblate of the planets has a figure of 0.108 (10.8%). Generally, the planets have an equatorial bulge and in the case of Saturn, its equatorial diameter is some 10.8% greater than its polar diameter.

Oblique rotator A star whose magnetic axis does not coincide with its rotational axis, but is inclined at an angle to it. The magnetic field strength therefore appears to vary as the star rotates. These fluctuations may be accompanied by small brightness variations, as in Alpha2 Canum Venaticorum stars. The Blazhko effect in certain RR Lyrae stars is thought to arise because they are oblique rotators.

Obliquity of the ecliptic (symbol e) The angle between the Earth's equator and the ecliptic. It is the same as the Earth's axial tilt. The mean obliquity, corresponding to the mean equator, is currently just over 23° 26', but is decreasing slowly because of planetary perturbations of the Earth's orbit. The true obliquity at any epoch is the sum of the mean obliquity and the nutation in obliquity

Observatory A place from which astronomical observations are made; sometimes the name is also used for an administration building completely separate from the observing site. An observatory contains instruments for detecting electromagnetic radiation from celestial objects at various wavelengths. Before the invention of the telescope, observatories such as those of Ulugh Beg and Tycho Brahe had only simple equipment for naked-eye sightings of positions. The development of optical telescopes in the 17th century led to the need for a permanent protective building, usually with a dome that could be opened to view the sky. To ensure the best possible observing conditions, astronomers began in the late 19th century to site their telescopes on high mountains, above the densest parts of the atmosphere. This trend has continued by placing telescopes in orbit, where they can also observe wavelengths such as X-rays, ultraviolet, and infrared that do not penetrate the atmosphere.

Occultation The passage of one astronomical body in front of another, usually the obscuration of a star by the Moon; strictly speaking, a solar eclipse, when the Moon passes in front of the Sun, is a special form of occultation. Precise timings of lunar occultations of stars help to refine our knowledge of the Moon's orbit. Stars can be occulted by asteroids or planetary satellites, which can provide improved diameters for the occulting bodies. Perhaps the best-observed such event was the occultation of the star 28 Sagittarii by Saturn's satellite Titan in 1989. The Galilean satellites of Jupiter are regularly occulted by the planet. Jupiter's satellites may occult one another when Earth lies close to their orbital plane.

Occulting bar An obstruction placed in the field of view of an eyepiece to hide or occult bright objects so that fainter objects may be seen more easily.

Occulting disk A small disk placed centrally in the field of view of an eyepiece, or at the focal point of a telescope, to hide or occult a bright object so that fainter objects can be seen more readily. An occulting disk may be used, for example, to reveal faint satellites of a planet, or to hide the Sun's disk in a coronagraph.

Ocean of storms Landing site of Apollo 12 on the Moon on Nov. 19, 1969; Oceanus Procellarum.

Oceanic plates The crust of the Earth is broken into plates. The plates are enormous chunks of rock that float atop the soft mantle. The plates are moving at a speed that has been estimated at 1 to 10 cm per year. Oceanic plates (those that are under the ocean) are thinner, younger, and denser than continental plates. These underwater plates are about 75 kilometers thick and are made of basalt rock. They are relatively young since plate formation (seafloor spreading) occurs at the margins of oceanic plates.

Oceanus A very large lowland plain on the surface of a Solar System. The only named example is Oceanus Procellarum on the Moon.

Octahedrite The most common type of iron meteorite, containing 6-12% nickel. The meteorite that formed Meteor Crater was an octahedrite. In addition to having a greater nickel content than hexahedrites, octahedrites have an additional mineral, taenite, which is nickel-rich. Octahedrite meteorites have a structure which consists of plates of the mineral kamacite orientated parallel to the sides of an octahedron (hence their name), with the intervening spaces filled with taenite. This arrangement of the two minerals forms a Widmanstätten pattern. Octahedrites are classified according to the thickness of the kamacite plates.Coarse octahedrites have band widths of 1.5-3 mm; medium octahedrites have band widths of 0.5-1.5 mm; and fine octahedrites have band widths of 0.2-0.5 mm. Below a band width of 0.2 mm, three different types of octahedral structure are apparent. The finest octahedrites have band widths of less than 0.2 mm.

Octans The constellation containing the south celestial pole, representing an octant.The closest naked-eye star to the pole is Sigma Octantis. The brightest star in the constellation is Nu Octantis, of magnitude 3.8.

Ocular Another term for an eyepiece.

Of star A star of spectral type O which shows emission lines due to doubly ionized nitrogen (N III) at 464 nm and singly ionized helium (He II) at 468.6 nm, in addition to the normal absorption-line spectrum. They are thought to be O stars undergoing substantial mass loss (about 10^{-5} solar masses per year) that will affect their evolution and lead to them becoming Wolf-Rayet stars. These emission features are variable, as are the overall spectra of Of stars. It has been found that all O stars earlier than O5 are actually type Of, due to their very high temperatures and consequent mass loss. The brightest example of this group is the O5 star Zeta Puppis.

Off-axis guider A device for following a guide star which is just outside the normal field of view of a telescope. The telescope is positioned so that the guide star is over a small mirror or prism, in the light path close to the focal plane. The mirror or prism reflects the guide star's light to an eyepiece with an illuminated reticle. The advantages of an off-axis guider over a guide telescope are that it eliminates the risk of flexure between the guidescope and the main instrument, it is lighter, and fainter guide stars can be used because of the usually larger aperture of the main instrument. However, the object being used for guiding cannot be photographed at the same time.

OH line Dark absorption lines on an electromagnetic spectrum or the emission of photos by hydroxyl (OH) molecules (one atom of oxygen and one atom of hydrogen). At presents, four principal lines are known in the radio domain at frequencies of 1612, 1665, 1667, and 1720 MHz, or wavelengths of approximately 18 centimeters.

OH maser A maser source in which the hydroxyl (OH) molecule is excited to maser action. There are four OH spectral lines at frequencies of 1612, 1665, 1667, and 1720 MHz (a wavelength of about 18 cm). Other OH spectral lines at higher frequencies are also sometimes detected as masers. OH masers are found in starforming regions, circumstellar envelopes, and comets, and in active galactic nuclei in the form of megamasers.

OH-IR source A highly evolved star obscured by a thick cocoon of gas and dust which it has ejected. Most of the energy from the star is absorbed by the dust and reradiated at infrared (IR) wavelengths. The gas is rich in the hydroxyl (OH) radical, which is excited by infrared radiation to produce a powerful maser.

Ohm's law $V = IR$, Where V is the potential across a circuit element, I is the current through it, and R is its resistance. This is not a generally applicable definition of resistance. It is only applicable to *ohmic* resistors, those whose resistance R is constant over the range of interest and V obeys a strictly linear relation to I.

Materials are said to be *ohmic* when V depends linearly on R. Metals are ohmic so long as one holds their temperature constant. But changing the temperature of a metal changes R slightly. Therefore such a device as an electric light bulb increases its temperature as it warms up, which is why it glows slightly brighter for a very brief time just after it is turned on.

For non-ohmic resistors, R is a function of current and the definition $R = dV/dI$ is far more useful. This is sometimes called the *dynamic resistance*. Solid state devices such as thermistors are non-ohmic, and non-linear. A thermistor's resistance decreases as it warms up, so its dynamic resistance is negative.

Olbers' paradox The apparent contradiction between the simple observation that the night sky is dark and the theoretical expectation that an infinite, static Universe, filled more or less uniformly with stars and galaxies, should be as bright as the surface of a star. The first correct discussion of this paradox was published in 1744 by the Swiss astronomer.Philippe Loys de Chéseaux H. W. M. Olbers published his discussion of it in 1826. The paradox can be resolved by identifying its incorrect assumptions. Most important of these is that, as shown by the Big Bang theory, the Universe is not infinite, having come into being around 15 billion years ago. An additional, less important, effect is that the expansion of the Universe weakens the light from distant galaxies, but this on its own cannot fully explain the paradox.

Olbers, Heinrich Wilhelm Matthäus German physician and amateur astronomer. The method he developed to calculate the orbit of a comet he discovered in 1796 expelled from their heads by the Sun, thereby anticipating the discovery of radiation pressure. In 1802, a year to the day after its discovery, Olbers recovered the asteroid Ceres in the position predicted by C. F. Gauss. Olbers subsequently discovered Pallas and Vesta.In 1823 he first pointed out what has come to be called Olbers' paradox.

Old A planetary surface that has been modified little since its formation typically featuring large numbers of impact craters.

Old style date The date of a historical or astronomical event on the Julian calendar, as opposed to the modern Gregorian calendar. Thus the Russian revolution of 1917, whose anniversary is November 7 on the Gregorian calendar, was referred to in Russia as the October Revolution, since Russia still used the Julian calendar at the time and its Old Style date was 1917 October 25.

Olivine Mineral found in basalt; ranges from Mg_2SiO_4 to Fe_2SiO_4.

Oljato Asteroid 2201, a member of the Apollo group, discovered by the American astronomer Henry Lee Giclas in 1947, then lost until 1979. Its diameter is 2.8 km. Oljato's highly elliptical orbit is chaotic, its semimajor axis varying violently due to its many close approaches to Earth and Venus. Its current semimajor axis is 2.176 AU, period 3.20 years, perihelion 0.63 AU, aphelion 3.72 AU, and inclination 2°.5. Oljato has a unique reflectance spectrum that does not resemble that of any other known asteroid, meteorite, or comet; it could be a dormant or extinct cometary nucleus. Oljato is associated with several meteoroid streams.

Olympus mons The largest volcano in the Solar System, lying on Mars at +18° lat., 133° long. From Earth, it is visible as a white spot previously known as Nix Olympica. The summit of Olympus Mons is 27 km above the mean surface level of Mars, and contains a caldera 80 km across. From the caldera, several lava flows have issued down to a scarp about 4 km high which surrounds the volcano. The diameter of Olympus Mons at the scarp is about 620 km. Around the scarp are many ancient landslips, themselves the largest known in the Solar System, some of which have produced an apron of debris extending out another 600 km or more. The whole complex of volcano and scarps is in places 1600 km across. Olympus Mons is thought to be comparatively young, and was probably active within the last billion years.

Omega Omega (the last letter in the Greek alphabet) refers to ratio of the observed density of the Universe (how much mass there is per unit volume) to the critical density of the Universe (the density that would be necessary to stop the expansion of the Universe). If omega is greater than one, then the Universe will eventually stop expanding and begin to contract, and the Universe will end in a "Big Crunch." If omega is equal to one, the Universe will eventually stop expanding but will not collapse. If omega is less than one, the Universe will continue expanding.

Omega Centauri The brightest globular cluster in the sky, of apparent magnitude 3.7 and width 0°.6, in Centaurus. It is also known as NGC 5139. Omega Centauri is markedly elliptical in shape, with a short axis only 80% the length of its long axis. Its actual diameter is about 180 L.Y, and it is the most luminous globular known, with an absolute magnitude of -10.3. At a distance of 17000 L.Y, it is one of the closest globulars to us.

Omega nebula The diffuse nebula M17 in Sagittarius, also known as NGC 6618; alternative names are the Horseshoe Nebula and the Swan Nebula. The Omega Nebula is ¾° across at its widest, and is variously described as being shaped like a Greek letter omega, a horseshoe, a swan, or the figure 2. It lies about 5000 L.Y away.

Onizuka, Ellison S Ellison Shoji Onizuka (June 24, 1946 - January 28, 1986) was a Lieutenant Colonel in the United States Air Force, an aerospace engineer, and a NASA astronaut. Onizuka flew on two Space Shuttle missions; he died in the Challenger explosion on January 28, 1986.

Onsala space observatory A radio observatory at Onsala in Sweden, 40 km south of Gothenburg, founded in 1955. It is operated by Chalmers University of Technology, Onsala. Its main instruments are a 25.6-m antenna, opened in 1964, and a 20-m millimetre-wave telescope in a radome opened in 1976. Onsala Space Observatory jointly operates the Swedish-ESO Submillimetre Telescope at the European Southern Observatory in Chile.

On-the-fly imaging A form of single dish observing and subsequent imaging in which the antenna is slewed continuously across the sky and data are recorded on a time scale short compared to the time in which the antenna moves in angle by an entire beamwidth. The pointing position, possibly including pointing error estimates, is recorded with the data. Proper imaging requires use of the pointing data.

Oort cloud Vast spherical swarm of comets orbiting our Solar System from roughly 2,000 to 100,000 AU. The Oort cloud, while roughly spherical at the largest radius, is wedge-shaped where it merges with the outer planet region in the vicinity of the Kuiper belt of comets. The cloud is divided into different regions of dynamical stability: the Kuiper belt (35-50 AU; affected by planetary perturbations), a dynamically inert region (50-2000 AU; not affected by gravity of planets or stars), the inner Oort cloud (2,000-15,000 AU; affected by galactic tidal forces), and the outer Oort cloud (15,000-100,000 AU; affected by stellar perturbations). The Oort cloud has never been observed, rather its existence is inferred from the careful analysis of the orbits of comets which come in from the cloud. Recent discoveries of objects in the Kuiper belt have confirmed its existence.

Oort, Jan Hendrik Dutch astronomer. He studied with J. C. Kapteyn, whose work on star streams he continued, finding an overall net motion of the Sun with respect to other stars. In 1927 he showed that highvelocity stars appeared to be rotating about the galactic centre, and went on to estimate the Sun's distance from the centre and also the Galaxy's diameter and mass. He also suggested that the Galaxy had missing mass. In the 1950s he and others used the 21-cm line of interstellar hydrogen to map the Galaxy, revealing its spiral structure. In 1950 he suggested that comets originate in a region now called the Oort Cloud.

Oort's constants Two parameters defined by J. H. Oort to describe the major features of our Galaxy's differential rotation in the Sun's neighbourhood. They are usually expressed in units of kilometres per second per kiloparsec. The two parameters are given the symbols A and B. Subtracting B from A gives the angular velocity of the local standard of rest around the centre of the Galaxy, which corresponds to a period of about 200 million years.

Oosterhoff group A category in a classification of globular clusters, in which they are divided into Group I or Group II according to their abundances of heavy metals. Group I clusters have higher metal abundances, somewhat brighter horizontal branch stars, and slightly shorter periods for their RR Lyrae variables (0.55 days compared with 0.65 days) than Group II clusters. Omega Centauri is an Oosterhoff Group II cluster, the iron abundance of its stars being only a few per cent of the Sun's value. They are named after the Dutch astronomer Pieter Theodorus Oosterhoff

Ootacamund radio astronomy centre A radio astronomy observatory in Tamil Nadu state, southern India, founded in 1968, owned and operated by India's National Centre for Radio Astrophysics; its name is popularly abbreviated to Ooty. The main instrument is the Ooty Radio Telescope, a parabolic cylinder 530 m long and 30 m wide, opened in 1970. Uniquely, its long axis is aligned north-south on a hill with a slope equal to the latitude of the station (about 11° north). Radio sources are tracked by rotating the cylinder along its long axis.

Opacity The property of a medium which determines how it absorbs or scatters electromagnetic radiation. It depends on the composition of the medium, its temperature, its density, and—unless an averaged value is being taken, as in the Rosseland mean opacity—on the wavelength of the radiation. In astronomical contexts, in a gas at a temperature below a few thousand degrees, molecules, grains, and ices suspended in the gas may all be significant causes of opacity. Above about 8000 K, the opacity rises very steeply with temperature as the gas is ionized, then above 10^4-10^5 K it declines slowly reaching a plateau at around 10^6 K as electron scattering opacity becomes dominant. Calculations of opacity are important in determining stellar structure.

Open cluster A group of stars formed together in the spiral arms of a galaxy; sometimes called a galactic cluster. Open clusters are usually irregular in shape and contain anything from a few dozen to several hundred relatively young stars in a volume up to 50 L.Y across. The Hyades and the Pleiades are well-known examples. Open clusters are divided into various types according to the Trumpler classification. They are more loosely packed than globular clusters, but may still have a central star density of up to 10 000 times that in the Sun's neighbourhood. Over a thousand are known, all in the galactic disk. Their ages range from a few million to several billion years, and the youngest are still surrounded by traces of the nebula from which they formed. The less dense of them are gradually disrupted by gravitational interaction with the rest of the Galaxy.

Open universe An open universe is a model of the universe in which it will expand forever. In this model, there isn't enough matter (and its accompanying gravitational forces) to stop the current expansion. Consequently, space and time are infinite in this universe. Contrast with a closed universe.

Ophelia The second-closest satellite of Uranus, distance 53 790 km, orbital period 0.376 days; also known as Uranus VII. Its diameter is 30 km, and it was discovered in 1986 by the Voyager 2 spacecraft.

Ophiuchid meteors A weak meteor shower with ill-defined activity limits during May and June. The maximum ZHR is only around 5, with possible peaks around June 9 from a radiant at RA 17m 56m, dec. -23°, and June 19 from RA 17h 20m, dec. -20°.

Ophiuchus A large constellation on the celestial equator, representing a man holding a serpent (the constellation Serpens). It contains the second-closest star to the Sun, Barnard's Star. Although Ophiuchus is not officially a constellation of the zodiac, the Sun nevertheless passes through it in the first half of December. Its brightest star is Rasalhague.Rho Ophiuchi is a quadruple star of magnitudes 5.0, 5.9, 6.7, and 7.3, surrounded by the faint Rho Ophiuchi Nebula. 70 Ophiuchi is a double star with an orbital period of 88 years, magnitudes 4.2 and 6.0, colours yellow and orange, 16 L.Y away. RS Ophiuchi is a recurrent nova that has flared up to naked-eye brightness five times, in 1898, 1933, 1958, 1967, and 1985. NGC 6633 and IC 4665 are large open clusters. Ophiuchus contains numerous globular clusters, including M9, M10, M12, M14, M19, M62, and M107.

Öpik, Ernst Julius Estonian astronomer. He established the process by which meteoroids ablate on entry into the Earth's atmosphere. From a study of the orbits and perturbations of comets, he predicted in 1932 that the Solar System's comets reside in a cloud extending out to a radius of 60 000 AU, an idea later revived by J. H. Oort. He pioneered the measurement of meteoroid size distributions, and suggested that Apollo asteroids are 'burnt-out' comets.

Oppenheimer-Volkoff limit The maximum mass that a neutron star can have without it being overwhelmed by its own gravity.

Calculations put this between 1.6 and 2 solar masses, although the exact figure is uncertain. A neutron star with a mass greater than this is expected to collapse further into a black hole. The limit is named after the American physicist Robert and the Russian-born Canadian George Michael Volkoff.

Opposition The occasion when a body in the Solar System lies opposite the Sun in the sky, and hence has a celestial longitude of 180°. At opposition, a body lies due south at midnight and is visible all night. It is the best time for observing the outer planets, since they are then at their closest to Earth. The inner planets, Mercury and Venus, cannot come to opposition.

Optical axis The line passing through an optical system, around which all image-forming properties are symmetrical. Usually, the optical axis is at right angles to principal optical components (lenses or mirrors) and passes through their centres.

Optical binary A pair of stars which happen to lie close to one another on the celestial sphere because of a chance alignment. They are not physically associated with one another and lie at vastly different distances. Optical binaries are also known as visual binaries.

Optical depth (symbol t) A measure of how far light will travel through a partially transparent medium, such as the atmosphere of a star or a planet, before it is absorbed or scattered. A completely transparent medium has an optical depth of zero. A medium with a low optical depth is described as optically thin, whereas one with a high optical depth is optically thick. Absorption and scattering vary with wavelength, so optical depth is usually specified for a particular wavelength.

Optical double Two stars that happen to lie close to the same line of sight, and thus appear near each other on the sky. The components of an optical double are at different distances and are not related. An optical double is therefore distinct from a physical double, whose components are bound by gravity. Optical doubles are rarer than true binaries.

Optical flat A piece of glass with one surface that is flat to a high accuracy, used for testing the flatness of other plane mirrors.

Optical glass Glass manufactured so as to have a uniform refractive index throughout, and annealed slowly so as to be suitable for lenses.

Optical interforometor An instrument that combines two separate light beams from the same object to form an interference pattern. Examples include the stellar interferometer, for measuring the diameters of stars.

Optical libration An apparent wobble in the movement of a celestial body caused by the changing geometry between it and the body from which it is observed; also known as geometric libration. The Moon's librations are optical librations.

Optical pathlength The length of the optical path through a refracting medium, such as the filter in a photometer. This differs from the geometrical distance travelled because the light is refracted on entering and leaving the filter, so if a filter is placed in front of the focal plane of a telescope then the focus must be lengthened.

Optical pulsar A pulsar that exhibits pulsations in the visible part of the spectrum, as well as at radio and other wavelengths. The first to have its optical pulsations discovered, in 1969, was the Crab Pulsar, followed in 1977 by the Vela Pulsar. optical thickness

Optical sign conventions In introductory courses in physics a sign convention is used for objects and images in which the lens equation must be written $1/p + 1/q = 1/f$. Often the rules for this sign convention are presented in a convoluted manner. A simple and easy to remember rule is this: p is the *object-to-lens* distance. q is the *lens to image* distance. The coordinate axis along the optic axis is in the direction of passage of light through the lens, this defining the *positive* direction. Example: If the axis and the light direction is left-to-right (as is usually done) and the object is to the left of the lens, the object-to-lens distance is positive. if the object is to the right of the lens (virtual object), the object-to-lens distance is negative. It works the same for images.

For refractive surfaces, define the surface radius to be the directed distance from a surface to its center of curvature. Thus a surface convex to the incident light is positive, one concave to the incident light is n

Optical tube The optical tube is the main body or tube of a telescope. This optical tube holds the objective.

Optical wedge A strip of glass or other transparent material whose light transmission decreases evenly from one end to the other. The change in light transmission is accomplished by applying a coating whose thickness increases along the wedge, or by an increase in photographic density. The wedge is used at the eyepiece to reduce the intensity of a star in the field of view until it appears the same brightness as another, fainter, comparison star. The magnitude difference between the stars can then be measured.

Optics Any optical components, such as the objective lens or mirror, secondary mirrors, and eyepieces of a telescope.

Orange "soil" On the Moon, a mixture of very small dark orange and black glass balls which formed from quickly cooled lava droplets during a pyroclastic eruption.

Orbit The path of a body around another in space. Planets moving around the Sun, and large satellites moving around planets, follow orbits that are approximately ellipses, governed by Kepler's laws. Other possible orbital shapes are a parabola and a hyperbola. The size and shape of an orbit is defined by its elements. The changes in those elements with time due to perturbing forces such as the gravitational influence of other bodies can be predicted by celestial mechanics.

Orbital eccentricity Eccentricity is a measure of how an orbit deviates from circular. A perfectly circular orbit has an eccentricity of zero; higher numbers indicate more elliptical orbits. Parabolas have an eccentricity of 1. Neptune, Venus, and Earth are the planets with the least eccentric orbits in our solar system. Pluto and Mercury are the planets with the most eccentric orbits in our solar system.

Orbital family A group of asteroids that follows the same relative orbital path, velocity, and is usually seen close together; they are thought to have once been one asteroid that was broken apart due to a collision in the past.

Orbital inclination is the angle between the plane of an orbit and the plane of the ecliptic. Orbital inclination is abbreviated as "i".

Orbital period The length of time it takes one body to orbit another

Orbital speed Is the planets orbit the Sun, they travel at different speeds. Each planet speeds up when it is nearer the Sun and travels more slowly when it is far from the Sun (this is Kepler's Second Law of Planetary Motion).

Orbital velocity The velocity an object has at a given point in its orbit. Unless the orbit is circular, the object's velocity continually changes in accordance with Kepler's second law.The velocity of an object defines not only speed but direction. The circular velocity of an object in a circular orbit is always perpendicular to the radius vector. An object moving faster than circular velocity will enter an elliptical orbit. Escape velocity puts the object into a parabolic orbit so that it departs from the body it was orbiting, never to return.Hyperbolic velocity is a velocity greater than escape velocity.

Orbiting astronomical observatory A series of NASA astronomy satellites operating mainly in the ultraviolet. The first successful mission, OAO-2, was launched in 1968 December after an initial failure in 1966. It carried Celescope, a group of four 0.32-m telescopes, plus several other telescopes for ultraviolet observations. A third satellite in 1970 failed to achieve orbit. The last of the series, OAO-3, was launched in 1972 and was later named the Copernicus satellite.

Ordinary chondrite The most common of the three main classes of chondrite meteorites, and the most numerous of the observed falls. Ordinary chondrites consist of 10-15% of fine-grained matrix material, 65-75% of chondrules and less than 1% of inclusions. The presence of chondrules, with diameters of 0.3-0.9 mm, in the ordinary chondrites shows that they have not been melted since they formed. The ordinary chondrites are divided into three subgroups on the basis of their content of iron and related elements, such as nickel. The H or high-iron group contains 25-30% iron; the L or low-iron group contains 20-25%; and the LL or low-iron, low-metal group contains only 18-20%. A substantial fraction of all three groups are impact breccias, resembling lunar breccias. The distinct chemistry of the three groups indicates that they originated from separate parent bodies. Each of the three subgroups may be further divided into a number of types on the basis of texture and mineralogy (e.g. H3, L6, and LL5).

Organic molecule A molecule containing carbon atoms. Organic molecules are present in living materials, or have the potential to form the fundamental building blocks of living materials. Many such molecules have been detected in interstellar matter, in the atmospheres of the giant planets, in meteorites, in comets, and on the surfaces of some asteroids and satellites.

Organic Pertaining to carbon-containing compounds. Organic compounds can be formed by both biological and non-biological processes.

Orgueil meteorite A meteorite of carbonaceous chondrite type, which fell at Orgueil, near Montauban in southern France, on 1864 May 14. Twenty fragments totalling 11 kg were collected, and it was identified as of CI1 type. CI carbonaceous chondrites are the most chemically primitive of meteorites and are rare. The Orgueil meteorite was found to have a substantial content of organic compounds.

Orientale basin A major impact basin on the Moon, 930 km across, centred at -19° lat., 93°W long.; the central part is known as Mare Orientale. It is the bestpreserved and probably the youngest of the large lunar basins, formed 3.8-3.9 billion years ago. At least three concentric rings of mountains can be seen within it. Long chains of secondary craters, caused by ejecta from the impact, extend to distances of more than 1000 km. Volcanic lava and small shield volcanoes cover part of the basin's interior.

Orion A magnificent constellation on the celestial equator, representing a great hunter of Greek mythology. Orion is outlined by the prominent stars Betelgeuse, Bellatrix and A line of three stars, Alnilam,Alnitak and Mintaka make up the Belt of Orion. The famous Orion Nebula, M42, contains the multiple star known as the Trapezium. On the southern edge of the Orion Nebula is the double star Iota Orionis, magnitudes 2.8 and 6.9. North of the Orion Nebula is another bright nebula, NGC 1977, and further north again is the 5th-magnitude open cluster NGC 1981. This complex of nebulosity and clusters forms the Sword of Orion, hanging from Orion's belt. Sigma Orionis is an impressive multiple star of magnitudes 3.8, 6.6, 6.7, and 8.8. Eta Orionis is a close double, magnitudes 3.8 and 4.8. The dark Horsehead Nebula intrudes into a faint strip of nebulosity, IC 434, which runs south of Alnitak. The Orionid meteors radiate from the constellation every October.

Orion arm The local spiral arm of our Galaxy. The Sun lies near its inner edge, that is, towards the galactic centre. All the main naked-eye stars are part of the Orion arm, including those of Orion itself. The arm stretches towards Cygnus in one direction, and towards Puppis and Vela in the opposite direction. Features in the Orion Arm include the Gum Nebula, the Orion Nebula, the North America Nebula, the Cygnus Loop, and the Great Rift. The Orion arm was once thought to be a connecting link between larger structures in our Galaxy; hence it is sometimes also known as the Orion Spur.

Orion association A large association of very young stars of spectral types O and B, centred on the Orion Nebula, 1500 1.y. away. It is about 400 1.y. across and contains all the main stars of Orion, except Betelgeuse. The area is still partly shrouded in gas from which stars continue to form.

Orion molecular Clouds A group of large clouds of gas and dust 1500 1.y. away in the constellation Orion, the nearest examples of giant molecular clouds. Each cloud is over 100 1.y. across and contains more than 100000 solar masses of gas, mostly in the form of hydrogen molecules. The young stars of Orion formed from these clouds within the last 10 million years. New stars are still forming in dense molecular cores within the clouds, hidden from sight by the dust, but detectable as sources of infrared and molecular radiation. The best-known molecular cores are designated OMC-1 and OMC-2.

Orion nebula A large and bright nebula, 1500 1.y. away in the constellation Orion; also known as M42 or NGC 1976. The nebula covers more than 1° × 1° of sky and is faintly visible to the naked eye as a fuzzy patch around the multiple star Theta1 Orionis, also known as the Trapezium. The ultraviolet light from the Trapezium stars ionizes the nebula. The Orion Nebula is over 20 1.y. in diameter, and contains several hundred solar masses of ionized gas. It forms a hot blister on the near side of one of the massive Orion Molecular Clouds.

Orion spur Alternative name for the Orion Arm of our Galaxy, given when it was thought to be a branch of the Perseus Arm.

Orion variable Any of the different types of eruptive variable associated with nebulosity, also known as a nebular variable. The majority are irregular variables of the various IN subtypes but the FU Orionis stars and the UVN subtype of flare star are also included.

Orionid meteors A meteor shower showing moderate activity between October 15 and November 2. The activity pattern is complex, with several peaks and troughs, presumably reflecting the filamentary nature of the meteor stream laid down over several perihelion returns by Halley's Comet. A broad maximum, with ZHR reaching 30, is seen for a couple of days around October 21, from a radiant at RA 6h 24m, dec. +15° (between Betelgeuse and Gamma.Orionid meteors are very swift (geocentric velocity 66 km/s), and a high proportion leave persistent trains, but most are faint.

Orrery A working model of the Solar System which represents mechanically the movements of the planets. The name is from the fourth Earl of Cork and Orrery, Charles Boyle for whom one of the first was made. Popular in the 18th century, they ranged from 'grand orreries' a metre in diameter that reproduced the rotational and orbital motions of planets and satellites, down to small portable devices featuring just the Earth, Moon, and Sun.

Orthoscopic eyepiece A design of eyepiece that usually consists of a cemented triplet field lens and a planoconvex eye lens. This combination provides good correction for chromatic aberration, spherical aberration, coma and distortion, at reasonable cost. Orthoscopic eyepieces have a field of view of 35-50° and good eye relief. The design was introduced in 1880 by the German physicist Ernst Karl Abbe.

Oscillating universe A modification of the Big Bang theory, in which an expanding universe eventually reverses and collapses back to a Big Crunch before starting to expand again, forming one cycle of an endless series. In such a model, each cycle has a finite duration but the universe itself can last for an infinite time. It is not known whether our Universe is of this type, or whether it will expand forever.

Osculating elements The orbital elements at a given time of a body in an orbit that is perturbed by the gravitational pulls of other bodies, such as the planets. Osculating elements describe the orbit that the body would follow if those perturbations were to cease. However, the real orbit gradually changes because of perturbations, and so the osculating elements alter. Predictions based on osculating elements are accurate only for times close to the epoch of osculation. The way in which osculating elements change with time is a useful way of demonstrating the effects of perturbations on a body's orbit.

Osculating orbit The orbit that one celestial body would pursue about another if no other celestial object or forces existed. At any given time the osculating orbit touches the real orbit at a point where the body would have the same position and velocity in both orbits. An osculating orbit would be a perfect ellipse; in practice, real orbits are not because of perturbations from the planets, so the real and osculating orbits gradually depart from each other with time.

OSO 3 Orbiting Solar Observatory 3

OSO 8 Orbiting Solar Observatory 8

O-type star A hot, massive blue star that emits strongly at ultraviolet wavelengths and has a surface temperature between about 28,000 to 40,000 kelvins

Outer planets The outer planets are those planets that orbit far from the Sun. They are: Jupiter, Saturn, Uranus, Neptune, and Pluto. They are mostly huge, mostly gaseous, ringed, and have many moons (the exception is Pluto which is small, rocky, and has one moon, Charon).

Outgassing The loss of gas from within a planet through processes such as volcanism. Outgassing is thought to have been primarily responsible for the atmospheres of the Earth and the other terrestrial planets.

Overcontact binary A close binary system in which each star has filled and expanded well beyond its Roche lobe. The system therefore consists of two stellar cores surrounded by a common, dumbbell-shaped convective envelope. Such a system is often an eclipsing binary of the W Ursae Majoris type.

Oversampling The circumstance in which a detector has more resolution elements (e.g. pixels) than are required to sample the finest details of the object or spectrum under study. For example, if an image is taken under seeing conditions of 1 ", but the CCD detector has a scale of 0".1 per pixel, then the image is said to be oversampled.

OVV quasar Abbr. for optically violently variable quasar.

Owens Valley radio observatory The radio astronomy observatory of the California Institute of Technology (Caltech), located at an altitude of 1235 m at Big Pine, California, and founded in 1956. It has two 27.4-m dishes opened in 1959, now used mostly for solar observation, and a 39.6-m dish opened in 1968. Six 10.4-m dishes movable along arms arranged in a T shape form a millimetrewave interferometer, the Caltech Millimeter Array, opened with three dishes in 1985 and enlarged in 1994.

Owl nebula The 11th-magnitude planetary nebula M97 in Ursa Major, also known as NGC 3587. It gets its name because in large apertures it shows two large dark patches like the eyes of an owl. It lies 1300 l.y. away.

Oxygen An element with atomic number 8; symbol: O. It is actually the most common element in the crusts and mantles of the inner planets and rocky moons, making up all silicate minerals. Along with hydrogen, carbon, and nitrogen, oxygen is essential to life.

Oxygen burning A set of nuclear reactions inside stars through which oxygen is converted to heavier elements such as silicon. It occurs in stellar cores (and sometimes later, when oxygen in the core is exhausted, in a surrounding shell) at temperatures over 2×10^9 K.

Ozone Ozone is a form of molecular oxygen (O_3); it consists of three connected oxygen atoms.

Ozone layer The ozone layer is a region of the stratosphere which contains most (about 90%) of the Earth's atmospheric ozone. It is about 10-25 miles (15-40 km) above the Earth's surface. The ozone layer shields the Earth from Ultraviolet B rays that come from the Sun. The ozone layer is becoming depleted, and there is an "ozone hole" over Antarctica.

Ozonosphere A level of the stratosphere at an altitude of 10-50 km where substantial amounts of ozone (O_3) are found. The highest concentrations occur between 20 and 25 km in the ozone layer. Stratospheric ozone is generated by the action of sunlight, which splits oxygen molecules into single atoms. Atomic oxygen (O) can combine with molecular oxygen (O_2) to produce the highly reactive triatomic form (O_3). The ozone layer is an important barrier to ultraviolet light from the Sun at wavelengths of 230-320 nm.

P

P Cygni line profile A feature in a spectrum that indicates a strong outflow of matter from a star. It is named after P Cygni, a Be star whose spectrum shows strong hydrogen and helium emission lines with absorption lines adjacent to their blueward side. These dual characteristics are caused by an expanding shell or wind of material that is being blown off the star either by radiation pressure or rapid rotation. Where the expanding shell is directly between us and the star, it produces absorption at a velocity which is blueshifted relative to the star. The other parts of the shell produce the emission lines. If material is falling on to the star, the absorption component can be redshifted relative to the emission lines, causing an inverse P Cygni line profile.

P magnitude The magnitude of a star measured with a filter with wavelength 12.2 μm and a bandwidth of 1.0 U+00μm. This infrared magnitude is now little used.

Pahoehoe Basaltic lava with a smooth, billowy, or ropy surface.

PAHS Polycyclic aromatic hydrocarbons (PAHs) are a class of very stable organic molecules. They are flat molecules made only of carbon and hydrogen atoms. PAH molecules are quite common and highly carcinogenic. They are one of the by-products of combustion from automobiles and airplanes, and some are present in charcoal broiled hamburgers.

Pair annihilation The mutual destruction of a particle and its antiparticle. The energy reappears as photons or other particle-antiparticle pairs. For example, an electron and positron combine to produce two 511 keV gamma-ray photons. Annihilation of a proton and antiproton forms pions.

Pair production The production of a particle and its antiparticle. This can happen in several ways. For example, an electron-positron pair can be produced from a gamma-ray photon of energy greater than 1.022 MeV in the electric field surrounding an atomic nucleus, or the pair may be produced by the collision of two photons whose total energy exceeds 1.022 MeV. The energies of photons in the early Big Bang were high enough to produce many varieties of pairs of particles. Protons and antiprotons could be produced during the first 2×10^{-6}s after the Big Bang when the temperature exceeded 10^{13} K, and positive and negative muon pairs until about 2×10^{-4}s. Electrons and positrons could be produced until the temperature fell below 10^{10} K, about 2s after the Big Bang. Pair production is also important in secondary cosmic-ray showers.

Palette In computer graphics, an ordered range of available elements, usually colors.

Palimpsest An impact crater whose topography has been smoothed by glacierlike flow of the surface. The term was coined for features on the icy surfaces of Ganymede and Callisto. Palimpsests there consist of bright circular features, sometimes at the centre of concentric rings.

Pallas Asteroid 2, the second asteroid to be discovered, by H. W. M. Olbers in 1802. Its mean diameter is 538 km, making it the second-largest main-belt asteroid, and its rotation period is 7.81 hours. Its shape is a triaxial ellipsoid, 559 × 532 × 525 km. Pallas is of B class, with a spectrum similar to that of the carbonaceous chondrite meteorites. Its mean magnitude at opposition is 8.0; only Vesta and Ceres can become brighter. Its orbit has a semimajor axis of 2.772 AU, period 4.62 years, perihelion 2.12 AU, aphelion 3.42 AU, and an unusually high inclination of 34°. 8.

Pallasite A class of stony-iron meteorite named after the German naturalist Peter Simon Pallas who found a 680-kg specimen near Krasnojarsk, Russia, in 1772. In pallasites, cavities in the nickel-iron metal are filled with crystals of the hard, glassy silicate mineral olivine; the proportions of metal and silicate are about the same.

Palomar observatory An observatory at an altitude of 1706 m, 80 km northeast of San Diego, California, owned and operated by the California Institute of Technology Pasadena. It was founded in 1934 but completion was delayed by World War II. Its main instrument is the 200-inch (5-m) Hale Telescope opened in 1948, named after the observatory's founder, G. E. Hale. Also in operation since 1948 has been the 1.2-m Oschin Schmidt Telescope, named after Samuel Oschin, a benefactor of the observatory, used for the Palomar Observatory Sky Survey a new corrector plate was installed in 1985. Also on the mountain are a 1.5-m reflector, opened in 1970, and a 0.46-m Schmidt, opened in 1936.

Palomar observatory sky survey (POSS) A photographic survey of the sky from the north celestial pole to declination -30°, made using the 1.2-m Oschin Schmidt Telescope at Palomar Observatory, California. The original survey, completed in 1958, consists of 936 pairs of plates 0.355 m square, each pair consisting of one plate sensitive to blue light and one to red light. The field of view of each plate is 6°.5. It was financed by the National Geographic Society. A second survey was begun in 1985 from the north celestial pole to the equator, using modern photographic plates. When complete, it will consist of 894 fields in blue, red, and infrared light, with greater overlap between fields than the original survey.

Palus A mottled area on a planetary surface; pl. paludes. The name, which means 'marsh' or 'swamp', was first used for features on the Moon in the 17th century. It is not a geological term, but is used in the nomenclature of individual features, for example Palus Putredinis on the Moon, or Oxia Palus on Mars.

Pan Pan is one of the smaller of the many moons of Saturn. Pan is located within the Encke Gap in Saturn's rings. Pan has a radius of about 10 km. Pan orbits at 133,583 km from Saturn. Its orbital period is 0.575 earth days. Pan was discovered by Mark R. Showalter/ Voyager 2 in 1990.

Pandora The fourth-closest satellite of Saturn, distance 141 700 km, orbital period 0.629 days; also known as Saturn XVII. It was discovered in 1980 by Voyager 1, between Saturn's F and G Rings. Pandora is irregularly shaped, 110 × 90 × 70 km in size. It acts as a shepherd moon to the F Ring.

Panel A segment of the reflecting surface of an antenna that is independently formed with its supporting rib structure, and is mounted on the *backup structure* as a unit.

Panels are mounted to the backup structure on screws so that their relative positions can be optimized. This allows the antenna gain and effective area to be maximized when an antenna is commissioned. It also allows the panels to be realigned later, either to compensate for changes in the antenna structure or to refine the initial adjustment using improved surveying techniques (such as holography).

In advanced antenna designs, the panels are mounted on calibrated *actuators*, so that their alignment may be adjusted continually under remote control. This permits active correction for modeled, or measured, deformations of the reflecting surface during an astronomical observation. In the GBT, such an *active surface* will be used to maximize the antenna gain and minimize pointing errors, especially at the higher observing frequencies.

Pangaea Pangaea was a supercontinent consisting of all of Earth's land masses. It existed during the Permian period through the Jurassic period. It began breaking up during the Jurassic period, forming the continents Gondwanaland (South America, Africa, India, Australia, and Antarctica) and Laurasia (North America, Europe, Asia, Greenland, and Iceland).

Parabola A type of curve whose 'arms' become parallel as they approach infinity, so that the curve never quite closes in on itself, defined mathematically as a conic section with an eccentricity of 1. It may be regarded as an ellipse in which the two foci are infinitely far apart, and is the limiting case between an ellipse and a hyperbola. Some comets have elliptical orbits that are so extended they are indistinguishable from parabolas.

Parabolic antenna A bowl-shaped aerial with a parabolic cross-section that reflects incoming radio waves to a focal point above the bowl; also known as a dish antenna. At the focal point is a collector known as a feed. High-performance parabolic antennas are often of the Cassegrain design, in which a secondary reflector directs radio waves to a focal point low down inside the bowl. The principle of a parabolic antenna is identical to that of the primary mirror in an optical telescope.

Parallactic angle The angle between the hour circle through a celestial object and the vertical circle through the object, at the time of an observation.

Note that the parallactic angle also describes the orientation of a fixed direction on a telescope, such as the direction of sensitivity of a linearly-polarized receptor on an antenna, relative to celestial co-ordinates. If an antenna is on an equatorial mount, the parallactic angle of any receptor will not change relative to any celestial object that it is tracking (unless the receptor is deliberately rotated during the observation).

Parabolic comet A comet whose orbit around the Sun has an eccentricity of exactly 1.0. Many cometary orbits are initially calculated on the basis that they are parabolic, since the observed arc close to perihelion is only a small part of the true ellipse, and a parabola represents the best first approximation.

Parabolic orbit An orbit having the shape of a parabola. A body pursuing a parabolic orbit about another makes only one approach to the other body, in theory coming from infinity and then receding to infinity.

Parabolic velocity The speed of an object that is moving at exactly the escape velocity of another body. The relation between parabolic velocity and circular velocity at any point in the orbit is that parabolic velocity equals $v^2 \times$ circular velocity.

Paraboloid A surface swept out by a parabola rotated 360° about its axis. The main mirrors in large telescopes are usually paraboloidal in cross-section, since such mirrors are free from spherical aberration.

Parallactic ellipse The path on the sky of the apparent position of a star as seen from the Earth, due to the Earth's annual motion around the Sun. The shape of the parallactic ellipse ranges from a circle for a star at the ecliptic pole to a line for a star on the ecliptic. Its size depends on the distance of the star, becoming smaller with increasing distance.

Parallactic inequality A variation in the Moon's orbital motion caused by the Sun's gravitational attraction. As a result of the Sun's attraction, the Moon is about 2' ahead of its expected orbital position at first quarter, and a similar amount behind at last quarter.

Parallactic motion parallactic motion is the apparent motion of the stars that is caused only by the motion of the Earth as it orbits the Sun.

Parallax (symbol p) The angular difference between an object's direction as seen from two points of observation, such as opposite sides of the Earth's orbit. It is thus a form of triangulation. Parallax can also be defined as the angular distance between two points as seen from a third point in space, such as the radius of the Earth's orbit as seen from a star. By extension, the term parallax can also be used to mean a star's distance, by whichever method it is determined.

Parallax, trigonometric The apparent shift of an object when viewed from two different directions. This shift may be used to gauge the distance of an object out to distances of about 1 000 light years or more. The object is observed six months apart and the amount it shifts relative to the background stars is measured in arcseconds. No star subtends an angle of one arcsecond, so the angles measured this way are extremely small. However, if the angle is measurable it is possible, by using trigonometry, to measure the distance to a star.

Parametric amplifier An amplifier used in radio astronomy for the frequency range of about 1-100 GHz. They are cooled to 15-20 K to reduce noise. Parametric amplifiers operate by transferring energy to the signal from a highly stabilized laboratory oscillator known as the pump.

Paraxial Describing light rays which are both close to and parallel to the optical axis of an optical system.

Parent body An object (e.g. asteroid) from which meteorites come.

Parent isotope An isotope that undergoes radioactive decay, its nuclei disintegrating spontaneously to form a daughter isotope (often of a different element). For example, rubidium-87 is the parent isotope of strontium-87, into which it decays with a half-life of 4.88×10^{10} years. The amount of the parent isotope in a sample decreases with time, while that of any daughter isotope increases; the relative proportions of the two can be used to derive the age of the sample

Parfocal Having the same focus position. A set of parfocal eyepieces, for example, are manufactured so that they can be interchanged without refocusing the telescope.

Parhelion An optical phenomenon resulting from the refraction and reflection of sunlight within ice crystals in cirrus cloud; also known as a mock Sun or sundog. Parhelia appear 22° either side of the Sun, and at the same altitude above the horizon, but disappear when the Sun's altitude exceeds 60°. Red light is refracted to the edge of the parhelion closer to the Sun, blue to the farther edge. Parhelia are often associated with an atmospheric halo.

Parkes observatory A radio observatory near the town of Parkes in New South Wales, Australia, owned and operated by the Commonwealth Scientific and Industrial Research Organisation.It consists of a 64-m diameter radio dish opened in 1961. As well as radio astronomy, it is used for tracking space probes. Since 1988 it has been part of the Australia Telescope National Facility.

Parkes telescope Parkes Telescope is a 64 meter radiotelescope in New South Wales, Australia.

Parsec (symbol pc) A basic unit of stellar distance, corresponding to a trigonometric parallax of one second of arc. In other words, it is the distance at which one astronomical unit subtends an angle of one second of arc. One parsec equals 3.2616 light years, 206 265 astronomical units, or 30.857×10^{12} km. For distances on galactic and intergalactic scales, kiloparsec and megaparsec are used.

Parselene An optical phenomenon resulting from the refraction and reflection of moonlight within ice crystals in cirrus cloud; also known as a mock Moon. It is the lunar equivalent of the parhelion. Parselenae appear 22° from the Moon and at the same altitude as the Moon above the horizon. They occur only when the Moon is bright, typically within five days of full. Being fainter than parhelia, they are usually colourless.

Partial eclipse A solar eclipse in which the Moon does not completely obscure the Sun's disk, which then appears as a crescent; or, a lunar eclipse in which the Moon is not entirely immersed in the Earth's umbra, so that some of the Moon remains sunlit.

Particle accelerator A machine that, using electromagnets, can accelerate moving charged particles, giving them more energy.

Pascal (symbol Pa) A unit of pressure equal to a pressure of one newton per square metre. One pascal equals 10^{-5} bar. Hence, the Earth's atmospheric pressure at sea level is approximately 10^5 pascal. It is named after the French mathematician and theologian Blaise Pascal.

Paschen series A sequence of absorption or emission lines in the near-infrared part of the spectrum, due to hydrogen. They are caused by electron jumps between the third energy level and higher levels. The Paschen-a line has a wavelength of 1875 nm, and the series limit is at 820 nm. The series is named after the German physicist Louis Carl Heinrich Paschen.

Paschen-back effect The splitting of lines in a spectrum when the source is located in a strong magnetic field and the magnetic splitting exceeds the normal multiplet splitting. The effect is named after the German physicists Friedrich Paschen and Ernst Emil Alexander Back.

Pasiphae Pasiphae is Jupiter's fifteenth moon. Pasiphae is 22 miles (36 km) in diameter and orbits 14,600,000 miles (23,500,000 km) from Jupiter. Pasiphae has a mass of 2×10^{23}kg. It orbits Jupiter in 735 (Earth) days and is in a retrograde orbit (orbiting opposite to the direction of Jupiter). Very little is known about Pasiphae. Pasiphae was discovered by P. Melotte in 1908.

Passband The range of wavelengths that are transmitted by a filter. It is often quoted as the wavelength limits within which the transmission exceeds 50% of the peak value.

Patera A crater on a planetary surface with irregular, fluted, or scalloped edges; pl. paterae. The name, which means 'shallow dish', is not a geological term, but is used in the nomenclature of individual features, for example Ulysses Patera on Mars, or Loki Patera on Io.

Paul Wild observatory A radio observatory at Culgoora, near the town of Narrabri in New South Wales, Australia. It is the site of the Compact Array of the Australia Telescope National Facility, and consists of a line of six 22-m antennas that can operate as one instrument by aperture synthesis. Five of these antennas can be moved along a 3-km stretch of rail track running east-west. A sixth antenna lies 3 km to the west of the main group, on its own short track. The array began operation in 1988. The observatory is named after the British-born Australian radio astronomer Paul Wild.

Pauli exclusion principle The physics principle that no two fermions can exist in identical quantum states. (In plain English, no two particles can occupy the same state at the same time.)

Pavo A constellation of the southern sky, representing a peacock. Its brightest star is called Peacock.Kappa Pavonis is a bright Cepheid variable that ranges between magnitudes 3.9 and 4.8 with a period of 9.1 days. NGC 6752 is a 5th-magnitude globular cluster.

Payload The payload is the cargo carried on the Space Shuttle or other spacecraft.

Payne-Gaposchkin, Cecilia Helena English astronomer who worked in America. She developed a temperature scale for the various spectral types of stars based on the strengths of their spectral lines. In 1925, in a study of the relative abundances of elements in stars of various ages, she established that hydrogen is the major constituent of stars. With the Russian-American astronomer Sergei Illarionovich Gaposchkin whom she married in 1934, she began a huge programme of measuring the magnitudes of variable stars from photographic plates, which resulted in a catalogue of variables published in 1938. A similar joint study of variables in the Magellanic Clouds appeared in 1971.

P-class asteroid A class of asteroid, fairly common in the outer main asteroid belt, with a distribution that peaks at a mean distance from the Sun of 4.0 AU. Members of this class have a featureless reflectance spectrum, flat to slightly reddish over the wavelength range 0.3-1.1 μm. Their spectra are intermediate between classes C and D. P-class asteroids are distinguished from the spectrally identical E and M classes by their lower albedos (0.02-0.06). Members of this class include (87) Sylvia, diameter 282 km, and (153) Hilda, diameter 222 km.

PCX An image file format, capable of representing 1, 4, 8 or 24 bits per pixel, originally introduced by ZSoft. Mild compression is achieved by storing adjacent replicated pixels as an intensity with a repeat count.

Peacock The star Alpha Pavonis. It is a B2 subgiant, magnitude 1.9, distance 360 1.y.

Peculiar galaxy A peculiar galaxy is an irregular galaxy that has an abnormal shape (neither elliptical, spiral, nor lenticular) and/ or has another unusual characteristic, like jets of gas spewing from the nucleus, unusual amounts of dust, low surface brightness, etc. They are probably formed as galaxies collide or have companion galaxies that influence them. Peculiar galaxies are generally smaller and optically dimmer than regular galaxies. They have the designation *p* or *pec* at the end of their name. In 1966, the astronomer Halton Arp published a catalogue of 338 of these galaxies in his "Atlas of Peculiar Galaxies." Arp suggested that peculiar galaxies create stars in intense bursts. Ring galaxies are a type of peculiar galaxy.

Peculiar motion The proper motion of a star due to its velocity relative to the local standard of rest. The peculiar motion is that part of a star's total proper motion which is not accounted for by parallactic motion.

Peculiar star A star that is known to be variable but cannot be assigned to any specific classification type, or one whose characteristics appear to be unique. Some peculiar stars probably represent objects in transition between one recognized type of variable and another.

Peculiar velocity The space velocity of a star relative to the local standard of rest.

Pedestal crater An impact crater formed in weak surface materials which have subsequently been eroded away, except where protected by the surrounding ejecta blanket. The term was coined for certain craters on Mars, which in some areas of the planet form prominent circular plateaux above the surrounding terrain.

Peekskill meteorite A meteorite that fell on 1992 October 9, damaging a parked car at Peekskill, New York. The fireball preceding the fall of the 12.6-kg H6 ordinary chondrite was observed over Virginia and New Jersey. Video and other records of its trajectory allowed the meteorite's orbit to be determined. It was found to have had an aphelion in the main asteroid belt.

Pegasus The seventh-largest constellation, lying in the northern sky, and representing the winged horse of Greek mythology. The stars Markab (Alpha Pegasi), Scheat (Beta Pegasi), and Algenib (Gamma Pegasi) form three corners of the Square of Pegasus, which is completed by the star Alpheratz (Alpha Andromeda) from neighbouring Andromeda. Enif (Epsilon Pegasi) is a double star. M15 is a 6th-magnitude globular cluster.

Pele An active volcano on Jupiter's satellite Io, at -19° lat., 258° long. It lies in a volcanic complex about 500 km across, which includes lava flows and tablelands. It was active in 1979 March during the Voyager 1 encounter, giving rise to the largest volcanic plume ever seen, 280 km high and 1000 km wide. Vaporized sulphur might be responsible for eruptions of this size, the heat being provided by an intrusion of molten silicate magma. The Pele eruption had ceased by the time Voyager 2 passed Io four months later.

Pelican nebula A diffuse nebula in Cygnus, so named because of its resemblance to a pelican on long-exposure photographs; also known as IC 5067 and 5070. The Pelican Nebula is about 1° wide. It lies next to the larger and brighter North America Nebula, and is part of the same cloud.

Pencil beam An antenna pattern with an extremely narrow main beam of circular cross-section and negligible side lobes. A radio telescope with a pencil beam has high angular resolution.

Penrose process A process for extracting energy from a rotating black hole (i.e. a Kerr black hole). The process requires sending a mass into a trajectory in the ergosphere around the black hole, against the direction of rotation of the black hole. While inside the ergosphere the mass splits into two parts, one of which enters the black hole while the other escapes. Given a suitable trajectory, the emerging fragment may possess a total energy (i.e. rest mass plus kinetic energy) greater than the total energy of the mass that went in. The extra energy has been extracted from the rotational energy of the black hole, which must therefore slow down slightly. The process is named after the English mathematician Roger Penrose who discovered it in 1969.

Penumbra 1. The outer region of the shadow cast in space by a planet or satellite; an observer within the penumbra would witness a partial eclipse. The penumbra surrounds the dark, narrower umbra (1), within which the Sun is completely obscured. The penumbra is darkest immediately adjacent to the umbra. During a lunar eclipse the Moon passes through the penumbra of the Earth's shadow before and after its umbral passage.

2. The lighter, outer region of a sunspot, with a temperature of around 5500 K. Small spots often do not have a penumbra, but in mature sunspots the penumbra is well developed and occupies about 70% of the total spot area. The penumbra consists of filaments radiating from the central umbra. These filaments are brighter than the umbra, but less bright than the surrounding photosphere. They consist of light grains lasting about an hour, which drift in towards the umbra where they become umbral dots, apparently smaller versions of photospheric granules.

Penumbral eclipse A lunar eclipse during which the Moon passes slightly to the north or south of the umbra, and only through the less intense outer regionof the Earth's shadow, usually showing only negligible dimming. When the Earth's atmosphere is heavily laden with the penumbra may be darker than normal.

Penzias, Arno Allan German-American physicist. In the early 1960s he and R. W. Wilson were using a low-noise horn antenna in connection with satellite communications, and picked up a mysterious, persistent signal. They consulted R. H. Dicke and colleagues, and concluded that they had detected the cosmic background radiation. This, the single most important piece of observational evidence for the Big Bang, gained Penzias and Wilson the 1978 Nobel prize for physics.

Perfect cosmological principle An extension of the cosmological principle which proposes that the Universe is not only the same in all places and in all directions, but also at all times. The principle is the cornerstone of the steadystate theory, but is incompatible with observations that show that the Universe is evolving with time.

Peri Prefix referring to the closest point in the orbit of one object around another, as in perihelion and perigee.

Periapsis The point in an orbit when two objects are closest together. Special names are given to this point for commonly used systems: see periastron, perihelion, and perigee. The opposite of apoapsis.

Periastron effect An increase in the brightness of a binary star that has a highly eccentric orbit, when the separation between the components is at a minimum. It is, in fact, an enhancement of the reflection effect at the time of periastron, and arises from the same cause: irradiation of one star by the other.

Periastron The point in an elliptical orbit around a star that is closest to the centre of the star.

Pericentre The point in an elliptical orbit that is nearest to the centre of mass of the orbiting system, such as a binary star or a planet and satellite.

Peridotite A course-grained igneous rock consisting largely of the mineral olivine.

Perigee For an object orbiting the Earth, the perigee is the point in each orbit which is closest to the Earth. The perigee varies a small amount from orbit to orbit. The closest perigee is called the proxigee.

Perihelion The point in its orbit where a planet is closest to the Sun. when referring to objects orbiting the Earth the term perigee is used; the term periapsis is used for orbits around other bodies. (opposite of aphelion)

Period of revolution The period of revolution of an orbiting body is the time it takes for it to make one revolution around its primary body. The period of revolution for the Earth is one year.

Period of rotation The period of rotation of an orbiting body is the time it takes for it to make one rotation around its axis. The period of rotation for the Earth is one day.

Period, orbital (symbol P) The time that a celestial body takes to make one orbit of another body. The period from one perihelion to the next is called the anomalistic period. The time the orbiting body takes to revolve through 360° is called the sidereal period. These two periods are different in duration if the orbit is perturbed by other bodies or forces.

Period-age relation A relationship between the inferred evolutionary age of a Cepheid variable and its period. High-mass stars evolve more rapidly and are therefore younger when they enter the Cepheid instability strip. At the same time, their high mass is reflected in a longer period of pulsation.

Period-colour relation A statistical relationship between the mean colour of a pulsating variable and its period. It reflects the fact that the period of pulsation depends on various factors, including a star's mass, radius, density, and temperature, with the temperature directly affecting the colour.

Period-density relation A relationship between the period of a pulsating variable and its mean density. The lower the density, the longer the period.

Periodic comet A comet with an orbital period less than 200 years; also known as a short-period coma. A period cannot reliably be established until a comet has reappeared, and the maximum of 200 years, chosen for historical reasons, is somewhat arbitrary. Most periodic comets have direct orbits with inclinations of less than 30°. The typical short-period comet has a period of 7 years, perihelion 1.5 AU, and inclination 13°. The names of periodic comets are prefixed by P/ (or D/ if they have disintegrated or disappeared), preceded by a number indicating the order in which their orbit was established (e.g. 1P/Halley, 2P/Encke, 3D/Biela, 109P/Swift-Tuttle). The periodic comet with the longest established period, 155 years, is Comet Herschel-Rigollet. Most periodic comets are thought to come from the Kuiper Belt.

Periodic orbit An orbit that is closed so that the orbiting body follows the same trajectory repeatedly. An elliptical orbit is a periodic orbit. The orbit of planets about the Sun and the major satellites about their planets are almost periodic, although the mutual perturbations of the planets prevent the orbits from being exactly periodic.

Periodic perturbation A disturbance in a body's orbit that is oscillatory in nature, having a certain period and amplitude. Such perturbations can occur in all six elements of the orbit and are usually termed longterm or short-term depending on whether the period is much larger than the body's orbital period, or of the same order as it.

Periodic table of the elements The periodic table of the elements is a tbale that shows the relative size and properties of the elements that make up the universe. The atomic number indicates the place of an element in the periodic table of the elements.

Period-luminosity relation A statistical relationship between the period of a pulsating variable and its luminosity. For classical Cepheids and W Virginis stars, the relationship is approximately linear: the longer the period, the greater the luminosity. Classical Cepheids show a greater colour dependence, so it is preferable to use a period-luminosity-colour relation. Mira stars have a nonlinear PL relation, but it is well-established and useful for determining distances within the Galaxy. The Cepheid PL relation is the basis of the traditional method of determining distances to nearby galaxies.

Period-luminosity-colour relation A relationship between the period and colour of a pulsating variable and its luminosity. It is an extension of the period-luminosity relation that takes differences in colour (i.e. temperature) into account.

Period-mass relation A relationship between the mass of a pulsating variable and its period. In approximate terms, the greater the mass of a star, the larger its radius, the lower its mean density, and thus the longer its period of pulsation

Period-radius relation A mathematical relationship between the period and mean radius of a pulsating variable. The exact equation depends on the type of variable.

Period-spectrum relation An approximate relationship between the mean spectral type of a pulsating variable and its period. Generally, the later the spectral type, the longer the period. The larger a star, the cooler its surface temperature and the longer it will take to pulsate. Period-spectrum relations are found in Cepheid variables and Mira stars. In Mira stars a period-spectrum-luminosity relation can be established.

Permian period The Permian period preceded the Triassic period, when dinosaurs and mammals evolved. It lasted from 280 to 248 million years ago.

Perseid meteors A meteor shower, one of the three most active of the year, detectable between July 25 and August 20. Peak activity normally falls on August 12, when the radiant lies at RA 3h 04m, dec. +58°, near the Double Cluster. Activity at maximum is high, but variable; in the 1980s the maximum ZHR ranged from 80 to 140. Perseid meteors are fast (geocentric velocity 60 km/s); many are bright, and a high proportion leave persistent trains. The Perseids are produced by debris from Comet Swift-Tuttle, whose return to perihelion in 1992 was accompanied by a new, sharp peak in the shower profile, some hours ahead of the regular August 12 maximum, caused by recently ejected meteoroids from the comet.

Perseus A prominent constellation of the northern sky, representing the mythical Greek hero who rescued Andromeda. Its brightest star, Mirfak (Alpha Persei), is surrounded by a loose open star cluster, Melotte 20. Beta Persei is Algol, the prototype eclipsing binary. Rho Persei is a semiregular variable red giant that ranges between magnitudes 3.3 and 4.0 with a period of about 7 weeks. NGC 869 and 884 are the famous Double Cluster, also known as h and Chi Persei. M34 is a 5th-magnitude open cluster. M76 is a planetary nebula known as the Little Dumbbell. NGC 1499 is the California Nebula, and NGC 1275 is a peculiar galaxy known as the radio source Perseus A, part of the Perseus Cluster of galaxies. The year's brightest meteor shower, the Perseids, radiates from the constellation every August.

Perseus arm The spiral arm of our Galaxy that is about 5000. L.Y farther from the centre than the local Orion Arm in which the Sun lies, and roughly parallel to it. The Perseus Arm can be traced from the constellation of Cassiopeia via Perseus to Auriga, Gemini, and Monoceros, behind stars in our local arm. It contains the Double Cluster, the Crab Nebula, and the Rosette Nebula.

Perseus cluster A cluster of galaxies about 250 million L.Y away, covering 4° of sky in the constellation Perseus; also known as Abell 426. At its centre lies the supergiant elliptical galaxy NGC 1275, also known as the radio souice Perseus A. The Perseus Cluster is dominated by elliptical galaxies.

Perturbation A change in the orbit of a body, usually due to the gravitational attraction of another body. Thus the planets in their elliptical orbits about the Sun mutually perturb one another, producing changes in their orbital elements. The orbits of satellites are perturbed by their mutual gravitational attractions and also by the Sun. Perturbations also arise in the orbits of satellites that lie close to their planets because the planets are not perfectly spherical in shape. The orbits of some artificial Earth satellites are also perturbed by atmospheric drag.

Pfund series A sequence of absorption or emission lines in the far-infrared part of the spectrum, due to hydrogen. They are caused by electron jumps between the fifth energy level and higher levels. Pfund-a is at a wavelength of 7.46 μm, and the series limit is at 2.28 μm. The series is named after the American physicist Herman Pfund.

Phaethon Asteroid 3200, a member of the Apollo group, discovered in 1983 by the Infrared Astronomical Satellite. Phaethon's orbit is very similar to that of the Geminid meteor shower, and it may be their parent body. Before the discovery of Phaethon, meteor showers were generally believed to be associated only with comets, but infrared observations suggest that Phaethon has a rocky surface, not the dusty crust expected for an extinct cometary nucleus. Phaethon is an F-class asteroid with a diameter of about 7 km. Its orbit has a semimajor axis of 1.271 AU, period 1.43 years, perihelion 0.14 AU, aphelion 2.40 AU, and inclination 22°.1. It can approach to within 0.026 AU (3.9 million km) of the Earth's orbit.

Phase 1. The proportion of the visible disk of the Moon or a planet that is illuminated as seen from Earth.

2. A measure of position along a cyclically varying quantity, such as a wave or a periodic vibration. Phase is measured as an angle, where one complete cycle is equivalent to a phase of 360° (or 2° radians), or occasionally as a number between 0.0 and 1.0. Two or more waves of the same frequency are said to be in phase when their maxima and minima occur at the same instants. Otherwise, they are said to be out of phase, or to have a phase difference. If their phases differ by exactly 180° they are said to be in antiphase.

Phase angle The angle between the Sun and an observer as seen from the centre of a body in the Solar System. At a phase angle of 0° the Sun and observer lie in exactly the same direction, and so the observer sees the object fully illuminated. At a phase angle of 180° the Sun and observer lie on opposite sides of the body, and the whole of its unilluminated side is facing the observer.

Phase defect The amount by which the illuminated disk of the Moon appears to depart from a complete circle at full Moon, because of the inclination of the Moon's orbit to the ecliptic. It is measured in degrees.

Phase diagram A graph showing the range of temperatures and pressures at which a substance can exist as a solid, liquid, or gas. These different physical states are known as phases, and the temperatures and pressures under which each phase exists are represented by an area on the diagram. A line between two areas defines the conditions under which two phases can exist in equilibrium. The triple point is the only point on the diagram where conditions are such that the solid, liquid, and gas phases can coexist in equilibrium.

Phase difference (symbol f) The amount by which two cyclical motions of the same frequency, such as electromagnetic waves, are out of step with each other. It can be measured in degrees, radians, or seconds of time.

Phase rotator A device used in radio astronomy to adjust the phase of an incoming signal. It is often used in interferometry to compensate for the slow turning of the baseline caused by the rotation of the Earth.

Phase tracking center In a correlating-type radio interferometer usually the fringe stopping center and the delay tracking center coincide. When this is the case, both are referred to as the *visibility phase tracking center*, sometimes called the *phase referencing center*.

Phases of the moon As the moon circles the Earth, the shape of the moon appears to change; this is because different amounts of the illuminated part of the moon are facing us. The shape varies from a full moon (when the Earth is between the Sun and the moon) to a new moon (when the moon is between the sun and the Earth). The phases of the Moon include: the new moon, waxing crescent, first quarter, waxing gibbous, full moon, waning gibbous, last quarter, waning crescent, and new moon again.

Phillips bands Features in the spectrum of the carbon molecule C2 which occur in the red and near-infrared. A prominent band lies at 1207 nm. Phillips bands are prominent in the atmospheres of carbon stars. They are named after the American astrophysicist John Gardner Phillips.

Phobos The inner satellite of Mars, 9378 km from the planet's centre and only 6000 km above its surface. Phobos orbits in 0.319 days, faster than Mars rotates on its axis, keeping one face turned permanently towards the planet. Phobos is irregularly shaped, 27 × 22 × 19 km in size, and is covered with impact craters, the largest, Stickney, being 11.5 km in diameter. There are also several families of parallel grooves or chains of craters, possibly secondary craters from large impacts on Mars. Phobos was discovered in 1877 by A. Hall, and may be a captured asteroid.

Phobos probes Two space probes to the planet Mars, launched by the former Soviet Union in 1988 July. Contact was lost with Phobos 1 en route, but Phobos 2 went into orbit around Mars in 1989 January, studying the planet and its larger satellite, Phobos, although it failed before it could drop its instrumented landers on to the satellite's surface.

Phocaea group A group of asteroids at a mean distance of about 2.36 AU from the Sun, with orbital inclinations of 23-25°. The group has been separated from the main asteroid belt by the gravitational effects of the major planets, notably Jupiter, and contains a broad selection of asteroid classes. The largest member of the group is the C-class (105) Artemis, diameter 126 km. Certain members, among them (323) Brucia, (852) Wladilena, (1568) Aisleen, and (1575) Winifred, may form a true family within the Phocaea group. The group is named after (25) Phocaea, an S-class asteroid, diameter 72 km, discovered in 1853 by the French astronomer Jean Chacornac.Phocaea's orbit has a semimajor axis of 2.402 AU, period 3.72 years, perihelion 1.79 AU, aphelion 3.01 AU, and inclination 21°.6.

Phoebe Phoebe is one of the smaller of the 18 moons of Saturn. Phoebe has an almost circular shape and a reddish color. It rotates on its axis every 9 hours, unlike the other moons of Saturn (except Hyperion) which always show the same face to Saturn. It has a radius of about 110 km and mass of 4 x 10^{18}kg. Phoebe orbits at 12,952,000 km from Saturn. Its orbital period is 550.48 earth days. Phoebe was discovered by W. Pickering in 1898.

Phoenicid meteors A weak southern-hemisphere meteor shower, showing low rates (maximum ZHR 5) between November 29 and December 9. At maximum on December 5, the radiant lies at RA 1h 00m, dec. -52° (northwest of Achernar). Unusually high activity (ZHR 100) occurred for a limited period around maximum in 1956.

Phoenix A constellation of the southern sky, representing the phoenix. Its brightest star is Alpha Phoenicis, magnitude 2.4. Beta Phoenicis is a close double, magnitudes 4.0 and 4.2. Zeta Phoenicis is an eclipsing binary that varies between magnitudes 3.9 and 4.4 in a period of 1.67 days; it has a magnitude 6.9 companion.

Pholus Asteroid 5145, an unusual object in the outer Solar System, discovered in 1992 by the American astronomer David Lincoln Rabinowitz.Pholus lies well outside the main asteroid belt, moving from near the orbit of Saturn out to beyond the orbit of Neptune. It was the second of the Centaur group to be discovered. Pholus is estimated to be 100-200 km in diameter. Its extremely red spectrum is suggestive of a comet but, unlike Chiron, Pholus seems inactive near perihelion. Its orbit has a semimajor axis of 20.30 AU, period 93.0 years, perihelion 8.68 AU, aphelion 31.91 AU, and inclination 24°.7.

Photino A hypothetical fundamental particle that, according to certain theories, is related to the photon. In so-called supersymmetry theories, all bosons have a partner which is a fermion; the partner of the photon (a boson) would be the photino (a fermion). Photinos may have been produced in sufficient numbers in the Big Bang to provide a candidate for the missing mass of the Universe.

Photoheliograph A telescope in which an image of the Sun in white light is recorded on a photographic plate. The first photoheliograph was designed by the English scientist Warren De la Rue in 1857. Daily solar photography with modern photoheliographs is carried out at observatories around the world.

Photometric parallax The distance of a star as inferred from its position on the lower main sequence, where colour and absolute magnitude are tightly correlated by the colour-magnitude relation. If the colour is known, the absolute magnitude may be inferred. The difference between the apparent magnitude, m, and the absolute magnitude, M, is related to the parallax, p, in seconds of arc by the formula $\log p = 0.2(M - m - 5)$.

Photometry The science of the measurement of light. The earliest estimates of star brightnesses were made with the naked eye, which formed a basis for the magnitude scale used in astronomy. Now, professional astronomers use photometers to compare objects under study with standard stars. Broad-band filters are used for the faintest objects to achieve sufficient signal, while brighter stars can be measured with intermediate-band or narrow-band filters. The filters are carefully chosen to reveal characteristics of the spectrum such as absorption lines in stars and galaxies, and emission lines and bands in nebulae. Magnitudes from different filters are usually compared to obtain a colour index. Satellites have extended astronomical photometry into the ultraviolet and infrared.

Photon A particle of electromagnetic radiation. A photon has zero rest mass, zero charge, and travels at the speed of light. The energy, E, of a photon is related to its frequency, f, by the formula $E = hf$, where h is the Planck constant. Hence a photon at a radio frequency is of much lower energy than a gamma-ray photon.

Photosphere The visible surface of a star, from which most of its energy is emitted in the form of visible and infrared radiation. The name means 'light sphere'. The Sun's photosphere is a thin layer about 500 km deep. Its temperature decreases steadily from about 6400 K at its base to 4400 K at the temperature minimum, where it merges with the chromosphere above. This drop in temperature with height causes limb darkening. The photosphere has a ricegrain texture called granulation, caused by rising convection cells of hot gas. Other photospheric features include sunspots, faculae, and filigree structures, all associated with strong magnetic fields. Almost all the features of the Sun's visible-light spectrum originate in the photosphere, including the dark Fraunhofer lines.

Photovisual magnitude (symbol m pv) The brightness of a star measured on a panchromatic (yellow-sensitive) photographic plate exposed through a yellow filter to mimic the wavelength response of the eye. Such magnitudes are similar to photoelectric V magnitudes, although not so accurate. The term is now obsolete; many astronomers are content to use the technique and call the resulting magnitude V.

Physical double A double star in which the components are physically linked by their mutual gravitational attraction and therefore form a true binary star, as opposed to an optical double.

Physical libration A real periodic variation in the rotation rate of a celestial object, as distinct from an optical libration. In addition to its familiar optical librations, the Moon also possesses a very small physical libration in its axial rotation, amounting to less than 2'. This small physical libration allows the moments of inertia of the Moon to be found.

Pi The constant equal to the ratio of the circumference of a circle to its diameter, which is approximately 3.141593.

Piazzi, Giuseppe Italian astronomer and monk. In 1789 he acquired a high-quality transit instrument made by the English optician Jesse Ramsden.From accurate positional measurements made with it, he compiled a star catalogue at the Observatory of Palermo, and in the process discovered the first asteroid, Ceres, on 1801 January 1.

Piazzi's proposed term 'planetoid' lost out to F. W. Herschel's suggestion of 'asteroid'.

Pic Du Midi observatory An observatory at an altitude of 2860 m in the Pyrenees mountains of southwestern France, owned by the French Centre National de la Recherche Scientifique and operated by Paul Sabatier University, Toulouse; its headquarters are in Bagnères-de-Bigorre. Pic du Midi Observatory was founded in 1878 primarily for meteorology, with solar observations starting in 1892. Its main instruments are the 2-m Bernard Lyot reflector, opened in 1979, a 1-m reflector opened in 1963, and several solar telescopes.

Pickering series A sequence ot emission and absorption lines in the visible part of the spectrum, caused by the singly ionized helium atom, He II. Such an ion is very similar to a hydrogen atom in that both contain only a single electron.

However, the helium nucleus is four times heavier and has twice the nuclear electric charge. For these reasons, alternating lines in the Pickering series almost align with lines in the hydrogen Balmer series: Pickering-ß, for example, is at 656 nm, close to the Ha line. The series limit is at 364.4 nm. Four times as much energy is needed to ionize a helium atom as a hydrogen atom, so He II lines are seen only in very hot stars (spectral type O) and accretion disks. The Pickering series is named after E. C.Pickering.

Pickering, Edward Charles American physicist and astronomer, elder brother of W. H. Pickering. As director of Harvard College Observatory, he produced a catalogue of brightnesses for 4260 stars, the Harvard Photometry, in 1884, and extended the work in 1908 with the Harvard Revised Photometry. He was an early exponent of astrophotography, and built up a massive library of photographic plates at Harvard. In 1903 he published the first photographic map of the whole sky. In the 1880s, Pickering began a programme to classify stellar spectra, discovering in the process the first spectroscopic binary, Mizar A. He invented a method for recording several spectra on one plate by placing a large, low-dispersion prism in front of the telescope. The routine work was largely carried out by a team of women assistants, including A. J.Cannon, W. P. Fleming, and Antonia Caetana de Paiva Pereira Maury.The programme yielded the Harvard classification and the Henry Draper Catalogue of stellar spectra.

Pickering, William Henry American astronomer, younger brother of E. C.Pickering. He assisted P. Lowell in establishing the flagstaff Observatory, but later disagreed with Lowell's exotic theories about life on Mars. From Harvard College Observatory's southern station in Peru, Pickering discovered Saturn's satellite Phoebe in 1898. His photographic lunar atlas was published in 1903. Pickering made many predictions of trans-Neptunian planets; a photographic search based on one of them recorded Pluto in 1919, but it was not identified as a planet.

Pico veleta The site in the Sierra Nevada mountains near Grenada, Spain, of the 30-m millimetre-wave telescope of the Institut de Radio Astronomie Millimétrique.

Pictor A constellation of the southern sky, representing a painter's easel. Its brightest star, Alpha Pictoris, is of magnitude 3.3. Beta Pictoris is surrounded by what may be a planetary system in the process of formation. Pictoris contains the fast-moving Kapteyn's Star.

Pinwheel galaxy A name sometimes applied to either of the Sc spiral galaxies M33 (NGC 598) in Triangulum or M101 (NGC 5457) in Ursa Major.

Pioneer A series of US space probes. The first Pioneers, launched in 1958-9, were intended as Moon probes, but none succeeded.

Pioneers 5-9, in 1960-8, were put into orbit around the Sun, monitoring solar activity and conditions in interplanetary space. Pioneers 10 and 11, the last and best-known of the series, were the first probes to reach Jupiter, photographing the planet and studying its environment. Pioneer 10, launched in 1972 March, flew past Jupiter in 1973 December. Pioneer 11, launched 1973 April, passed Jupiter in 1974 December and then became the first probe to reach Saturn, in 1979 September. Both Pioneers are currently on their way out of the Solar System. Pioneer 10 crossed the orbit of Pluto in 1983 June.

Pioneer venus Two NASA space probes to Venus. Pioneer Venus 1 (also known as Pioneer Venus orbiter) went into orbit around Venus, photographing the planet's clouds and mapping its surface by radar. Pioneer Venus 2 (also known as the multiprobe) ejected four sub-probes that analysed the planet's clouds and atmosphere during their descent. One of the sub-probes survived its impact with Venus, continuing to transmit from the surface for over an hour.

Pipe nebula A dark nebula in Ophiuchus, with the curving shape of a pipe. It is one of the largest dark nebulae in the Milky Way, extending for several degrees. Sections of the pipe's stem have the designations Barnard 59 and 65-7, while the bowl is Barnard 78.

Pisces A constellation of the zodiac, representing a pair of fishes. The Sun passes through Pisces between mid-March and the third week of April, and so is in the constellation at the vernal equinox. The constellation's brightest star is Eta Piscium, magnitude 3.6. Alrescha (Alpha Piscium) is a close double star. TX Piscium (also known as 19 Piscium) is a red giant that varies irregularly between magnitudes 4.8 and 5.2. M74 is a 9th-magnitude spiral galaxy seen face-on.

Piscis Australid meteors A meteor shower active during July and August. At maximum (ZHR 5), on July 31, the radiant lies at RA 22h 40m, dec. -30°, west of Fomalhaut. The shower is best observed from more southerly latitudes, and has received relatively little attention. There seems to be little to distinguish this shower from the Capricornid-Aquarid activity which emanates from roughly the same area around the same time.

Piscis Austrinus A constellation of the southern sky, popularly known as the Southern Fish. It contains the bright star Fomalhaut (Alpha Piscis Austrini) but little else of note.

Pistol star The Pistol star is the largest-known star; it is the most massive and the brightest star. It is 10 million times brighter than our Sun and 100 times more massive. This star is at the center of the Pistol Nebula and created it by expelling tremendous amounts of gas in violent eruptions.

Pitch angle A measure of how tightly the spiral arms of a galaxy are wound, defined as the angle between the tangent to a spiral arm and a circle about the galactic centre at the same radius.

Plage A brighter, hotter patch in the Sun's chromosphere, visible in Ha light and the calcium K line. Plages are the chromospheric equivalent of faculae on the photosphere, as can be seen when an active region is near the limb. They are regions of particularly strong magnetic field.

Planck era In the Big Bang theory, the fleeting period between the Big Bang itself and the so-called Planck time when the Universe was 10^{-43}s old and the temperature was 10^{34} K In this period, quantum gravitational effects are thought to have dominated. Theoretical understanding of this phase is virtually nonexistent. It is named after Max Planck.

Plane of the ecliptic The plane of the ecliptic is the plane of our solar system. In the course of a year, the sun traces a path in the sky along the ecliptic. The Earth's axis is tilted at a 23.5° from the ecliptic (which causes the seasons).

Planet An object orbiting a star that is not a brown dwarf but bigger than an asteroid. This is somewhat ambiguous. In our solar system it is really an historical definition which boils down to "a planet is a member of the set {Mercury, Venus, Earth, Mars, Jupiter, Saturn, Uranus, Neptune, Pluto}". All the known planets orbiting other stars are large gas planets so no ambiguity yet arises there. But when we begin to discover terrestrial planets and then smaller objects it will become necessary to make some further arbitrary definition.

Planet "X" Planet "X" was the temporary name given to the then-unknown planet beyond Neptune that disturbed the orbits of Uranus and Neptune. Percival Lowell calculated the rough location of Planet "X's" orbit, but died in 1916 before it was found. This planet was eventually found by C. Tombaugh in 1930 and named Pluto). Planet "X" is also the name given to any as-yet-undiscovered tenth planet. There may be another Planet "X" beyond Pluto.

Planet symbols Each of the planets in our Solar System has been given a symbol; the symbols are pictured above.

Planetarium A device for projecting a simulated view of the heavens on to the inside of a dome; or the building that houses such a projector.

Planetary aberration The angle between the geometric direction to an object in the Solar System at the instant of observation and its apparent direction as seen by a moving observer. It is the combined effect of aberration due to the observer's motion, and the movement of the planet during the time that its light takes to reach the observer.

Planetary alignment Planetary alignment is when many of the planets in the solar system are roughly in a line. A large planetary alignment occured on May 5, 2000, when Mercury, Venus, Earth, Mars, Jupiter, and Saturn were (very roughly) lined up, with all the planets on the other side of the Earth from us. Also, the Moon was almost between the Earth and Sun. This type of alignment happens every few hundred years and has little or no significance.

Planetary nebula A bright cloud of glowing gas and dust surrounding a highly evolved star. A planetary nebula forms when a red giant ejects its outer layers at speeds of about 10 km/s. The ejected gas is then ionized by ultraviolet light from the hot core of the star. As mass is lost this core is progressively exposed, and it ultimately turns into a white dwarf. Planetary nebulae are typically 0.5 L.Y across, and the amount of ejected material is 0.1 solar mass or more. Because the core is so hot the gas in the nebula is highly ionized. The planetary nebula exists for up to 100 000 years, during which time a sizable fraction of the star's mass is returned to interstellar space. Planetary nebulae were so named because they appeared to early observers to resemble a planetary disk. In fact the detailed shapes of planetary nebulae revealed by modern telescopes cover many different types, including ring-shaped dumbbell-shaped, or irregular.

Planetary precession The perturbation of the Earth's orbital plane by the attractions of the planets on the Earth's centre of mass. It causes the equinox to move eastwards along the celestial equator (i.e. in the opposite direction to lunisolar precession) at a rate of about 0'.12 per year. Its effect is much smaller than that of lunisolar precession.

Planetesimal A 0.1-100 km body of rock and/ or ice that is presumed to have formed in the early history of the Solar System. The planets are thought to have grown from the accumulation of planetesimals. Most planetesimals left over from planetary accretion were ejected by perturbations of the planets into the Kuiper Belt and Oort Cloud beyond Neptune. planetoid An obsolete name for an asteroid.

Planetesimal hypothesis The planetesimal hypothesis is a theory about the formation fo the Solar System. It was proposed by Thomas Chrowder Chamberlain (Sept. 25, 1843 - Nov. 15, 1928), an American geologist and teacher. In this theory, a star is supposed to have passed near the Sun, pulling matter away from the Sun. Later, this matter is to have condensed into larger masses, forming the planets.

Planetoid A planetoid is a very small planet. The larger asteroids are sometimes called planetoids.

Planetologist A planetologist is an astronomer who studies the physical features of planets.

Planetology The study of the planets, including their surfaces, interiors, and atmospheres; also known as planetary science.

Planisphere A circular map of the stars covered by a rotating mask that has a window or opening in it. The map and mask are the same size and have time and date markings on the outside. Lining up the time/date marks then expose a given area of the map through the mask so that the stars visible for a given time can then be seen. Planispheres are usually devised so that they can only be used within a certain latitude range.

Planitia A large, low plain on a planetary surface; pl. planitiae. The name is not a geological term, but is used in the nomenclature of individual features, for example Guinevere Planitia on Venus or Sarandib Planitia on Enceladus.

Plaskett's star A 6th-magnitude spectroscopic binary star in Monoceros named after J. S. Plaskett, who discovered in 1922 that it was the most massive binary known; it is also known as V640 Mon. According to current measurements, the individual stars are both blue supergiants of 51 and 43 solar masses. Plaskett's Star lies about 5000 L.Y away.

Plasma A plasma is an extremely hot gas that is composed of free-floating ions (atomic nuclei stripped of some electrons - making the ions positively charged) and free electrons (negatively charged). A plasma behaves much differently than a neutral gas, and is considered the fourth state of matter. A plasma conducts electrical currents. Stars are composed of plasma.

Plasmapause The outer boundary of the plasmasphere, marked by a sharp decrease in plasma density.

Plasmasphere A region within the Earth's magnetosphere containing relatively cool, low-energy plasma at a temperature of 2000 K extending to a distance of about 4 Earth radii (25 000 km), bounded by the inner Van Allen Belt. The particles in the plasmasphere are believed to originate in the ionosphere.

Platonic year The time taken for the Earth's poles to describe one complete circle on the celestial sphere as the result of precession: 25 800 years. PLC relation Abbr. for period-luminosity-colour relation.

Pleiades A prominent open cluster in Taurus, popularly termed the Seven Sisters, and also known as M45. The cluster spans over 1.5 of sky and contains about 100 stars, the brightest of which is 3rd-magnitude Alcyone. Several other members are visible to the naked eye. Its true diameter is about 13 L.Y, although there may also be scattered outliers. The cluster lies 410 L.Y away and is about 80 million years old. Its stars are embedded in a reflection nebula, which is now thought to be the result of a chance encounter rather than the nebula being the remains of the gas cloud from which the stars formed.

Plerion A rare form of supernova remnant in which radiation is being emitted from the central region as well as from the expanding shell; also known as a filled-centre supernova remnant. Plerions tend to be relatively young remnants containing a pulsar, the energy from which makes central regions glow by synchrotron radiation. The best-known example is the Crab Nebula.

Plough Popular name for the shape formed by the stars Alpha, Beta, Gamma, Delta, Epsilon, Zeta, and Eta Ursae Majoris, which resembles the outline of an old horse-drawn plough. An alternative name for the same shape is the Big Dipper. PL relation Abbr. for period-luminosity relation.

Plume A cloud of gas, liquid, aerosols, or particles originating from within a planet. The term is most commonly used for the gas and aerosols emitted by a volcano that drift downwind. The plumes above the erupting volcanoes of Io are the largest and most spectacular such examples, some rising to 280 km altitude and depositing material 500 km from the vent. The plumes seen rising from the surface of Triton by Voyager 2 appear to be from geyser-like eruptions.

Plutino A Plutino (meaning "little Pluto") is a main Kuiper belt objects, ones that have a 2:3 resonance with Neptune (as Pluto is); a 2:3 resonace means that the Plutino completes 2 orbits around the Sun in the same time that it takes Neptune to complete 3 orbits around the Sun. About one-quarter of the Kuiper Belt objects (also called transneptunian objects) are Plutinos.

Pluto (symbol E) The ninth planet from the Sun, and the smallest planet in the Solar System. Its orbit has the greatest inclination of any planet, 17°.1 to the ecliptic, and is the most elliptical of all the planets (eccentricity 0.25). At aphelion Pluto lies 7375 million km from the Sun, but only 4425 million km at perihelion, inside the orbit of Neptune; it last reached perihelion in 1989. Its mean opposition magnitude is +15. Pluto was discovered in 1930 by C. W. Tombaugh. Its rotation axis is tilted at 123° to the upright, so that its rotation is retrograde, and it presents its poles and its equator alternately towards the Sun and the Earth as it moves around its orbit. Its axial rotation period, 6.387 days, is the same as the orbital period of its satellite, Charon, so that Pluto always keeps the same face towards Charon. As seen from Earth, Pluto and Charon underwent a series of mutual occultations in 1985-90. Pluto has an extremely thin atmosphere with a surface pressure of about 10 μbar.

Pogson step method A visual method of estimating the magnitude of variable stars, based on training the eye to recognize the difference between the variable and a comparison star in steps of 0.1 magnitude. Consistent estimates by this method are generally possible only after considerable experience.

Pogson, Norman Robert English astronomer. He discovered eight asteroids, the first, (42) Isis, in 1856. It was in connection with the light curves of variable stars, his other main interest, that he introduced a mathematically rigorous definition of stellar magnitudes based on a logarithmic scale Others, notably the German scientist Carl August von Steinheil had proposed similar scales, but Pogson's advocacy led to its adoption as a universal standard.

Poincaré, (Jules) Henri French mathematician. From 1889 he worked on the three-body problem in both its general and restricted forms, and showed that there are no exact solutions to the n-body problem. He expanded the field of celestial mechanics, for example demonstrating the possibility of chaotic orbits. Poincaré's mathematical studies of special relativity were the first of importance.

Polar cap A bright surface layer of ice or similar substances surrounding the poles of a planetary body, usually varying in size during the year. On Earth, the polar caps are made of water ice, whereas on Mars they consist mostly of carbon dioxide with a central portion of water ice that does not melt. The Martian polar caps can extend down to latitude 60° in winter, but retreat back to 85-87° latitude at the height of summer, revealing the permanent water-ice cap. Of the other bodies in the Solar System only Triton has a definite polar cap, which appears to consist of nitrogen and may vary seasonally.

Polar motion The displacement within the Earth of the axis of rotation. The principal components are an annual term, and a 14-month term known as the Chandler wobble.

Polar orbit An orbit with an inclination close to 90°, passing over, or almost over, the poles of a planet. A satellite in polar orbit will in due course pass over every point on the planet below as the planet rotates on its axis.

Polar ring galaxy A rare type of galaxy, almost always a lenticular galaxy, that has a luminous ring of stars, gas, and dust orbiting over the poles of its disk. Hence, the rotation axes of the ring and disk are almost at right angles. Such a system may be the result of a collision, tidal capture, or merger of a gas-rich galaxy with the lenticular galaxy.

Polaris Polaris (alpha UMi) is the current pole star for the Northern Hemisphere; it is 1 degree from the exact Northern celestial pole. In 1780, Sir William Herschel discovered that Polaris was a double star with a faint companion star. Polaris is a blue-green Cepheid variable star (its size brightness changes periodically, with period of 3.969778 days; it varyies between mag 1.92 and 2.07). Polaris has a relatively dim companion star (9th magnitude). Polaris' distance from Earth has been estimated to be from 360 to 820 light years. At its brightest, Polairs is about 6,000 to 10,000 times brighter than our Sun. It is the larger star at the end of the handle of the Little Dipper (Ursa Minor). Polaris is also called the Lodestar or the Cynosure.

Pole star The naked-eye star nearest the north or south celestial pole. The current north pole star is Polaris, and the south pole star is Sigma Octantis. However, the position of the celestial pole changes with time due to the effect of precession.

Poliakov, Valeri Valeri V. Poliakov is a Russian physician and cosmonaut who broke many space endurance records, and holds the second-longest record for being in space: 678 days over two flights (Sergei Avdeyev beat his record in 1999, having been in space for 758 days). Dr. Poliakov holds the record for the single longest stay in space, 437 days, 17 hours, 58 minutes. Poliakov was on the Russian space station MIR missions Soyuz TM6 and TM7 (240 days, 22 hours, 26 minutes in 1988), and Soyuz TM18 (437 days, 17 hours, 58 minutes in 1994-1995).

Pollux The star Beta Geminorum, magnitude 1.14. It is a KO giant 35 L.Y away.

Population 1 stars Population 1 stars are relatively young stars. They are found mainly in the disk of galaxies.

Population 2 stars Population 2 stars are relatively old stars. They are found mainly in the halo of galaxies.

Populations, stellar There are essentially two types of star regions. I - in which the brightest stars are hot and white; and II - in which the brightest stars are old Red Giants. This is an odd distinction because the first stars to have populated a given region would now be classified as Population II! In our own galaxy Population II stars are more numerous toward the centre of the galaxy while Population I stars are more numerous in the outer regions of the galaxy and spiral arms.

Pore A small, dark area on the Sun's photosphere, from which a sunspot may develop. Pores last less than an hour, and are up to 2000 km across. In an alternative usage, a pore is a sunspot without a penumbra.

Porrima The star Gamma Virginis. It is a binary, consisting of a pair of F-type dwarfs each of magnitude 3.5, which orbit each other in a period of 169 years. Together they appear to the naked eye as a star of magnitude 2.8. Porrima lies 31 L.Y away.

Portia The seventh-closest satellite of Uranus, distance 66090 km, orbital period 0.513 days; also known as Uranus XII. Its diameter is 108 km, and it was discovered in 1986 by the Voyager 2 spacecraft.

Position angle The direction in which one object lies relative to another on the celestial sphere, such as the two components of a double star or the axis of rotation of the Sun or a planet. Position angle is measured in degrees from north via east, although for the axial inclination of the Sun it is measured positive (+) to the east and negative (-) to the west.

Position circle A circle on the Earth's surface on which the observer is situated at some point. The centre of the circle is directly under a celestial body, and its radius is equal to the zenith distance of the star. The position of the circle on the Earth is calculated by measuring the altitude of the celestial body at a particular time. In practice it is possible to select a small part of the circle, called the position line, on which the observer is situated. If a second position line is found from another observation (either of the same object at a different time, or of a different object at the same time) the intersection of these lines identifies the observer's location.

Position-angle effect An error affecting visual estimates of variable stars. It causes a difference in the perceived brightness of a star, relative to another, depending on its position in the field of view. Generally, the star that is either lower in the field or closer to the observer's nose appears relatively brighter. The effect is caused by the variation of sensitivity across the retina, and is independent of variations caused by colour differences such as the Purkinje effect.

Post-nova The stage following a nova outburst, when the star has returned to a quiescent state. Some decades after the eruption the spectrum remains characteristic, with H I and He I emission, accompanied by high-excitation lines such as He II, C III, and N III. Such a spectrum is sometimes encountered in a star for which no outburst has been observed (a quiescent nova).

Poynting-Robertson effect An effect in which dust particles in space are slowed down in their orbits and spiral into the Sun. It occurs because the particles absorb solar radiation and then re-radiate it in all directions, which slows them down. The effect is particularly marked for small (micrometre-sized) particles. However, for the smallest particles (under about a micrometre), radiation pressure is greater. The effect is named after the English physicist John Henry Poynting and the American mathematician and cosmologist Howard Percy Robertson.

PPM star catalogue A catalogue listing the positions and proper motions of 181731 stars north of declination -2.5° for epoch 2000.0, published in 1991 by the Astronomisches Rechen-Institut, Heidelberg. It is based on the same selection of stars as the AGK3, and supersedes both the AGK3 and the Smithsonian Astrophysical Observatoty Star Catalog. PPM is the first large star catalogue of positions and proper motions to be based on the FK5 reference system (the AGK3 and SAO catalogues used the FK4 system). A southern hemisphere extension, containing 197179 stars, was published in 1993.

Preceding Referring to the side of an object, or member of a group of objects, that leads in motion across the sky or across the face of a rotating body. Examples are the preceding component of a double star or a group of sunspots; the preceding side of a planetary feature; the preceding limb of a planet; or the preceding side of a telescopic field of view. The trailing side is said to be following.

Precession This is the slow movement of the celestial poles tracing out large circles on the celestial sphere. It is caused by a slow wobble in the Earth's axis due to the gravitational effects of the Sun and Moon on the Earth's equatorial bulge. A reasonable analogy is that of a spinning top - As the top slows down it will start to wobble. This 'Earth wobble' is very slow, one 'wobble' takes about 26 000 years to complete. The other effect is that the equinoxes are slowly moving or precessing, at the rate of about 50 arcseconds a year. The Vernal Equinox is also called the First Point of Aries but due to the effects of precession the Vernal Equinox is now in the constellation of Pisces and moving towards Aquarius.

Precession of the equinoxes The motion of the equinoxes along the ecliptic, arising from the combined motion of the equator (lunisolar precession) and the ecliptic (planetary precession); also known as general precession.

Pre-nova The precursor to a nova, which is expected to be a form of cataclysmic binary. To date, no object that has been studied in detail has subsequently become a nova, but some novae have been identified on earlier photographs. Of the few pre-outburst light-curves that can be constructed, some appear to show a rise in brightness of 1-4 mag. (perhaps with oscillations), over a period of 1-5 years before the explosion.

Pressure broadening The broadening of spectral lines from a star due to the high pressure of gas increasing the number of collisions between atoms in a star's atmosphere. High pressures are usually due to a strong surface gravity, as occurs on white dwarfs, which have extremely broad hydrogen lines because of this effect.

Pressure ridges Long, narrow wavelike folds in the surface of lava flows; formed where lava may have buckled up against slower moving or stationary lava downstream.

Príbram meteorite The first meteorite fall to be photographed by a camera network, enabling its trajectory and orbit to be determined and the meteorite to be recovered. Cameras operated by the Ondrejov Observatory in the Czech Republic photographed a brilliant fireball on 1959 April 7. Nineteen fragments, totalling 9.5 kg, of an H6 ordinary chondrite meteorite were subsequently found at the impact site, close to the town of Príbram, near Prague. The largest fragment weighed 4.3 kg. Its calculated aphelion was in the outer main asteroid belt.

Primary A primary is a planetary body around which a moon or satellite orbits.

Primary cosmic ray An atomic particle moving close to the speed of light, with extremely high energy.On collision with atoms in the Earth's atmosphere, a primary cosmic ray produces showers of secondary cosmic rays. These secondary particles decay into electrons, positrons, and neutrinos. Deceleration of the electrons and positrons by the atmosphere produces a flash of light which can be observed from the ground with a special telescope, and gives information about the primary cosmic ray.

Prime meridian A fixed circle of longitude on the surface of a body, used as a reference from which other longitudes can be measured west or east. The prime meridian on Earth passes through the transit circle established by G. B. Airy at Greenwich, and defines the zero of terrestrial longitude. On other bodies the prime meridian is defined with respect to a known surface feature, such as a small crater.

Prime vertical The great circle that passes through the observer's zenith, perpendicular to the meridian. It intersects the horizon at the west and east points. The parts of the circle going from zenith through east and west are known as the prime vertical east and the prime vertical west.

Primordial fireball An alternative term for the radiation era in the Big Bang theory, during which the Universe was hot and dense and was dominated by the effects of radiation. The existence of such a phase is strongly suggested by the observed properties of the cosmic background radiation.

Primordial galaxies Galaxies that formed very early in the history of the Universe and which would be observed today at very high redshift. Objects apparently similar to present-day galaxies have been found with redshifts greater than 3, but it is not yet clear how these objects are related to the population of galaxies we observe in our neighbourhood.

Principia The short form of the title of Isaac Newton's treatise Philosophiae naturalis principia mathematica ('Mathematical Principles of Natural Philosophy', 1st edition 1687, 2nd edition 1713). One of the most important scientific books ever written, it deals with the celestial mechanics of the Solar System and for the first time weaves together mathematical analysis and physical observation. It sets out Newton's law of gravitation and Newton's laws of motion, presents derivations of Kepler's laws, and shows how the inverse-square law of gravitation produces motion in an ellipse. There are mathematical treatments of many other phenomena, including the tides and cometary orbits. principle of equivalence.

Prismatic astrolabe A horizontal telescope for measuring the instant at which a star reaches a specific altitude on the celestial sphere. A prism is mounted with one face vertical immediately in front of the telescope's objective lens. Below the prism is a horizontal mirror formed by a dish of mercury. The star's image, and its reflection in the mercury mirror, are reflected internally in the prism. The two images coincide in the focal plane of the telescope when the star's zenith distance is half the angle at the leading edge of the prism (30° for a 60° prism).

Probe A probe is an unmanned spacecraft that is sent to collect data from space. Some probes return to Earth, others are not meant to.

Procyon Procyon (also called Alpha Canis Minoris, Antecanis, and the Little Dog Star) is the brightest star in the constellation Canis Minor and the eighth brightest star in the sky. This yellowish star is also part of the Winter Triangle. Procyon is a relatively close star; it is only about 11.4 light years from us. Procyon is about as seven times more luminous than our Sun.This sub-giant star is near the end of its life; it's hydrogen core is almost spent (it has been converted into helium). Procyon has a companion, a bersly-visible white dwarf called Procyon B (it is only about the size of the Earth and is about 14.9 astronomical units from Proycon A), and this small star makes Procyon wobble from our perspective. The existence of Procyon B was predicted in 1840 by Arthur Julius Georg Friedrich von Auswers (1838-1915) but wasn't seen until 1896 (by John M. Schaeberle (1853-1924)).

Prometheus Prometheus is one of the smaller of the 18 moons of Saturn. Prometheus is a shepherd satellite for the inner edge of Saturn's F Ring. It has a radius of about 74x50x34 km. Prometheus orbits at 139,353 km from Saturn. Its orbital period is 0.613 earth days. Prometheus was discovered by S. A. Collins and D. Carlson/Voyager 1 in 1980.

Prominence spectroscope A spectroscope by which solar prominences can be observed. This is usually done by observing the Ha line and opening the spectroscope slit to reveal any prominences present on the Sun's limb.

Prominence, solar A loop of cooler gas seen above the sun's surface, which sometimes erupts outwards into space.

Promontorium A headland or cape extending from a larger upland on the Moon; pl. promontoria. The name is not a geological term, but is used in the nomenclature of individual features, for example Promontorium Heraclides.

Proper motion The progressive change in position of a star due to its motion relative to the Sun. Absolute proper motion is the change in position relative to an inertial reference frame such as that defined by extragalactic objects. Relative proper motion is obtained by measurements relative to selected reference stars at different epochs. The average absolute proper motion of these reference stars is known as the reduction from relative to absolute proper motion. Relative proper motions can be used to separate members of a star cluster from field stars in the same line of sight. Proper motions are usually listed in star catalogues as changes in right ascension and declination per year or century.

Proplyds Proplyds (short for protoplanetary disks) are disks of dust and gas in space that surround newborn stars. They appear as fuzzy blobs, even when seen through the Hubble Space Telescope. Proplyds may be very young solar system, consisting of a rotating disk of gas that will eventually coalesce into a solar system.

Proportional counter A detector for X-rays and low-energy gamma rays, consisting of a chamber filled with a noble (i.e. unreactive) gas such as argon or xenon. Ionization of the gas by an X-ray or gamma-ray photon in an electrical field produces an avalanche of electrons. Since the size of the electrical pulse is proportional to the energy of the photon, the detector is able to give spectral information. The instrument can also be used as an imaging device when a wire grid arrangement is incorporated into the construction.

Proteus Proteus is the second-largest moon of Neptune. It was discovered using NASA's Voyager 2 mission in 1989. It orbits 117,650 km from the center of Neptune and is about 218 x 208 x 201 km in size. This dark moon has a circular orbit just over the cloud-tops of Neptune. Proteus is dotted with impact craters and is irregularly shaped.

Protogalaxy The precursor of a galaxy. In the Big Bang model, galaxies are thought to form from large gas clouds which collapse under their own self-gravity. A protogalaxy is such a cloud which is just beginning to collapse. No protogalaxy has yet been unambiguously identified from astronomical observations, suggesting that galaxy formation occurs at very high redshifts.

Protoplanet A body presumed to have formed in the early history of the Solar System, from which the major planets grew. Protoplanets are believed to have formed from the aggregation of planetesimals. They eventually became the major planets by sweeping up other bodies that crossed their orbits, in a process of accretion.

Protoplanetarty disk A protoplanetary disk is a rotating disk of dust that surrounds the central core of a developing solar system. This disk eventually coalesces into planets that orbit the star (which forms from the central core).

Protoplanetary nebula 1. An early stage in the formation of a planetary nebula. In this phase the central star has shed its outer layers, exposing the hot stellar core. Ultraviolet light from the core starts to ionize the surrounding cloud of gas and dust, and for a brief phase the circumstellar envelope contains both hot ionized material near the star and cool molecular material far from the star.

2. A cloud from which planets are formed around a newborn star, as in the solar nebula.

Protostar A star in the earliest phase of its life, condensing out of a cloud of gas and dust, before the onset of nuclear burning. Its mass grows over about 100 000 years as material falls on to it from the surrounding cloud. A protostar is not visible at optical wavelengths because the infalling material obscures it, but it is bright at infrared wavelengths.

Protosun The Sun in the initial stages of its formation, nearly 5 billion years ago. The protosun had less mass than the Sun has today, and was larger, with a radius comparable to that of the orbits of the inner planets. Over about a million years the protosun's mass increased through the addition of gas from a surrounding disk, the solar nebula, while the weight of this infalling material made it shrink to only a few times its present size.

Proxigean spring tide The Proxigean Spring Tide (also called the proxigean tide) is a rare,

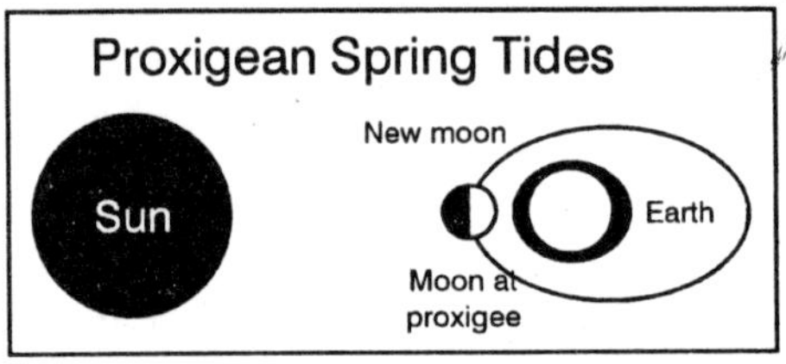

Fig. Proxigean spring tide

unusually high tide. This very high tide occurs when the moon is both unusually close to the Earth (at its closest perigee, called the proxigee) and in the New Moon phase (when the Moon is between the Sun and the Earth). The proxigean spring tide occurs at most once every 1.5

Proxigee For an object orbiting the Earth, the proxigee is the closest perigee. The perigee is the point in each elliptical orbit which is closest to the Earth, but the perigee varies a small amount from orbit to orbit.

Proxima centauri The closest star to the Sun, 4.2 L.Y away, the third member of the Alpha Centauri system. It is an M5.5 dwarf of magnitude 11.2, but is also a flare star and undergoes sudden brightness increases of up to 1 mag. lasting several minutes. Intrinsically it is 20 000 times fainter than the Sun. It is thought to be gravitationally bound to the other two members of Alpha Centauri, which lie 0.1 L.Y farther away. Its orbital period around them must be a million years or more.

Ptolemaic system The ancient Greek geocentric model of the Solar System, as described by Ptolemy. It may be traced back through the work of, for example, Hipparchus, Apollonius, Callippus, and Eudoxus. The Earth is placed at the centre of the Universe, and around it revolve the Moon, Mercury, Venus, the Sun, Mars, Jupiter, and Saturn. Beyond Saturn is the sphere of the fixed stars. In the basic model, each body moves along the circumference of a small circle, the epicycle, whose centre in turn follows the circumference of a larger circle, the deferent, centred on the Earth. In later refinements, Ptolemy introduced two points equally spaced on either side of the Earth: the eccentric and the equant. The centre of the epicycle revolved around the eccentric, not the Earth, and the orbiting body moved uniformly with respect to the equant.

Ptolemy Egyptian astronomer and geographer. He produced the Almagest, a compendium of contemporary astronomical knowledge, drawing on writers, such as Plato and Hipparchus, whose works were kept in the great library at Alexandria. His Ptolemaic system was a geocentric model of the Universe. Highly contrived as it now appears, it successfully accounted for the observed apparent motions of the planets, and remained largely unquestioned until the 16th century, when it was challenged by N. Copernicus. Ptolemy's Geography enjoyed a similar period of dominance (it convinced Columbus that he could sail westwards to India); his Tetrabiblos was an astrological treatise.

Pulsar A radio source from which is received a highly regular train of pulses. More than 600 pulsars have been catalogued since the first was discovered in 1967. Pulsars are rapidly spinning neutron stars, 20-30 km in diameter. The stars are highly magnetized (about 10^8 tesla), with the magnetic axis inclined to the spin axis. The radio emission is believed to arise from the acceleration of charged particles above the magnetic poles. As the star rotates, a beam of radio waves sweeps across the Earth and a pulse is seen, much like the beam from a lighthouse. Pulse periods are typically 1 s, but range from 1.56 ms (millisecond pulsars) to 4.3 s. The pulse periods are lengthening gradually as the neutron stars lose rotational energy, but a few young pulsars are prone to abrupt disturbances known as glitches. Precise timing of pulses has revealed the existence of binary pulsars, and one pulsar, PSR 1257+12, has been shown to be accompanied by objects of planetary mass.

Pulse broadening A phenomenon in which irregularities in interstellar matter broaden the pulses from a pulsar by scattering the radio waves. Pulse broadening is more pronounced at lower radio frequencies, and takes the form of a characteristic 'tail' to the pulse. It is closely related to interstellar scintillation.

Pulse profile The variation with time of the signal strength from a pulsar, usually plotted against pulse phase where one complete rotation is equal to 360 of longitude. Pulse profiles vary erratically from pulse to pulse, but the integrated pulse profile, the average of a large number of pulse profiles, is characteristic of each individual pulsar. The shape of the profile is believed to reflect the structure of the pulsar's emitting region.

Pulsar glitch A sudden change in the pulsar period due to a sudden shift in the crust of the neutron star (a "starquake").

Pulsar, millisecond As the name implies, these are pulsars with periods measured in terms of milliseconds (thousandths of a second). The shortest have a period of about one and two milliseconds. Millisecond pulsar periods are very constant. They don't slow down much implying a weak magnetic field. Most millisecond pulsars are found in binary systems. It is believed that millisecond pulsars are old pulsars that have had their spin rates increased through the accretion of angular momentum-containing matter from the other star in the binary. Mass transfer has spun up these pulsars back to very fast spin rates. The idea that the millisecond pulsars are actually old systems gains support from the presence of millisecond pulsars in globular clusters.

Pulsating variable Any intrinsic variable in which the flow of energy from the interior varies in a more or less rhythmic manner, causing large-scale motion in the outer layers and resulting in brightness changes. The pulsation may be essentially symmetrical (radial pulsation), or take the form of waves travelling across the star's surface (non-radial pulsation), or combine both forms. As stars evolve they all pass through one or more pulsating stages.

Pulsation mode The way in which pulsations occur in a star. It may be described in terms of either the frequency of pulsation, or the physical motions of the star's surface. Pulsation at the lowest (forcing) frequency is known as the fundamental mode; at twice the frequency, the first overtone mode; and so on. A star may pulsate either with approximately spherical symmetry (radial pulsation), or as a series of waves running across the surface.Pulsation may occur in a single mode or in multiple modes, depending on the type of star.

Puppid-velid meteors A complex southern-hemisphere meteor shower, producing moderate activity from a number of radiants in Puppis, Vela, and Carina between late November and January. The peak ZHR may be as high as 15 at the two principal maxima on December 9 and 26, on which dates the main radiant lies at RA 9h 00m, dec. -48° and RA 9h 20m, dec. -65°, respectively.

Puppis A significant constellation of the southern sky, representing the stern of the ship Argo Navis. Its brightest star is Naos. Xi Puppis is an optical double, magnitudes 3.4 and 5.3. L Puppis is another optical double, one star of magnitude 4.9 and an unrelated semiregular red-giant variable that ranges between magnitudes 2.6 and 6.2 in a period of about 140 days. V Puppis is an eclipsing binary, ranging from magnitude 4.4 to 4.9 in a period of 1.45 days. M46, M47, NGC 2451, and NGC 2477 are all large, bright open clusters.

Purkinje effect An error affecting visual estimates of variable stars that arises when comparing stars of different colours, especially when one is red. It is caused by the differing sensitivity of the rods and cones in the retina to light of various wavelengths. It varies between observers, and also depends on the aperture of instrument used (i.e. on the amount of light gathered by the system), because the cones of the eye—unlike the rods—do not function at low light levels. It is named after the Czech physiologist Johannes Evangelista Purkinje.

Q

Q-class asteroid A rare class of asteroid whose members are distinguished by moderately high albedos and reflectance spectra that display a strong absorption feature at wavelengths shorter than 0.7 μm and a modest absorption feature near 1 μm. The spectra are interpreted as being similar to ordinary chondrite meteorites. At present, only Apollo and a few other near-Earth asteroids have been identified as belonging to this class.

QSO(quasi stellar object QSOs are objects that, at first glance, appear as normal stars. Upon closer inspection, however, QSOs have very large redshifts (i.e. the light they emit is strongly displaced toward the red end of the spectrum). Although their exact nature is controversial, they are commonly considered to be extremely distant, unusually bright nuclei of galaxies. If so, then the light we see from them would have been emitted when the universe was a fraction of its present age.

Quadrant An obsolete navigational instrument used for measuring angles. It consisted of an engraved arc of a quarter of a circle, with a plumb line suspended from the centre of the circle. Stars were sighted along one arm and their elevation was read off the scale against the plumb line.

Quadrantid meteors A meteor shower active between January 1 and 6, but with marked activity confined mainly to a 12-hour interval around January 3/4. At maximum, the radiant lies at RA 15h 28m, dec. +50, a few degrees northwest of Tau, Phi, and Nu Herculis in what used to be the constellation of Quadrans Muralis. The ZHR at maximum can be as high as 120, placing the Quadrantids among the three most active regular annual showers. Quadrantid meteors are moderately fast, with a geocentric velocity of 41 km/s. The stream orbit oscillates up and down relative to the ecliptic over the centuries. As a result, the stream will fail to encounter the Earth after about AD 2200. Attempts to identify a parent body are also confounded by this rapid motion of the orbit, although a connection with Comet 96P/Machholz 1 has been suggested.

Quadrature The position of the Moon or another planet when it is at right angles to the Sun as seen from Earth. An example would be when the Moon is seen at First or Last Quarter phases.

Quantum cosmology The study of the very earliest stages of the Universe after the Big Bang, including the properties of the initial singularity and of the Planck era in general. This field is highly speculative since it requires an understanding of quantum gravitational effects, and there is as yet no satisfactory theory of quantum gravity.

Quantum efficiency The efficiency of a detector in recording radiation. It is the ratio of the number of photons usefully detected to the number (of a given frequency) that strike the detector. An ideal detector would have a quantum efficiency of 1. Photographic emulsion records only a small proportion of incoming photons, and thus has a low quantum efficiency of 0.001-0.01. Electronic detectors, such as CCDs and photomultipliers, have quantum efficiencies of 0.6-0.8 and hence can record faint objects during short exposures. The quantum efficiency of the human eye is about 0.01-0.05. Quantum efficiency is frequently expressed as a percentage.

Quantum gravitation A theory in which gravitational interactions between bodies are described by the exchange of hypothetical elementary particles called gravitons. The graviton is the quantum of the gravitational field. Gravitons have not been observed, but are presumed to exist by analogy with photons of light.

Quaoar Quaoar (also known as 2002 LM60) is a newly-discovered object located in the far reaches of our Solar System, about a billion kilometers beyond Pluto (in the Kuiper Belt). Its designation is Kuiper Belt Object 2002 LM60. Quaoar is over half the diameter of Pluto (Quaoar is about 800 miles=1300 km across). It has an orbital period of about 288 years, with a nearly circular orbit. Michael Brown and Chadwick Trujillo (of Cal Tech) discovered Quaoar in June, 2002, using the Palomar 48-inch telescope. This comet-like object is made up of low-density ices mixed with rock. Quaoar was named for a Native American (Tongva) god.

Quark star A hypothetical star intermediate in density between a neutron star and a black hole. Such a star would consist of free quarks. The forces between the quarks balance the gravitational forces. Quark stars are unlikely to exist in reality, but some models for the cores of neutron stars suggest that the neutrons break up into a broth of quarks.

Quarter moon A quarter moon looks like half a circle. This moon has completed one quarter of an orbit around the Earth from its full or new position, and one quarter of the moon's surface is visible from Earth. It is also called a half moon.

Quasar An object with a high redshift which looks like a star, but is probably the very luminous active nucleus of a distant galaxy. The name is a contraction of quasistellar, from their star-like appearance. The first quasars discovered were strong radio sources (quasi-stellar radio sources, or QSSs), but many more are now known which are relatively radio-quiet (quasi-stellar objects, or QSOs). At the distance implied by the redshift, the nucleus must be up to 100 times brighter than the whole of a normal galaxy. Yet some quasars vary in brightness on a time-scale of weeks, indicating that this huge amount of radiation originates in a volume only a few light weeks across. The source may therefore be an accretion disk around a black hole with 107 or 108 solar masses. Some quasars show little change in their light output, while others are much more variable: for example, 3C 279 has varied by a factor of nearly 500 in four months.

Quasi-stellar radio source A quasar which is a strong radio source. This was the first type of quasar to be discovered. It has been found that the intrinsically brightest quasars are the most likely to be strong radio sources, while only a few per cent of lower-luminosity quasars show strong radio activity.

Quiescent prominence A long-lived solar prominence that lasts up to several months, changing little in appearance. Quiescent prominence are arch-shaped, several hundred thousand kilometres long, a few thousand kilometres thick, and up to 50 000 km high. When seen against the solar disk they appear as darker features called filaments. Quiescent prominences are found to the poleward side of active regions. Over the course of the 11-year sunspot cycle they gradually migrate to higher latitudes, eventually forming a polar crown at sunspot minimum. Quiescent prominences are most frequent on the rising portion of the sunspot cycle.

Quiet sun The Sun when it is at or near minimum activity in its 11-year cycle, and the numbers of sunspots and active regions are at their lowest. At such times, activity is still present in the form of small X-ray bright points, prominences, and some coronal features.

R

R association A stellar association containing a number of reflection nebulae (hence the R). In such associations the stars are less than a million years old and are still embedded in the nebula from which they formed. Because of this they may be obscured, and their light is visible only after reflection from nearby dust.

R Canis Majoris star A binary star with components of widely differing masses, and somewhat ambiguous spectral features. The type star is a semidetached Algol star, with a late-A-type primary (1.5 solar masses) and a G-type subgiant of very low mass (about 0.1 solar mass) and evidence of circumstellar material. The stars in this group have periods of the order of 1 day that are variable, and thus imply mass transfer or mass loss.

R Coronae Borealis star A type of eruptive variable that undergoes occasional fades of variable amplitude, with a sudden onset and generally slower recovery, often with major fluctuations; abbr. RCB. The fades have depths of 1-9 mag., and last from months to years. The onset, depth, and duration of the fades are unpredictable. They occur when clouds of carbon particles in the outermost atmospheric layers are ejected in the direction of the observer, and absorb the light from the underlying photosphere. When the photosphere is obscured the star enters a chromospheric phase in which the residual light originates in the deep overlying chromosphere. The stars are high-luminosity supergiants of spectral types B-M and R, hydrogen-deficient but rich in helium and carbon. The majority also appear to be pulsating variables with amplitudes of up to a few tenths of a magnitude, and periods of 30-100 days.

R magnitude The magnitude of a star measured with an R filter. The filter used has several different definitions. The central wavelength and bandwidth of the R filter in Kron-Cousins RI photometry are 638 and 138 nm; in Johnson photometry they are 720 and 220 nm; in the six-colour system, 719 and 180 nm; and in RGU photometry, 638 and 50 nm. The R most frequently used nowadays is that of the Kron-Cousins system. R is usually measured photoelectrically, except in the RGU system, which is photographic.

Radar astronomy The study of the bodies in the solar system by radio waves reflecting off of them. A radio telescope is used to transmit a powerful pulse of radio waves, and then to detect the faint echo that returns a few seconds later.

Radar scatterometer A radar scatterometer is a device that emits and detects reflected microwave radiation The scatteromenter measures the radar echo (or radar backscattering cross-section, called "sigma-0") of the Earth's surface. Different terrains (like flat land, rough surfaces, and water) scatter microwaves differently. A radar scatterometer has been used to map Earth's polar regions.

Radiant The region of sky from which a meteor shower appears to emanate, as a result of perspective. In reality, meteoroids in a meteor stream pursue parallel orbits around the Sun, and enter the atmosphere along parallel trajectories. As an observational convenience, a radiant diameter of 8° is adopted. A meteor shower is named for the constellation in which its radiant lies (e.g. the Perseids emanate from Perseus). Because of the Earth's orbital motion, a radiant appears to drift eastwards by about 1° each day. Photographic and telescopic studies can reveal true radiant diameters and a degree of structure; splitting of a meteor stream by planetary perturbations, for instance, can result in showers with multiple sub-radiants.

Radiation 1. Energy propagating in the form of electromagnetic waves or photons

2. A stream of particles, most commonly protons and electrons.

Radiation belt A toroidal region within a planet's magnetosphere where charged particles become trapped for long periods, spiralling backwards and forwards between 'mirror points' at opposite ends of magnetic field lines. The Van Allen Belts surrounding the Earth are examples. Extensive radiation belts also surround Jupiter. Energetic particles in planetary radiation belts can present a hazard to spacecraft, causing damage to electronic components.

Radiation era The period from about 10^{-43}s (the Planck era) to 30000 years after the Big Bang. During this time, the expansion of the Universe was dominated by the effects of radiation or high-speed particles (at high energies, all particles behave like radiation). The lepton era and the hadron era are both subdivisions of the radiation era. The radiation era was followed by the matter era, during which slow-moving particles dominated the expansion of the Universe.

Radiation temperature The calculated surface temperature of a celestial body, assuming that it behaves as a black body. The radiation temperature is the same as the effective temperature but is usually measured over a narrow portion of the electromagnetic spectrum, such as the visible range; this gives the optical temperature.

Radiative zone The region of a star in which radiation, and not convection, is the main process by which energy is transferred outwards. Such zones occur in the cores of low-mass stars and the envelopes of high-mass stars.

Radio astronomy The study of the Universe in the radio part of the electromagnetic spectrum. There is a wide radio window covering wavelengths from about 1 mm to 30 m, almost all of it accessible from ground-based observatories, both day and night. Emission mechanisms are either thermal (black-body radiation, free-free transition) or non-thermal (mainly synchrotron radiation). Maser sources are also common. Study of emission and absorption lines from molecules provides information about conditions in interstellar space. The 21-cm hydrogen line has proved to be a particularly valuable probe of the structure of our Galaxy and others.

Radio astronomy began in the 1930s with the pioneering work of K. G. Jansky and G. Reber, but it was not until after World War II that major research groups were set up. The first sources to be identified were the Sun, the Milky Way, and supernova remnants such as the Crab Nebula. As radio telescopes improved in sensitivity and resolution, radio galaxies and quasars were identified. New understanding of the late stages of stellar evolution came with the discovery of pulsars. The counting of radio sources of different brightnesses (and hence presumably different distances) has helped astronomers understand how the Universe has evolved, and the discovery of the cosmic background radiation has provided direct support for the Big Bang theory of cosmology.

Radio bubble The "radio bubble" is a sphere centered on Earth that is expanding at the speed of light (the speed of radio transmissions). The leading edges of the bubble correspond with the first artificial radio signals that escaped the Earth's atmosphere. The radio bubble now extends beyond Alpha Centauri.

Radio emission Emissions of the sun in radio wavelengths from centimeters to dekameters, under both quiet and disturbed conditions.

Type I. A noise storm composed of many short, narrow-band bursts in the metric range (300 - 50 MHz).

Type II. Narrow-band emission that begins in the meter range (300 MHz) and sweeps slowly (tens of minutes) toward dekameter wavelengths (10 MHz). Type II emissions occur in loose association with major FLAREs and are indicative of a shock wave moving through the solar atmosphere.

Type III. Narrow-band bursts that sweep rapidly (seconds) from decimeter to dekameter wavelengths (500 - 0.5 MHz). They often occur in groups and are an occasional feature of complex solar active regions.

Type IV. A smooth continuum of broad-band bursts primarily in the meter range (300 - 30 MHz). These bursts are associated with some major flare events beginning 10 to 20 minutes after the flare maximum, and can last for hours.

Radio galaxy A galaxy that is an unusually powerful emitter of radio waves. The output of a radio galaxy can be up to 10^{38} watts, a million times greater than a normal galaxy such as our own. Radio galaxies have a compact radio nucleus coincident with the core of the visible parent galaxy, a pair of opposed jets emerging from the nucleus, and a pair of lobes (1) far outside the visible confines of the galaxy. The galaxy is almost always a giant elliptical, which may be the result of the collision and merger of two or more smaller galaxies. The source of the radio galaxy's energy is believed to be a massive black hole in the galactic nucleus from which the jets emerge, delivering energy to the lobes. Notable radio galaxies include Centaurus A, Cygnus A, and Virgo A.

Radio meteor A meteor detected by forward scatter, in which its ionization trail momentarily reflects the signal from a transmitter beyond the observer's horizon. Raw counts of reflections can be processed to give an index of meteor shower activity, even in cloudy or daylit conditions. As with radar meteors, most events detected by forward scatter are produced by much smaller meteoroids than those which cause visual meteors. Small particles produce short, sharp signals, while larger meteoroids give longer-lasting reflections.

Radio source Any cosmic source of radio waves. Major celestial radio sources include the Sun, Jupiter, flare stars, novae, pulsars, supernova remnants, H II regions, the Milky Way, the galactic centre, radio galaxies, quasars and other active galaxies, and the cosmic background radiation. Radio sources are designated by a number of methods. The first sources were named after the constellation in which they lay (e.g. Sagittarius A, Virgo A), but later they were labelled with a serial number and a prefix denoting the survey in which they were catalogued (e.g. 3C for the Third Cambridge Survey). Modern practice is to identify sources by their coordinates, either right ascension and declination, or galactic coordinates. Radio source counts can yield important information about the early Universe.

Radio star An obsolete term for a cosmic source of radio waves. It is now known that the vast majority of so-called 'radio stars' are galaxies and other non-stellar objects, and for them the term radio source is now used instead. However, some stars, such as novae and flare stars, do emit radio waves.

Radio window A region of the electromagnetic spectrum in which the Earth's atmosphere is relatively transparent to radio waves. It extends over a wavelength range of approximately 1 cm to 30 m. At the short-wave end of the window, radio waves are blocked by water vapour, and at the long-wave end by the Earth's ionosphere. Most of the radio spectrum is occupied by broadcasting and telecommunications services and only a few narrow bands are kept free for radio astronomy, notably those coinciding with the position of important spectral lines, such as the 21-cm hydrogen line at 1420 MHz.

Radiosonde A radiosonde is a small device that measures many atmospheric conditions high in the atmosphere, like temperature, wind speed and direction, air pressure, and humidity while suspended from a weather balloon. The device has a radio transmitter and sends its data back to ground recorders. A radiosonde can ascend to about 100,000 feet (30,000 m), giving a 3-dimensional model of the weather. The radiosonde was developed in the 1930's.

Rampart crater A crater surrounded by a ring of low hills. Such craters are thought to be formed by the impact of a meteorite on a planet with an atmosphere or a wet or muddy surface. The impact produces a splash pattern and surface flow within the ejecta blanket, forming a rampart at its edge. An example is the 18-km Martian crater Yuty.

Ramsden disk The disk of light seen through the eyepiece of a telescope pointed at bright sky, as observed from some distance behind the eyepiece. It is actually the image of the objective, reduced in size. The magnification given by an eyepiece can be estimated by dividing the diameter of the objective lens by that of the Ramsden disk when the telescope is focused on infinity.

Ranger A series of NASA Moon probes, originally intended to photograph the Moon as they approached and then to eject an instrument package on to the surface. After a sequence of failures the Rangers were changed into photographiconly missions. Three probes were successful, sending back increasingly detailed pictures before they crashed into the Moon. The Ranger pictures of the lunar surface assisted the design of the later Surveyor soft landers.

Rasalgethi The star Alpha Herculis, an M5 supergiant or bright giant that is one of the largest stars known, with an estimated diameter some 500 times the Sun's. It is a semiregular variable that ranges between magnitudes 2.7 and 4.0 with a period of 100 days or so, although there are also longer-period variations lasting about 6 years. It is a binary, with a companion of magnitude 5.4 that appears greenish by contrast with the orange primary; this star is itself a spectroscopic binary. The orbital period of the visual pair is estimated at around 3600 years. Rasalgethi lies 630 L.Y away.

Ra-shalom Asteroid 2100, the second-largest of the Aten group, diameter 2.4 km. It was discovered in 1978 by the American astronomer Eleanor Kay Helin, née Francis. Its name comes from the Egyptian Sun-god Ra, and Shalom, the Hebrew greeting meaning 'peace'. It is of C class. Ra-Shalom's orbit has a semimajor axis of 0.832 AU, period 0.76 years, perihelion 0.47 AU, aphelion 1.20 AU, and inclination 15U+00B0.8.

Rays, crater Bright streaks radiating from young impact craters. They are prominent on the

Moon. Rays appear to consist of crushed surface rock thrown out by the impact. Clusters of secondary impact craters are often found inside rays. Some ray systems, such as those from the large lunar craters Tycho and Copernicus, stretch for hundreds of kilometres. Similar rays are found on Mercury and on the satellites Ganymede, Callisto, and Oberon.

R-class asteroid A rare class of asteroid, distinguished by a moderately high albedo and a reflectance spectrum with a strong absorption feature at wavelengths shorter than 0.7 μm and a fairly strong absorption feature near 1 μm. At present only (349) Dembowska, diameter 164 km, has been allocated to this class.

Reber, Grote Grote Reber (1911-2002) was a radio engineer and pioneering amateur astronomer who was the first person to follow up Karl Jansky's discovery of radio waves coming from space. Reber built a 9-meter parabolic reflector dish radio antenna in his yard in Illinois - it was tte first radio telescope used for astronomy. He detected the first signals (at a frequency of 160 megahertz, about 2m wavelangth) in 1939, using his third receiver. Reber's work led to many developments in radio astronomy; he made the first radio maps of the sky and showed that the brightest areas corresponded to the center of the Milky Way. Reber started the field of very long-wavelength/low-frequency (1-2 MHz, 150-300 m wavelength) radio astronomy, moving to Tasmania (an island off the southeastern coast of Australia) where these radio waves can be received (because the long-wavelength radio waves can get through the Earth's ionosphere over that part of the globe due to a hole in the ozone layer).

Recombination epoch The era reached about 300000 years after the Big Bang when the Universe cooled to a temperature of a few thousand degrees, allowing electrons and protons to combine to form hydrogen atoms. The occurrence of recombination is closely associated with the decoupling of matter and radiation, so observations of the cosmic background radiation provide a means of studying the Universe when it was roughly this age.

Recurrent nova A cataclysmic binary that has undergone repeated outbursts similar to those found in novae. Five or six stars are assigned to this class but only two, T Coronae Borealis (the Blaze Star) and RS Ophiuchi have much in common, giant secondaries and long orbital periods (227.6 and 230 days respectively). These two stars appear to be similar to an RR Telescopii star or symbiotic nova.

Recycled pulsar A pulsar of unusually low magnetic field (1-100 tesla) and low spin-down rate, frequently of very short pulse period, and often found in a binary system. Recycled pulsars are believed to be ordinary pulsars which have lost energy and faded out, and then been spun up again by accretion of gas from a companion star in a binary system. A high proportion of recycled pulsars are found in the cores of globular clusters, where the high density of stars makes the capture of an old neutron star into a binary system more likely. The first recycled pulsars to be discovered had very short pulse periods and are known as millisecond pulsars, but later examples had much longer periods.

Red dwarf A cool, faint, low-mass star lying at the lower end of the main sequence. Red dwarfs have masses and diameters less than half those of the Sun. They are red because of their low surface temperatures, less than 4000 K, and are of spectral type K or M. Red dwarfs are the most common type of star, and also the longest-lived, with potential lifetimes greater than the current age of the Universe. Because of their low luminosity, no more than 10% of the Sun's, they are inconspicuous. Barnard's Star and Proxima Centauri are nearby examples. Many red dwarfs are flare stars.

Red giant A cool, large, and highly luminous star. Red giants are stars that have left the main sequence, having exhausted the hydrogen fuel in their cores, and are powered by nuclear reactions between elements heavier than hydrogen. They owe their colour to their cool surface temperatures (below 4000 K, spectral type K or M). Because of their large diameters, 25 times that of the Sun or more, they are highly luminous, many hundreds of times that of the Sun. Massive mainsequence stars may evolve into red giants after passing through a blue-giant phase, while lower-mass stars, such as the Sun, will evolve directly into a red giant. Many red giants are pulsating variable stars.

Red planet Popular name for the planet Mars, arising from its distinctive ruddy colour.

Red shift The red shift is an increase in the wavelength of the light that is emitted from an object that is moving away from us. This increase in wavelength makes the object appear to be redder than it actually is. For example, when a star is travelling away from Earth, its light appears redder (the light waves are elongated, lengthening the wavelength). The expansion of the universe was discovered when E. Hubble observed that the light from almost all other galaxies was red-shifted. Compare with blue shift.

Red supergiant A cool, massive star near the end of its life that has expanded to a size from a hundred to a thousand times the diameter of the sun

Redshift, cosmological A redshift caused by the expansion of space. The wavelength of light increases as it traverses the expanding universe between its point of emission and its point of detection by the same amount that space has expanded during the crossing time.

Reference star A star whose position and/or proper motion are known, so that it can be used to define a local frame of reference for relative positions or proper motions of other stars in the same area of sky.

Reflectance spectrum A spectrum which shows the proportion of sunlight reflected from the surface of a planet, planetary satellite, or asteroid over a range of wavelengths. Various minerals on the surface reflect sunlight in characteristic ways, so the reflectance spectrum of a body provides information about its mineralogical composition.

Refracting telescope A refracting telescope uses two lenses which magnify what is viewed; the large primary lens does most of the magnification. The first refracting telescope was invented by Hans Lippershey in 1608.

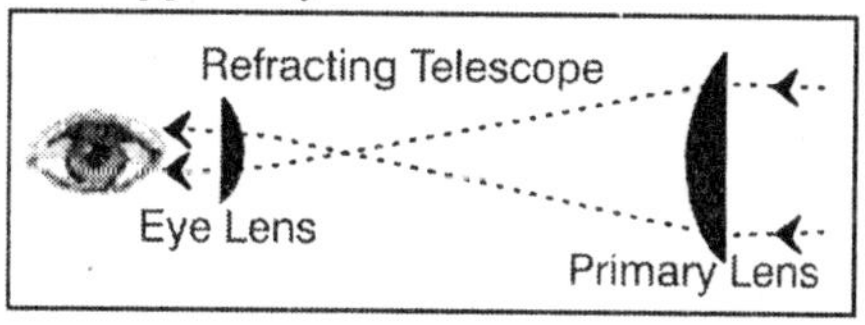

Fig. Refracting telescope

Reflection nebula A cloud of interstellar gas and dust that appears bright because it reflects or scatters starlight. The light from a reflection nebula has the same spectral lines as the starlight it reflects, but it is usually bluer and may be polarized. Reflection nebulae often occur together with emission nebulae in regions of recent star formation. The Pleiades cluster is surrounded by a reflection nebula.

Reflection variable A type of close binary system in which the variations arise largely from non-uniform surface brightness caused by the reflection effect; abbr. R. The amplitude may be 0.5-1.0 mag. even when (as in many cases) eclipses are absent.

Regolith A layer of loose and broken rock and dust on a planetary surface. It is analogous to a soil, but has no biological content. The lunar regolith, 5-15 m deep, is formed by meteoritic and micrometeoritic bombardment. Mercury is thought to have a regolith produced by the same process, whereas on Mars the regolith may include deposits from the erosional effects of wind and ice, and volcanic ash.

Regular galaxy cluster These are great groupings of galaxies into huge spherical distributions that have large numbers of galaxies concentrated in their centers. The tend to contain thousands of galaxies and to have many bright elliptical and S0 type galaxies.

Relative orbit The orbit of a celestial object with respect to the body it is orbiting. The Moon and Earth perform elliptical orbits about their centre of mass, a point about 1600 km below the Earth's surface. The Moon's relative orbit about the Earth is of the same shape, but with the Earth's centre at a focus of the ellipse. Similarly, the planets have relative orbits about the Sun and each component of a binary star has a relative orbit with respect to the other.

Relative sunspot number (symbol R) A measure of the number of spots on the Sun. It was introduced in 1848 by J. R. Wolf, and was formerly known as the Wolf number or Zürich relative sunspot number. The relative sunspot number is calculated by taking into account both the total number of individual spots (f) and the number of sunspot groups (g), plus a factor k that represents the efficiency of the observer and telescope: $R = k(10g + f)$. The factor k is equal to 1 for Wolf's own observatory at Zürich. The relative sunspot number used nowadays is based on observations from a network of 25 observing stations; the observatory at Locarno, Switzerland, acts as a reference station to preserve continuity with the Wolf numbers. Values are published by the Sunspot Index Data Centre in Brussels.

Relativity Albert Einstein formulated the Theories of Special and General Relativity. Special relativity supplanted Newtonian mechanics, yielding different results for very fast-moving objects. The Theory of Special Relativity is based on the idea that speed has an upper bound; nothing can pass the speed of light. The theory also states that time and distance measurements are not absolute but are instead relative to the observer's frame of reference. Space and time are viewed as aspects of a single phenomenon, called space-time. Energy and momentum are similarly linked. As a result, mass can be converted into huge amounts of energy, and vice versa, according to the formula E=mc2. General Relativity expands the theory of special relativity to include acceleration and gravity, both of which are explained via the curvature of space-time.

Resolution (spectral or frequency) Similar to spatial resolution except that it applies to frequency, spectral resolution is the ability of the telescope to differentiate two light signals which differ in frequency by a small amount. The closer the two signals are in frequency while still allowing the telescope to separate them as two distinct components, the higher the spectral resolution of the telescope.

Resolving power The ability of a telescope to image two closely spaced objects as separate.

Retardation The difference between the times of moonrise on successive nights, due to the Moon's motion in its orbit. It has an average value of 50'.4, but the actual figure can vary widely depending on the Moon's declination. On about September 23, the autumnal equinox, the full Moon rises only 18 min later on successive evenings. This is known as the harvest Moon. A similar effect in October is called the hunter's Moon.

Reticulum The seventh-smallest constellation, lying in the southern sky. Reticulum represents the reticle used in an eyepiece for measuring star positions. Its brightest star is Alpha Reticuli, magnitude 3.4. Zeta Reticuli is a wide double of stars similar to the Sun, magnitudes 5.2 and 5.5.

Retrograde motion The movement of a body such as a planet from east to west on the celestial sphere; or the movement of a body in its orbit in a clockwise sense as seen from above the Sun's north pole; or the rotation of a body on its axis in a clockwise sense as seen from above the Sun's north pole. Motion in the opposite direction is termed direct or prograde motion. Objects with retrograde motion have an orbital or axial inclination greater than 90°. The outer planets appear to move in a retrograde direction temporarily when the Earth catches them up and overtakes them; they are said to describe a retrograde loop.

Reversing layer A supposed region in the atmosphere of a star, particularly the Sun, where the dark Fraunhofer lines of the star's spectrum were once presumed to originate. At the end of the 19th century the solar chromosphere was thought to be the reversing layer (its spectrum being the reverse of the Fraunhofer spectrum), but it is now known that the Fraunhofer lines are produced in the photosphere, along with the continuum radiation, so that a specific reversing layer does not exist.

Revolution The movement of one body in orbit around another, or around a common centre of mass, such as the monthly revolution of the Moon around the Earth or the yearly revolution of the Earth around the Sun. For artificial satellites, revolutions are counted relative to a given point on the planet's surface. Because the Earth is rotating from west to east, a satellite in a west-to-east orbit takes slightly longer to complete one revolution of the Earth than to complete one orbit relative to a fixed point in space. Hence a spacecraft such as the Space Shuttle may make 16 orbits of the Earth in a day but only 15 revolutions.

Rhea The second-largest satellite of Saturn, 1530 km in diameter; also known as Saturn V. Rhea orbits Saturn in 4.518 days at a distance of 527040 km. Its axial rotation period is the same as its orbital period. It was discovered in 1672 by G. D. Cassini. Rhea has an icy surface, saturated with impact craters.

Rho cassiopeiae star One of a small group of yellow, cool, hypergiant, pulsating variables. The type star is itself a semiregular variable that has undergone shell episodes, with low-velocity ejection and corresponding fades. Stars in the group have periods of 60-400 days, and they may all exhibit non-radial pulsation. The stars have masses slightly below 40 solar masses.

Richter scale The Richter scale is a logarithmic scale that measures the intensity of an earthquake. It was developed in 1935 by Charles F. Richter. The magnitude of an earthquake is calculated from the logarithm of the amplitude of waves recorded by seismographs. Beno Gutenberg also contributed to the more general application of the Richter scale. A magnitude 2.0 or less earthquake is called a microearthquake and is not felt by people. A magnitude 4.5 or more earthquake can be measured by seismographs all over the world.

Richter, Charles F Charles Francis Richter (April 26, 1900- April 30, 1985) was a geologist who developed the Richter scale, a logarithmic scale that measures the intensity of an earthquake. He developed it in 1935 at the California Institute of Technology.

Right ascension (RA) Celestial coordinate which is equivalent to the longitude of an object on Earth. The starting point for longitude on Earth is at Greenwich England, whereas in the celestial coordinate system it is at a point in the sky called the vernal equinox. The two great circles defined by the celestial equator and the ecliptic, which is defined as the plane of the solar system, cross each other at 2 points: at the vernal equinox (first day of spring) and the autumnal equinox (first day of autumn). The vernal equinox is the starting point for the RA coordinate system.

Rille A well-defined, long, narrow valley on the Moon with steep walls and roughly parallel sides. There are three main types: straight (or linear) rilles which appear to be graben, where the surface has dropped between two parallel faults; arcuate rilles, which are similar but curved in plan view, and often concentric with a circular mare basin; and sinuous rilles, which are different in origin.

Rima A long, well-defined narrow furrow on a planetary surface; pl. rimae. The name, which means crack or fissure, was first used for features on the Moon in the 17th century. It is applied to features of various origins. For example, Rima Hadley on the Moon is a sinuous rille, formed as a lava channel in a vast lava flow, whereas the Rimae Hippalus are graben surrounding a large impact basin.

Ring, Planetary A disk of matter encircling a planet, consisting of numerous particles in orbit ranging in size from dust grains up to objects tens of metres across. Only Saturn has rings bright enough to be seen at visible wavelengths from Earth; they are composed mainly of ice. In 1979 the Voyager probes detected a ring around Jupiter. Rings were found around both Uranus and Neptune by stellar occultations, and confirmed when Voyager 2 visited them. The various planetary rings have different proportions of ice and dust, and different distributions of particle sizes. Planetary rings may consist of debris from impacts on the planets' satellites.

Ring galaxy An unusual galaxy with a well-defined luminous ring around a bright core. The ring can appear smooth and regular, or knotted or warped, and may include gas and dust in addition to stars. Such an arrangement may result from a collision of a disk galaxy with another smaller galaxy, which passes through the disk. An example is the Cartwheel Galaxy.

Ring nebula A 9th-magnitude planetary nebula about 2000 L.Y away in Lyra, also known as M57 or NGC 6720. The hot 15th-magnitude central star is surrounded by a shell of luminous gas and dust some three-quarters of a light year in diameter. In small telescopes the shell looks like an elliptical smoke ring 70" × 150" across. The ring is expanding at 19 km/s.

Rise time The time taken for a variable star to rise from minimum to maximum. In an eclipsing binary it is the time either from mid-eclipse or from third contact to maximum brightness. In novae and supernovae, it is usually taken to be the time to any pre-maximum halt, rather than to the extreme maximum.

Rising The moment at which a celestial body appears above an observer's horizon. For bodies with an observable disk, particularly the Sun and Moon, rising is taken as the moment when their upper limb lies exactly on the observer's horizon. Refraction in the Earth's atmosphere means that an object appears to rise before it actually does, and this effect must be taken into account when calculating observed rising and setting times.

Roche limit The distance from the centre of a planet within which any large satellite would be torn apart by tidal forces. The Roche limit lies at approximately 2.46 times the planet's radius, if the planet and satellite have similar densities. Within the limiting distance no major satellite could form, although small, strong objects such as artificial satellites can survive well within this distance. All four major planets have ring systems within the Roche limit. It is named after the French mathematician Édouard Albert Roche.

Roche lobe The volume around a star in a binary system in which, if you were to release a particle, it would fall back onto the surface of that star. A particle released above the Roche lobe of either star will, in general, occupy the 'circumbinary' region that surrounds both stars. The point at which the Roche lobes of the two stars touch is called the inner Lagrangian or L1 point. If a star in a close binary system evolves to the point at which it 'fills' its Roche lobe, theoretical calculations predict that material from this star will overflow both onto the companion star (via the L1 point) and into the environment around the binary system.

Römer, Ole Christensen Danish astronomer. While observing Jupiter's satellites from the Paris Observatory, intending to base an international time standard on their eclipses, he noticed that the times of eclipses disagreed with predictions by G. D. Cassini, varying by 22 minutes over a six-month period. Römer deduced that light took longer to reach Earth when Jupiter was farther away. In 1676 he calculated the speed of light from his timings, using the contemporary value of the astronomical unit, finding it to be about 200 000 km/s. This is two-thirds of the true value, but was the first demonstration that light had a finite speed. He also built the first transit instrument.

Ronchi test A method of testing the shape of a telescope mirror. It is based on the Foucault test, but instead of a knife edge it uses a fine grating with about 40 dark lines per centimetre. When this apparatus is used, the mirror appears to be crossed by lines. Any curvature of these lines indicates deviations from a spherical figure. It is named after the Italian physicist Vasco Ronchi.

Roo-Sastry type A category in a classification scheme for clusters of galaxies, based on the type and distribution of the ten brightest members. cD clusters contain a cD galaxy; B clusters have a binary pair of supergiant galaxies less than ten diameters apart; L clusters have their brightest members approximately in line; F clusters are highly flattened; C clusters have a core-halo structure with four or more of the brightest members near the core and fainter members in a halo; I clusters are irregular with no well-defined centre. The classification was introduced in 1971 by the American astronomer Herbert Jesse Rood and the Indian astronomer Gummuluru Narasimha Sastry.

Rosat A joint German-UK-US soft X-ray and extreme ultraviolet astronomy satellite; its name is a contraction of Röntgen Satellit. Its name derives from Wilhelm Konrad von the German physicist who discovered X-rays. In the first six months after its launch in 1990 June it made the first imaging survey of the X-ray sky, with a German 0.8-m telescope, and the first-ever survey of the whole sky in the EUV, with the UK's Wide Field Camera.Subsequently, the mission was devoted to detailed observations of individual objects. The grazing-incidence X-ray telescope has a choice of two detectors, either a position-sensitive proportional counter or microchannel plate detector, covering the energy range 0.1-2.4 keV (0.52-10 nm). The 0.525-m aperture WFC grazingincidence telescope has a microchannel plate detector with a series of thin metal foil filters, mounted on a wheel, used to select EUV wavelengths from 6 to 73 nm.

Rosetta A planned ESA probe to a comet, planned for launch in 2003 January. On its way it will look at asteroids (3840) Mimistrobell and (2703) Rodari before arriving at Comet 46P/Wirtanen in 2011 August. It will fly alongside the comet for two years as it approaches perihelion, and release a lander to sample the nucleus.

Rossi x-ray timing explorer A NASA satellite launched in 1995 December. It carries the largest array of proportional counters ever flown, nearly a metre square, to study X-rays of 2-200 keV (0.006-0.6 nm). RXTE uses its Proportional Counter Array and High Energy X-ray Timing Experiment to study the variability of the thousand or so brightest X-ray sources at time-scales down to 10 μs. Another instrument, the All-Sky Monitor scans the sky every 100 min. It is named after the Italian-born American astronomer Bruno Benedetto Rossi.

Rossiter effect An irregularity in the radial-velocity curve of an eclipsing spectroscopic binary immediately before and after eclipse; also known as the rotation effect. In most binaries the stars orbit in the same direction as they rotate on their axes. Immediately before maximum eclipse, the small segment of the star that remains uncovered therefore has a radial velocity that is directed away from the observer. The opposite applies as the star emerges from eclipse. The effect shows up as a short-lived rise and fall on the radial-velocity curve. It is named after the American astronomer Richard Alfred Rossiter.

Rotation The turning of a body on its axis, such as the daily rotation of the Earth. It is usually measured relative to the stars, and termed the sidereal period of axial rotation.

Rover A rover (or microrover) is a remote-controlled robotic vehicle. Sojuner rover was the first robotic explorer to explore Mars; this 25-pound, six-wheeled microrover explored an ancient Martian flood plain (Ares Vallis) in 1997.

Royal astronomical society An organization for the promotion of astronomy and geophysics, founded in 1820 as the Astronomical Society of London. It became the Royal Astronomical Society in 1831. Its headquarters are in London. The RAS publishes Monthly Notices, Astronomy & Geophysics, and Geophysical Journal International.

Royal Greenwich observatory An observatory owned and operated by the UK's Particle Physics and Astronomy Research Council. It was founded in 1675 at Greenwich, southeast London. After World War II it relocated to Herstmonceux Castle, Sussex, where it operated the Isaac Newton Telescope from 1967 to 1979. The original building at Greenwich is now maintained as a museum. In 1990 the RGO moved again, this time to Cambridge. The RGO contains the office of The Nautical Almanac. The Time Department, which formerly maintained and distributed Greenwich Mean Time, was closed in 1990 before the move to Cambridge.

RR lyrae star A type of yellow giant pulsating variable, with a period of 0.2-1.2 days and an amplitude of 0.2-2.0 mag.; abbr. RR. The stars belong to Population II, lying in the galactic halo and globular clusters (hence their earlier name of cluster variables). Two subtypes are now recognized: RRAB, which are pulsating in the fundamental mode; and RRC, pulsating in the first overtone. The RRAB type has an asymmetric light-curve, with a steep rise and gentler decline, whereas light-curves of type RRC are approximately sinusoidal and have amplitudes of less than 0.8 mag. Periodic modulation of the periods and amplitudes (the Blazhko effect) is common. The RRAB type all have approximately the same absolute magnitude (+0.5), which makes them valuable distance indicators. The similarity of the RRAB light-curves to those of Cepheid variables gave rise to the other obsolete name of short-period Cepheid.

RS canum venaticorum star A type of binary star, formally classified as an eclipsing variable but more correctly described as an eruptive variable, with strong emission lines of Ca II; abbr. RS. The low-amplitude light-curve resembles a sine-wave, but also exhibits slow changes in amplitude and phase which arise from chromospheric activity and starspots, coupled with differential rotation of the star's surface. The systems are detached binaries, and many undergo eclipses. Superimposed fluctuations of up to 0.2 mag. with periods of years are indications of changes in activity akin to the solar cycle.

Rubin-ford effect An apparent anisotropy in the expansion of the Universe on a scale of around 100 million L.Y, as revealed by a study of the motions of a sample of spiral galaxies by the American astronomers Vera Cooper Rubinand William Kent Ford, Jr.The reason that there is not a purely isotropic Hubble flow is because the Universe is not homogeneous on these scales.

Runaway star A young star moving at high velocity, perhaps hundreds of kilometres per second, suggesting that it has been ejected by some violent event from its birthplace. Such stars were probably once part of a binary, but were ejected either when their companion exploded as a supernova, or else through a close encounter with another binary.

Russell, Henry Norris American astronomer. By 1910 he had accumulated data on enough stars for him to plot a diagram of their absolute magnitude against their spectral type, finding that red stars fell into two groups, giants and dwarfs—unaware that E. Hertzsprung had done the same in 1906. This diagram, further developed by Russell in 1913, is now known as the Hertzsprung-Russell diagram. Russell also evolved a method of calculating the masses of binary stars. In 1928 he established the composition of the Sun's atmosphere from its spectrum, and went on to suggest that all stars contain a high proportion of hydrogen, a conclusion originally reached by C. H. PayneGaposchkin.

RV tauri star A highly luminous, yellow supergiant pulsating variable; abbr. RV. Spectra are F-K at maximum, K-M at minimum. The light-curves have overall amplitudes of 3-4 mag.; all have double maxima (normally with slightly different amplitudes) and an intervening secondary minimum. The amplitudes of the maxima and secondary minimum vary, and the secondary minimum may become as deep as the primary minimum. The formal period is the double interval from one deep minimum to the next, and generally lies in the range 30-150 days. In the RVA subtype the mean magnitude remains constant with time.

The RVB subtype exhibits a superimposed secondary wave, with amplitudes up to 2 magnitudes and periods of 600-1500 days.

RW aurigae star A widely used, but now obsolete, term for a young irregular variable. The designation and its subgroups were based solely on the form of the light-curve, whereas the current system of classification also takes into account the presence of nebulosity, speed of variation, and spectral type. RW Aurigae itself is now classified as of type IsT, with rapid variations and a T Tauri spectrum.

Ryle, Martin English radio astronomer. After World War II he began to establish Cambridge as a centre for radio astronomy, opting for radio interferometers rather than parabolic receivers. He developed the technique of aperture synthesis and commenced a series of surveys of radio sources that were published as the Cambridge catalogues, the best known of which is the third (objects in it are preceded by the designation 3C). The large number of faint radio sources discovered helped discredit the steady-state theory. Ryle and A. Hewish shared the 1974 Nobel prize for physics for their work on aperture synthesis and pulsars.

S

S star A red giant similar to spectral type M except that the dominant molecular bands in its spectrum are caused not by titanium oxide but by zirconium oxide (ZrO); hence they are sometimes also called zirconium stars or heavy-metal stars. Oxides of lanthanum, yttrium, and barium are also found in place of the usual scandium and vanadium. These elements must have been created in the core of the star by nucleosynthesis during its later evolutionary stages. The majority of S stars are irregular or long-period variables. Related subtypes are MS stars (M stars with zirconium oxide in their spectra) and SC stars, which are intermediate between S stars and carbon stars.

S vulpeculae star A yellow giant or supergiant semiregular variable (type SRD) that shows intervals of reasonable periodicity, interspersed with unpredictable episodes of irregular variations

Sabik The star Eta Ophiuchi. It appears to the naked eye of magnitude 2.4, but it is in fact a close binary consisting of two A-type dwarfs of magnitudes 3.0 and 3.5 that orbit with a period of 85 years. It lies 63 L.Y away.

Sachs-Wolfe effect A phenomenon in which irregularities in the distribution of mass in the early Universe give rise to localized temperature variations in the cosmic background radiation. It occurs because photons are gravitationally redshifted by regions where the density of matter is higher than average. The ripples detected by the Cosmic Background Explorer atellite in 1992 are thought to be a manifestation of this effect. It is named after the American astrophysicists Rainer Kurt Sachs and Arthur Michael Wolfe.

Sagittarius A constellation of the zodiac, representing a centaur with a bow, popularly known as the Archer. The Sun passes through Sagittarius from the third week of December to the third week of January, and hence is in the constellation at the winter solstice. Its brightest star is Epsilon Sagittarii.Beta Sagittarii is a naked-eye double, consisting of unrelated stars of magnitudes 4.0 and 4.3. Sigma Sagittarii is Nunki. RY Sagittarii is a variable star of the R Coronae Borealis type, which periodically dips from 6th to 14th magnitude. The constellation contains dense Milky Way starfields that lie towards the centre of our Galaxy; the exact centre is believed to be marked by the radio source Sagittarius A. M8 is the Lagoon Nebula; M17 is the Omega Nebula; M20 is the Trifid Nebula. M22 is a rich globular cluster of 5th magnitude, the third-best in the heavens.

Sagittarius arm The spiral arm of our Galaxy that is about 5000 L.Y closer to the centre of the Galaxy than the local Orion Arm, and roughly parallel to it. The Carina Arm may be an extension of this arm. The galactic centre lies behind the Sagittarius Arm as viewed from Earth, but is highly obscured at visible wavelengths by dust in the Galaxy's disk. The Sagittarius Arm can be traced behind the stars of our local arm from Serpens via Scutum to Scorpius, Sagittarius, Centaurus, and Carina. Among the objects it contains are the Eagle Nebula, the Trifid Nebula, the Lagoon Nebula, the Jewel Box cluster, and Eta Carinae.

Salyut 1 The first space station, Salyut 1 (also called DOS 1 and Zarya), was launched on April 19, 1971 from Baikonur. This small cylindrical-shaped Soviet space station had room for a crew of three; it was 13.1 meters long and about 4.2 meters across. The crew (ladimir Shatalov, Alexei Yeliseyev and Nikolai Rukavishnikov) unsuccessfully docked with the space station on June 7, 1971 (the docking mechanism was damaged during the docking process). They couldn't enter the space station, but spent 21 days orbiting the Earth. The next crew (Georgi Dobrovolski, Vladislav Volkov and Viktor Patsayev) spent 24 days on the station; they died during the return trip to Earth (on June 30). The space station orbited for 179.93 days but was never used again.

Sandage, Allan rex American astronomer. In 1950 he became an assistant to E. P. Hubble, whose work on measuring the distances of galaxies and refining the value of the Hubble constant he continued, finding a value of around 50 km/s/Mpc. He also calculated an early value for the deceleration parameter. In 1960 he and the Canadian astronomer Thomas Arnold Matthews made the first optical identification of a radio source (3C 48) that would turn out to be a quasar. In 1965 Sandage found the first radio-quiet quasar, and went on to identify many more quasars by their high redshifts.

Satellite A small body that orbits a larger one, particularly the natural satellites of the planets. A natural satellite is also known informally as a moon. All the planets have at least one natural satellite, except Mercury and Venus. A spacecraft sent into orbit around the Earth or another body is an artificial satellite.

Saturn The sixth planet from the Sun. Its mean opposition magnitude is +0.7, the faintest of the five planets known since antiquity. Saturn is the most flattened in shape of all the planets, with an equatorial diameter of 120 540 km and a polar diameter of 108 730 km. It is also the least dense of all the planets (0.7 g/cm^3) and the only planet less dense than water. The rotation period of the visible surface ranges between about 10h 14m near the equator and 10h 40m at 60° south. The rotation of the solid interior, derived from radio observations, is 10h 39m 22s.

Saturn has a thick atmosphere composed of about 96% hydrogen and 4% helium with traces of methane, ammonia, ethane, ethyne, and phosphine. The temperature near the top of the atmosphere is around -195 °C. Internally, Saturn is thought to possess a rocky high-temperature core, perhaps containing iron, about 20 000 km across.

Scatter ellipse The elongated area on the ground within which fragments of a meteorite fall are scattered. The long axis of the ellipse is parallel to the meteorite's direction of travel. The sizes of the fragments increase along the ellipse, because air resistance causes smaller fragments to fall to the ground sooner.

Schiaparelli,Giovanni Giovanni Schiaparelli (1835-1910) was an Italian astronomer (and director of the Milan Observatory) who first mapped Mars (in 1877) and brought attention to the network of "canali" (Italian for canals or channels) on Mars. These "canals" were later found to be dry and not to be canal at all. A Martian impact crater (Crater Schiaparelli, 461 km = 277 mi in diameter) and a hemisphere of Mars have been named after

Schiaparelli. His late niece Elsa was an influential fashion designer.

Schmidt-Cassegrain telescope A Schmidt-Cassegrain telescope (SCT) is a wide-angle reflecting telescope with a correcting lens that minimizes spherical aberration and a concave mirror that receives light and focuses an image. A second mirror reflects the light through a gap in the primary mirror, allowing the eyepiece or camera to be mounted at the back end of the tube. The Cassegrain telescope (named for the French sculptor Sieur Guillaume Cassegrain) was developed in 1672; the correcting plate (a lens) was added in 1930 by the Estonian astronomer and lens-maker Bernard Schmidt (1879-1935)

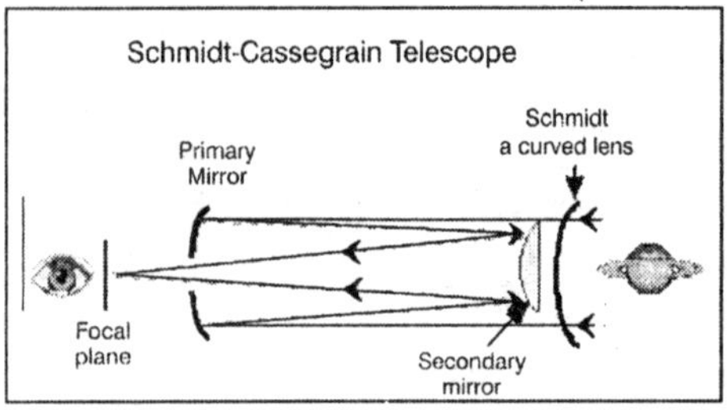

Fig. Schmidt-Cassegrain telescope

Schröter effect A phenomenon in which the observed phase of Venus appears less than the predicted phase. As a result, dichotomy occurs early at evening elongations, and late at morning elongations. The time difference between calculated and observed dichotomy is about a week. The effect was discovered by the German astronomer Johann Hieronymus Schröter. The cause may be that the region near the terminator is darker than the rest of the illuminated disk.

Schwarzschild radius The Schwarzschild radius is the radius of the event horizon of a black hole. This is the distance from a black hole at which nothing can escape, bot even light. Within the Schwarzschild radius, the escape velocity from the black hole is greater than the speed of light. The size of the Schwarzschild radius may be proportional to the mass of the black hole. For a typical black hole with a mass 10 times that of the Sun, the Schwarzschild radius would be roughly 18.6 miles (30 km). The Schwarzschild radius is named for the German astronomer Karl Schwarzschild, who predicted the existence of collapsed stellar bodies that cannot emit radiation, in 1916.

Schwarzschild black hole A black hole described by solutions to Einstein's equations of general relativity worked out by Karl Schwarzschild in 1916. The solutions assume the black hole is not rotating, and that the size of its event horizon is determined solely by its mass.

Schwarzschild, Karl German astronomer. He established a method for determining stars' brightnesses from photographs, comparing their visual and photographic magnitudes to obtain their colour index. In 1905 he obtained ultraviolet photographs of a solar eclipse, and went on to study energy transfer in the Sun, deducing that its outer regions had a layered structure. In 1916 he showed that, in the general theory of relativity, a sphere of material collapsing under its own gravitational field past its Schwarzschild radius Would cease to radiate energy (i.e. it would become a black hole). His son, Martin Schwarzschild,became a naturalized American and studied stellar evolution.

Scintillation 1. The rapid fluctuation in brightness of an astronomical object ('twinkling') due to scattering of electromagnetic waves by irregularities in the medium between the source and the observer. It is most pronounced in sources of small angular size, so at optical wavelengths it affects stars more than planets, while at radio wavelengths it affects point sources such as pulsars. Twinkling of stars at visible wavelengths is caused by convection currents in the atmosphere. Its cause is the same as that of bad seeing. At radio wavelengths, it occurs in the Earth's ionosphere (ionospheric scintillation), in the solar wind (interplanetary scintillation), and in interstellar matter (interstellar scintillation).

2. A phenomenon in which high-energy particles emit flashes of visible light (scintillations) on collision with atoms in certain materials (scintillators). It is the principle of the scintillation counter used in gamma-ray astronomy.

Scintillation counter A device for detecting high-energy X-rays and gamma rays. It makes use of the effect of scintillation, in which the incident radiation is converted into a flash of light by interaction with a sensitive material such as caesium iodide or sodium iodide.The intensity of the flash is proportional to the energy of the radiation, and it can be converted into an electronic signal by a photomultiplier tube or, in an imaging telescope, by a microchannel plate detector or charge-coupled device.

S-class asteroid A class of asteroid, relatively common in the inner main asteroid belt, having a moderate albedo (0.10-0.28), a reddish spectrum at wavelengths shorter than 0.7 μm, and moderate to non-existent absorption features at longer wavelengths. The proportion of S-class asteroids decreases outwards through the main belt. The S is for silicaceous, since their composition is thought to include silicate minerals such as olivine and pyroxene. They could be the parent bodies of the stony-iron meteorites. Members of this class include (15) Eunomia, diameter 260 km, and (29) Amphitrite, diameter 219 km.

Scorpius A constellation of the zodiac, representing a scorpion. The Sun passes through Scorpius for a week at the end of November. Its brightest star is Antares. Beta Scorpii is a double star of magnitudes 2.6 and 4.9; the stars are unrelated. Another optical double is the naked-eye pair Zeta Scorpii, magnitudes 3.6 and 4.7. Nu Scorpii is a notable quadruple star. M4, a 6th-magnitude globular cluster nearly ½° wide, is 6800 L.Y away, one of the closest globulars to us. M6 is a 4th-magnitude open cluster ¼° wide, and M7 is a 3rd-magnitude open cluster more than 1° across. Scorpius X-1 is the brightest X-ray source in the sky.

Scorpius X-1 The brightest persistent X-ray source in the sky, and the first known celestial X-ray source apart from the Sun. It was discovered in 1962 during a sounding-rocket flight. Scorpius X-1 is a low-mass X-ray binary with an orbital period of 18.9 hours. The optical counterpart is a 13th-magnitude blue star known as V818 Sco, 2300 L.Y away. The X-rays arise from the transfer of material from this star on to a companion neutron star via an accretion disk.

Sculptor A faint constellation of the southern sky, representing a sculptor's workshop. Its brightest star, Alpha Sculptoris, is of magnitude 4.3. The constellation contains the Sculptor Dwarf Galaxy, a member of our Local Group, and the south galactic pole lies in Sculptor.

Sculptor group The nearest group of galaxies to the Local Group, about 12 million L.Y away. It is a loose association containing the late-type spirals NGC 45, 55, 247, 253, 300, 7793, and possibly IC 5332.

Scutum The fifth-smallest constellation, lying in the equatorial region of the sky, representing a shield. Its brightest star, Alpha Scuti, is of magnitude 3.9. Delta Scuti is the prototype of a class of pulsating variables with small amplitudes, the Delta Scuti stars. M11 is the Wild Duck Cluster.

Sea of serenity One of the maria on the Moon's nearside; Mare Serenitatis.

Sea of tranquility The Sea of Tranquility (Mare Tranquillitatis) is a large plain on the moon. The first moon landing, Apollo 11, in 1969, was to this sea.

Seashell galaxy An unusually shaped galaxy about 280 million L.Y away in Centaurus, interacting with the galaxy NGC 5291. The Seashell has no NGC number of its own. Both galaxies are part of a cluster, the brightest member of which is IC 4329.

Secchi classification An early attempt to classify the spectra of stars, developed by P. A. Secchi in the 1860s. Secchi grouped stellar spectra into five types, on the basis of direct visual observation. Type I was the equivalent of today's spectral types B and A; Type II encompassed F, G, and K stars; Type III was the equivalent of the M stars; Type IV was the equivalent of N; and the rare Type V described spectra with unusual bright lines, such as from Wolf-Rayet stars and planetary nebulae. Secchi's scheme was superseded by the photographic Harvard classification system.

Secchi, (Pietro) Angelo Italian astronomer and priest. He was a pioneer of astronomical photography. By photographing a total solar eclipse in 1860 he, and independently the English scientist Warren De la Rue settled the controversy over whether prominences belonged to the Sun or the Moon. Secchi also made the first classification of solar prominences into quiescent and eruptive types. He and W. Huggins were the first to approach astronomical spectroscopy systematically. Secchi made the first spectroscopic survey of stars, which yielded a catalogue of over 4000 spectra, published 1863. He suggested a spectral classification for stars from which the modern system of spectral types evolved.

Second contact At a total solar eclipse, the moment when the leading limb of the Moon just touches the Sun's eastern limb, and totality begins; or, at a total lunar eclipse, the point at which the Moon's trailing limb becomes completely immersed in the Earth's umbra. Second contact in a total solar eclipse may be immediately preceded by the appearance of Baily's beads or the diamond ring. At an annular eclipse, second contact refers to the instant when the Moon's trailing (western) limb leaves the Sun's western limb, so that the whole of the Moon's body is silhouetted against the Sun.

Second generation star Second generation stars do not just burn hydrogen, they also burn heavier elements, like helium and metals, and were formed from supernova explosions (the debris of exploded population II stars). Our Sun is a seond or third generation star.

Second of arc (arcsec) (symbol ") A very small unit of angular measure, equivalent to one-sixtieth of an arc minute or 1/3600 of a degree. The resolving power of telescopes, the separation of double stars, and the apparent diameter of celestial objects are usually given in seconds of arc.

Secondary cosmic ray One of the shower of atomic particles produced when a primary cosmic ray enters the Earth's atmosphere. The interaction first produces pions, which subsequently decay rapidly into muons, some of which may also aecay into electrons. The bulk of the secondary cosmic rays present at sea level are muons..

Secondary crater A small crater formed by ejecta thrown out from a larger impact crater. Secondary craters tend to cluster in a ring around the main crater, the greatest number lying slightly more than about one crater diameter away on the Moon, but rather closer on Mercury because of its higher surface gravity. Ejecta falling nearer than about half the crater diameter is travelling too slowly to form craters, and piles up as an ejecta blanket. Secondary craters may be widely dispersed over the planet, if it has no atmosphere. Those far from the primary crater tend to be fairly circular, whereas those nearby may be very irregular in shape, a result of the lower velocities of the nearer ejecta.

Secondary minimum 1. The shallower of the two minima in the light-curve of an eclipsing binary. It occurs when the secondary (the star of lower surface brightness) is eclipsed by the brighter primary. Depending on the relative brightness of the components, the secondary minimum of an Algol star may be either barely detectable or nearly as deep as primary minimum. In a Beta Lyrae star it is shallower, and in a W Ursae Majoris star it may be nearly as deep as the primary eclipse. 2. The dip between the double maxima of an RV Tauri star.

Secular In astronomy, secular mean that a phenomenon takes a tremendous amount of time to unfold, and occurs gradually. Secular change, for example, is a long-standing, continuous (and nonperiodic) change to a system.

Secular parallax The angle subtended at a star by a baseline equal to the distance travelled by the Sun in a given time, such as a year or century. It can be used to estimate the average distance of a sample of stars which has a common motion relative to the Sun, such as the stars which define the local standard of rest. Care must be taken to distinguish other populations, such as RR Lyrae variables, which have different speeds and directions of motion.

Secular perturbation A perturbation in a body's orbit that is cumulative, causing an orbital element to grow or decrease. The gravitational attractions of the planets upon each other cause secular changes in the longitude of the ascending node, longitude of perihelion, and times of perihelion passage, as well as causing periodic perturbations.

Secular variable A variable star that has supposedly undergone an increase or decrease in brightness on a time-scale of hundreds or thousands of years. Although a number of possible candidates have been advanced, the poor quality of early magnitude estimates means that such changes cannot be confirmed.

Sedana Sedna is a reddish-colored, planet-like body located in the Oort Cloud, approximately 13 billion kilometers (8 billion miles) away (three times farther from Earth than Pluto). Sedna is abour 800-1100 miles in diameter (this is about three- fourths the size of Pluto). Sedna turns on its axis once every 40 days (it's day is 40 Earth days long); this is a very slow rotation. Sedna was discovered by Mike Brown, Chad Trujillo, and David Rabinowitz on Nov. 14, 2003. The name Sedna is from an Inuit (a North American Indian tribe) goddess of the ocean.

Seeing The movement or distortion of a telescopic image as a result of turbulence in the Earth's atmosphere. It is thought to be caused by undulations between separate layers of air, which disturb the path of light. The scale of these undulations is often regarded as giving cells of seeing, typically being around 100-150 mm across at sea level, and larger at higher altitude. A small telescope may look through individual cells, whose movements give rise to a sharp but wandering image, while a larger one may look through several at once, producing multiple images. The 'boiling' seen at the limb of the Moon or Sun is another manifestation of bad seeing. The Antoniadi scale is widely used by amateur astronomers to evaluate seeing.

Segmented mirror A telescope mirror that consists of several adjacent components which bring light to a common focus like one larger aperture. Typically the components are hexagonal in shape, and interlock like tiles. Each segment has a separate mounting with position sensors and actuators to maintain its alignment. Each of the two Keck Telescopes, for example, has 36 components 1.8 m across, forming a combined hexagonal aperture of 9.82 m. Segmented mirrors can be very light for their aperture, and can be made larger than individual mirrors.

Semidetached binary A close binary star in which one component fills its Roche lobe. Mass transfer occurs from this star to the other. In some systems the gas stream falls directly on to the surface of the accompanying star, but in most cases it forms an accretion disk. All active cataclysmic binary systems are semidetached.

Semiregular variable A type of giant or supergiant pulsating variable, with intermediate or late spectra, whose variations show a greater or lesser degree of regularity; abbr. SR. The amplitudes (average 1-2 mag.) are less than those of the Mira stars, and the mean periods range from 20 days to more than 2000 days. Four subtypes are defined. SRA are giants with reasonably persistent periodicity, although amplitudes (less than 2.5 mag.) and the form of the light-curves vary. SRB are giants with occasional periodicity interspersed with intervals of slow, irregular variations (sometimes with larger amplitude) or even constant brightness. SRC are supergiants with moderate periodicity and low amplitudes (about 1 mag.). These three types have M, C, S, or Me, Ce, Se spectra. SRD are giants and supergiants with F, G, or K spectra, sometimes with emission lines, and amplitudes of 0.1-4 mag.

Serpens A constellation of the equatorial region of the sky, representing a large snake. It is unique in being split into two halves, which lie on opposite sides of Ophiuchus: Serpens Caput, the serpent's head, which is the larger half; and Serpens Cauda, the serpent's tail. The two are regarded as a single constellation. The brightest star is Alpha Serpentis (Unukalhai), magnitude 2.7. M5 is a 6th-magnitude globular cluster, and M16 is a 6th-magnitude star cluster in the Eagle Nebula.

Serrurier truss A tube framework for telescopes. It consists of eight struts arranged in triangles. The struts run between a ring which forms either the top or bottom end of the tube and an open box at the centre of gravity of the assembly. This box is attached to the declination axis of the mounting (or the altitude axis in an altazimuth mounting). The triangular framework ensures that strains and loads are carried evenly around the tube, whatever the orientation of the telescope, and that sag is identical at both the top and bottom of the tube. The system was devised by the American engineer Mark Serrurier.

SETI Abbr. for 'search for extraterrestrial intelligence'. The first attempt to detect possible incoming radio signals from alien civilizations around other stars was Project Ozma in 1960. Various other searches have since been made for artificial signals from space, mostly at radio wavelengths, but without success.

Sextant A navigational instrument for measuring the altitude of celestial objects above the horizon. It consists of a graduated 60° arc (one-sixth of a circle) with a movable arm and sighting devices. An index mirror attached to the movable arm reflects the chosen object to another mirror, silvered across half its width, so that its image can be seen alongside the horizon. The object's altitude can then be read off the scale. A telescope and dark filters are provided to improve accuracy and allow observation of the Sun. Because the objects observed are reflected, the range of angles which can be measured is 120°, double that of the graduated scale.

Seyfert galaxy A type of galaxy with a small, bright nucleus which shows broad, strong emission lines in its spectrum. The first galaxies of this type were described in 1943 by C. K. Seyfert. Nearly all known Seyfert galaxies are spirals or barred spirals, and Seyfert-type activity probably occurs in a small percentage of all spiral galaxies. Seyfert galaxies are classified according to the relative widths of the emission lines in their spectra. Class 1 Seyferts (such as NGC 5548) have broad emission lines of hydrogen but narrow forbidden lines of heavier elements. In Class 2 Seyferts (such as NGC 1068) the hydrogen lines and forbidden lines both have the same width, which is broader than the forbidden lines in Class 1 Seyferts but not as broad as the latter's hydrogen lines. Seyferts have active galactic nuclei that produce strong radiation, probably from an accretion disk around a massive black hole. The radiation excites gas around the central regions, giving rise to the observed emission lines.

Seyfert, Carl Keenan American astrophysicist. In 1943 he found several spiral galaxies with very faint arms and unusually bright nuclei. Their spectra showed emission lines from hot ionized gases moving at high velocities. These so-called Seyfert galaxies have since been shown to belong to the wider class of objects with active galactic nuclei. In 1951 he observed the group of galaxies now known as Seyfert's Sextet.

Seyfert's sextet A group of six galaxies in Serpens; also known as NGC 6027. One of the group is a small spiral with a redshift that indicates a recession velocity some 15000 km/s greater than the rest of the group. The high-redshift member is probably a distant background galaxy, but the sextet has sometimes been cited as evidence of the existence of redshifts not caused by the overall expansion of the Universe.

SGR Soft Gamma Repeater (SGR) is probably a magnetar (a highly-magnetic neutron star) that emits bursts of soft (low-energy) gamma rays at irregular intervals. The gamma ray bursts may be caused by starquakes on the surface crust of the neutron star. (The starquakes are caused by aberrations in the star's strong magnetic field.) These SGR bursts are different from gamma ray bursts, which are one-time events. SGRs were discovered in 1979.

Shadow bands An optical phenomenon seen fleetingly around the onset of totality in a solar eclipse, comprising slow-moving, low-contrast waves of light and dark, perhaps only a few centimetres broad, seen on light-coloured surfaces on the ground. Shadow bands are presumed to originate from differential refraction of the last remaining segment of sunlight in the cooling upper atmosphere around the path of the total eclipse.

Shell burning The nuclear reactions in a shell around a star's core that continue after the fuel in the core itself has been exhausted. As the fuel, whether hydrogen or helium, is progressively exhausted, the shell moves outwards until it enters regions too cool for the reactions to continue.

Shell galaxy An elliptical galaxy surrounded by faint arcs or shells of stars, lying at right angles to its major axis. One to twenty shells may be seen, nearly concentric but incomplete. They are 'interleaved' in such a way that successive shells usually occur on opposite sides of the galaxy. About 10% of bright ellipticals show shells, most of them in regions of low density of galaxies. No spirals are known with such shell structure. Shell galaxies may be the result of a giant elliptical digesting a low-mass companion galaxy.

Shell star A star whose spectrum contains prominent absorption lines that originate in a shell of material surrounding the star. If variable, the star is classified as a Gamma Cassiopeiae star, in which ejection of a shell is accompanied by a temporary fade.

Shepherd moon A satellite of a ringed planet whose gravitational field has a significant effect on the configuration of planetary rings near it. Shepherd moons were first discovered on Voyager images of Saturn. Saturn's F Ring is kept very narrow by two shepherd satellites, Pandora and Prometheus, which orbit either side of it. Uranus's outer ring, the Epsilon Ring, is kept in place by the shepherd moons Ophelia and Cordelia.

Shepherd satellite A shepherd satellite is a moon of a planet that orbits along side a ring of that planet; the gravitational forces of the moon confines the ring and giving it a sharp edge. Examples of shepherd satellites include Saturn's moons Prometheus and Pandora which shepherd its narrow, outer F ring. Peter Goldreich and Scott Tremaine proposed the idea of shepherding moons in 1979 to explain why Uranus' rings were so narrow.

Shergottite A very rare type of achondrite meteorite, named after the meteorite that fell at Shergotty, India, in 1865, the first known fall of this type. The shergottites are divided into two subtypes: the basaltic or pyroxene-plagioclase shergottites, and the Iherzolitic or olivine-pyroxene shergottites. The basaltic shergottites, of which the two earliest recorded shergottite falls, Shergotty and Zagami, are both examples, consist primarily of pyroxene (pigeonite and augite), together with maskelynite (a plagioclase glass formed by shock metamorphism) and a trace of olivine. The Iherzolite shergottites, of which the Antarctic find ALHA 77005 was the first example, contain less pyroxene and maskelynite than the basaltic shergottites, but much more olivine. Shergottites belong to the class of SNC meteorites. Most shergottites share a formation age of 170 million years and an exposure age of about 3 million years.

Shoemaker-levy 9, comet A comet discovered on 1993 March 25 by the American astronomers Eugene Merle Shoemaker his wife Carolyn Shoemaker, née Spellmann and David Howard Levy formerly designated 1994 X. The comet was in a 2-year orbit around Jupiter, having been captured in, or before, 1929. A close approach to Jupiter (21 000 km) in 1992 July had disrupted the nucleus into at least 21 fragments. These fragments hit Jupiter over the week of 1994 July 16-22. The impacts produced prominent dark spots, visible even in small amateur telescopes, at latitude 44°S, distributed in longitude; they subsequently merged into a dark belt which persisted for 18 months.

Shooting star A shooting star is not a star; it is a meteor (made of rock and/or iron) which is burning up in the Earth's atmosphere.

Sickle Popular name for the asterism formed by the stars Epsilon, Mu, Zeta, Gamma, Eta, and Alpha Leonis. The Sickle makes up the head and chest of Leo, the lion.

Sidereal day The interval of time between successive transits of the mean equinox, equal to 23h 56m 04s. Because of the precession of the equinoxes, the mean equinox is not a completely fixed sidereal point. As a result the sidereal day is 0.0084 s shorter than the Earth's rotation period relative to the stars.

Sidereal month The average period of revolution of the moon around the earth in reference to a fixed star, equal to 27 days, 7 hours, 43 minutes in units of mean solar time.

Sidereal year A sidereal year is measured relative to the stars (the period between successive conjunctions with any star). One sidereal year, equal to 365 days, 6 hours, 9 minutes, and 9.5 seconds, is the period during which the earth completes one revolution around the Sun (this is the same as the time it takes to come into alignment with a particular star).

Sigma octantis The south pole star, being the closest naked-eye star to the south celestial pole. It is an F0 giant of magnitude 5.4, lying 300 L.Y away. It is about 1° from the south celestial pole, and the distance is slowly increasing as a result of precession.

Sikhote-alin meteorite A major shower of iron meteorites that fell in the western part of the Sikhote-Alin mountain range in southeast Siberia on the morning of 1947 February 12. A total of 383 impact sites were found, spread over 1.6 km^2. The largest crater was 27 m in diameter. The parent body broke up at about 5 km altitude, producing about 70 tonnes of metal fragments, containing both hexahedrite and octahedrite material. The largest single fragment weighed over 1.7 tonnes. A dark dust train, marking the path of the fireball, remained in the sky for several hours.

Singularity In astronomy, a term often used to refer to the center of a black hole, where the curvature of spacetime is maximal. At the singularity, the gravitational tides diverge; no solid object can even theoretically survive hitting the singularity. Mathematically, a singularity is a condition when equations do not give a valid value, and can sometimes be avoided by using a different coordinate system.

Singularity theorem A theorem that shows that a singularity must exist under certain circumstances - in particular, that the universe must have started with a singularity.

Sinope Sinope is Jupiter's sixteenth and outermost moon. Sinope is 17.5 miles (28 km) in diameter and orbits 14,700,000 miles (23,700,000 km) from Jupiter. Sinope has a mass of 8 x 10^{16}kg. It orbits Jupiter in 758 (Earth) days and is in a retrograde orbit (orbiting opposite to the direction of Jupiter). Very little is known about Sinope. Sinope was discovered by S. Nicholson in 1914.

Sinuous rille A long, winding valley, usually with steep walls, on a planetary surface. Sinuous rilles were first seen on the Moon. They have similar characteristics to lava channels and tubes on Earth, but are distinctly different from rivers. For example, they usually start in a crater, become narrower instead of larger as they flow downslope, have abrupt breaks, and sometimes have distributaries (where the main channel splits in two, as in a river delta) but not tributaries. After the Apollo 15 astronauts visited Hadley Rille it became clear that the sinuous rilles are indeed vast lava channels, formed during the eruption of the mare lavas 2.0-3.9 billion years ago. Sinuous rilles have since been identified on Mars and Venus.

Sinus An indentation in the edge of high ground on a planetary surface; pl. sinus. The name, which means 'bay', is not a geological term, but is used in the nomenclature of individual features, for example Sinus Iridum on the Moon. It is also used for a protrusion from a dark area in the naming of Martian albedo features, for example Sinus Meridiani.

Sio maser A maser source in which the silicon monoxide molecule is excited to maser action. SiO has many maser lines, which occur in groups at frequencies near 43, 86, 129 GHz, and so on. SiO masers are commonly found in circumstellar envelopes, close to the surface of the red giant where mass-loss originates. SiO masers are occasionally found in star-forming regions.

Sirius Sirius (meaning "scorching" in Greek), also known as the dog star, is the brightest star in the sky (except for the sun). It is in the constellation Canis Major (The Great Dog). Sirius is a main sequence star that is about 70 times more luminous than the sun. It is about 8.6 light-years from Earth. It has an apparent magnitude of -1.46 and an absolute magnitude of +1.4. Sirius has a companion star (called the Pup), which is a white dwarf.

Sky The sky on Earth appears blue because our atmosphere scatters the blue colors from the sunlight (which consists of all colors of light). The sky on planets with a different atmospheric composition would appear to be a different color.

Skylab Skylab was America's first orbiting space station. It was made from Saturn and Apollo equipment. Skylab was launched May 14, 1973 and orbited at an altitude of 270 miles (435 km). It orbited the Earth every 93 minutes. Three three-man crews (SL-2: Charles Conrad, Paul J. Weitz, Joseph Kerwin; SL-3: Alan L. Bean, Jack R. Lousma, Owen Garriott; SL-4: Gerald P. Carr, William R. Pogue, Edward Gibson) lived in Skylab for a total of 171 days and 13 hours; astronauts traveled to and from Skylab via Apollo spacecraft. These astronauts carried out astronomical observations and experiments on how humans react to a microgravity environment in space. Skylab re-entered the Earth's atmosphere and fell to Earth on July 11, 1979 scattering debris over the Indian Ocean and parts of Western Australia.

Small astronomy satellite A NASA series of three small satellites devoted to X-ray and gamma-ray astronomy. SAS-1, also known as Explorer 42, was renamed Uhuru after launch in 1970. It was the first X-ray satellite, and performed the first X-ray sky survey. SAS-2, otherwise known as Explorer 48, was a gamma-ray mission, launched in 1972 November. The last of the series, SAS-3, or Explorer 53, was launched in 1975 May to study individual X-ray sources.

Small magellanic cloud The smaller of the two irregular galaxies that accompany our Galaxy; also known as the Nubecula Minor. It is about 9000 L.Y across and 190 000 L.Y away, visible to the naked eye as a misty patch about 3° across in Tucana. Its visible mass is less than 2% of our Galaxy's, and it contains relatively more gas and less dust than the Large Magellanic Cloud (LMC), but fewer clusters and nebulae. Its structure may be elongated along the line of sight to Earth. Like the LMC, the SMC shows evidence for star formation early in its history, followed by a lull, and then a more recent burst.

Smithsonian astrophysical observatory A research centre of the Smithsonian Institution, founded in 1890 at Washington, DC, but which moved to Cambridge, Massachusetts, in 1955. The SAO operates the International Astronomical Union's Central Bureau for Astronomical Telegrams and Minor Planet Center. In 1973 the SAO and the neighbouring Harvard College Observatory established the joint Harvard-Smithsonian Center for Astrophysics. Since 1982 the SAO has operated the Oak Ridge Observatory at Harvard, Massachusetts, founded by HCO in 1933 and still owned by them. Instruments here include a 1.5-m reflector opened in 1933, and a 25.6-m radio telescope opened in 1965. SAO owns and operates the Fred Lawrence Whipple Observatory on Mount Hopkins, Arizona, and jointly owns and operates the Multiple Mirror Telescope, also on Mount Hopkins. The SAO, in conjunction with the Academia Sinica of Taiwan, is constructing the Smithsonian Submillimeter Array of eight 6-m dishes on Mauna Kea, Hawaii, due for completion in 1999.

SNC meteorites A small group of rare and unusual meteorites, consisting of three main subgroups: the shergottites, nakhlites, and chassignites. The SNC meteorites are igneous rocks that solidified from a cooling magma near the surface of their parent body, probably Mars. All but one are relatively young (less than 1.3 billion years old); the single ancient SNC meteorite found to date was formed about 4.5 billion years ago. The proportions and isotopic abundances of noble gases trapped in one shergottite resemble the composition of the Martian atmosphere as analysed by the Viking landers. SNC meteorites were probably ejected from Mars by impacts, and entered orbits around the Sun before falling to Earth.

Sojuner rover Sojuner rover was the first robotic explorer to explore Mars. Weighing only 25.4 pounds (11.5 kilograms), Sojuner was launched aboard NASA's Pathfinder mission on Dec. 4 1996. The 25-pound, six-wheeled microrover explored an ancient Martian flood plain (Ares Vallis) in 1997. Each wheel has a diameter of 5 inches (13 cm). The rover has a maximum speed of 1.3 feet per minute (0.4 meters per min.). It was powered by a solar panel.

Sol A local day on a planet, defined by its period of rotation. On Mars, for example, a local day lasts 24 hours and 37 minutes. During the Mars Pathfinder mission, scientists began counting Sols with the successful landing of the spacecraft on July 4, 1997 (Sol 1). Images from the Mars Pathfinder mission always contain a reference to the Sol on which it was taken.

Solar activity The collective term for all active phenomena on the Sun, including sunspots, faculae, active regions, plages, active prominences, and flares. Solar activity is strongly associated with magnetic fields, which are thought to arise from a dynamo action within the Sun. Solar activity increases and decreases in a cycle lasting approximately 11 years.

Solar antapex The direction in the sky towards which a sample of stars tends, on average, to be moving because of the motion of the Solar System relative to the sample. This direction cannot be defined precisely because of the random motions of the stars themselves, and it differs according to spectral type because of varying population mixtures, but the approximate position is RA 6h. dec. -30°.

Solar apex The direction towards which the Solar System is moving relative to the local standard of rest. Its position is approximately RA 18h, dec. +30°. As a result of this motion, stars seem to be converging towards a point in the opposite direction, the solar antapex.

Solar atmosphere The region around the Sun including the photosphere, chromosphere, transition region, and corona. The temperature of the solar atmosphere varies with height, at first decreasing to the temperature minimum region, and then increasing, very strongly so in the transition region. The temperature rise is thought to be due to the dissipation of waves, both sound waves (in the low chromosphere) and magnetohydrodynamic waves (in the rest of the chromosphere and corona).

Solar constant The amount of solar energy, in all wavelengths, that falls on a given area at the top of the Earth's atmosphere. In fact, the solar 'constant' varies by 0.3% over the 11-year solar cycle, and perhaps by more over longer periods, so a better term is solar irradiance. Modern spacecraft measurements give the solar irradiance as 1368 W/m^2 solar cycle.

Solar coordinates Central Meridian Distance (CMD). The angular distance in solar longitude measured from the central meridian.

Solar cycle The solar cycle is the roughly 11-year, quasi-periodic variation in the frequency or number of sunspots, solar flares, and other solar activity.

Solar day A solar day is the amount of time that passes between two subsequent times when the Sun reaches its highest position above the horizon (passing through the meridian). The solar day varies greatly throughout the year, so the mean solar day is used instead. This is calculated as the average of all of the solar days in one year. A solar day is exactly 24 hours (since this is how we define hour, as 1/24 of a day); a solar day is slightly longer than a sidereal day. A solar day is not the time it takes for the Earth to rotate on its axis (that is the sidereal day).

Solar dynamo The action within the Sun whereby the kinetic energy of the hot, highly ionized gas of the solar interior is converted into the magnetic field that gives rise to solar activity. In the picture due to H. W. Babcock, magnetic field lines under the photosphere running from pole to pole (the poloidal field) are twisted parallel to the equator (the toroidal field) by the Sun's differential rotation. Active regions, including sunspots, are thought to be generated as the distorted magnetic field lines rise through the photosphere. Convection currents gradually turn the toroidal field into a poloidal field of the reverse direction for the next cycle.

Solar eclipse The passage of the Moon across the Sun's disk. Solar eclipses occur only at new Moon when the Moon lies close to the node of its orbit around the Earth. They do not occur each month because the new Moon is usually either north or south of the node, due to the inclination of the Moon's orbit. Total solar eclipses are rare at any particular place on Earth, since the Moon's shadow falls on only a limited area. The theoretical maximum duration of totality is 7m 31s, but is usually no longer than 3-4 min. Around totality, a number of interesting phenomena occur, including shadow bands and Baily's beads. During totality, solar prominences and the corona can be seen. Solar eclipses around lunar apogee may be annular eclipses. Either side of the ground track of a total or annular eclipse, a partial eclipse is visible. Partial eclipses produce little obvious diminution in the level of sunlight unless their magnitude exceeds about 0.9 (90%)

Solar halo A solar halo is a luminous ring that is sometimes seen surrounding the Sun. Some parts of the halo are very bright, others are not very bright. Sometimes, only a part of the ring is visible. Within the solar halo, on opposite sides of the Sun, there can be two very bright spots called "sun dogs" or "mock suns." The halo is produced as sunlight is reflected and refracted through tiny, flat ice crystals in the atmosphere. Sun dogs and halos always are at an angle of 22° away from the Sun, due to the hexagonal structure of the ice crystals. The diameter of the halo is about an eighth of the sky.

Solar interior That part of the Sun lying below the deepest visible layer of the photosphere. The solar interior cannot be directly observed, but its structure can be deduced from standard solar models. Such models can be checked against observations of global oscillations, where the agreement is fairly satisfactory; and against the numbers of solar neutrinos, where there is a considerable disagreement, the models predicting more neutrinos than are actually observed.

Solar mass A solar mass is the amount of mass in our Sun; it is also the unit in which the masses of other stars, galaxies, and other large celestial bodies are expressed. A black hole, for example has anywhere from a million to a billion solar masses.

Solar maximum A solar maximum is the highest level in solar activity (like flares, prominences, sunspots, soronal holes, etc.), and occurs between consecutive solar minima.

Solar maximum mission A NASA satellite that observed solar activity at wavelengths from gamma rays to white light. SMM, also known as Solar Max, was launched in 1980 February. A failure of its pointing system after nine months led to a repair mission in 1984 April by Space Shuttle astronauts. SMM continued operations until it re-entered the Earth's atmosphere in 1989 December. The mission led to improved understanding of the energy-release mechanisms of solar flares and related phenomena.

Solar minimum The month(s) during the solar cycle when the 12-month mean of monthly average sunspot numbers reaches a minimum. the most recent minimum occurred in September 1986.

Solar motion The linear velocity of the Solar System in space relative to the local standard of rest. It is usually given in the form of three components along either galactic or equatorial coordinate axes; alternatively, it may be expressed as a speed towards the solar apex, about 19.5 km/s or 4 AU/year.

Solar nebula The cloud of gas and dust from which our Solar System formed some 5 billion years ago. The cloud is thought to have been shaped like a flattened disk and was dispersed by the T Tauri wind from the young Sun. Comets, asteroids, and meteorites provide important clues to the composition of the solar nebula. Similar disks of gas and dust have been detected around several nearby young stars, notably Beta Pictoris.

Solar neutrino unit A measure of the flux of neutrinos from the Sun reaching the Earth. The neutrinos are produced in nuclear reactions at the Sun's centre. High-energy solar neutrinos are detected by a tank of fluid containing chlorine, situated deep underground in the Homestake gold mine, South Dakota. A tiny proportion of the chlorine atoms interact with solar neutrinos to produce argon atoms. The number of neutrino interactions is measured by the solar neutrino unit (SNU), 1 SNU equalling

1 interaction per second per 10^{36} chlorine atoms. The predicted neutrino rate from current solar models is about 8 SNU, but the measurements from the Homestake experiment indicate a rate of only about 2 SNU. More recent measurements, both from Homestake and from the gallium detectors SAGE (in Russia) and Gallex (near Rome, Italy), which detect neutrinos from the main proton-proton fusion reaction, indicate a smaller discrepancy, with the neutrino rate about half that expected. solar oscillations.

Solar parallax The angular width of the Earth's equatorial radius as seen from the centre of the Sun, when the Earth is at a distance of 1 AU. Its value is 8".794148. Historically, the solar parallax was the prime quantity giving the linear scale of the Solar System, and was derived from observations of bodies in the Solar System as seen from widely spaced observatories, for example of transits of Mercury or Venus across the Sun. The solar parallax is now derived from direct measurements of distances in the Solar System by radar and transmissions from spacecraft.

Solar plume A solar plume is a long, feathery jet of high-speed electrified gas that is expelled from the Sun's corona. Solar plumes emanate near the Sun's poles and travel over 13 million miles (21 million km) into space.

Solar system The collective name for the Sun and all the bodies that orbit it. It includes the nine major planets and their 61 known satellites, plus countless asteroids, comets, and meteoroids. The aphelion of the orbit of Pluto, over 7.3 billion kilometres from the Sun, marks the outer limit of the known planetary system, but many Kuiper Belt objects lie beyond this and some long-period comets travel perhaps half-way to the nearest star.

Solar time Time with respect to the Sun; technically, the hour angle of the Sun plus 12 hours, added to make the day begin at midnight rather than noon. The true Sun is the basis of apparent solar time, as shown on a sundial, but this runs irregularly because the Earth's orbit is not circular and the Sun's path (the ecliptic) is inclined to the celestial equator. For accurate timekeeping purposes mean solar time is used, based on the hour angle of the fictitious mean sun. Solar time, which is the basis of all civil timekeeping, loses about 4 min a day against sidereal time, so that a star will rise about 4 min earlier each night.

Solar wind The flow of atomic particles from the Sun's corona. The solar wind consists of electrons, protons, and (to a lesser extent) the nuclei of elements such as helium. As the gas expands into interplanetary space it carries with it magnetic field lines, which form into a spiral pattern as the Sun rotates. The velocity of the solar wind increases from about 50 km/s at a few solar radii from the Sun to several hundred kilometres per second at the distance of the Earth. The particle densities at the Earth's distance are extremely low (about 5 million particles per cubic metre). The flow speeds vary widely. The low-speed solar wind has a speed of about 300 km/s. At times of solar minimum, streams are observed with speeds as high as 750 km/s. They flow from coronal holes, and pass the Earth once every 27 days as the Sun rotates. High-speed streams continuously emanate from each of the solar poles.

Solar year The solar year (also called the tropical year) is the amount of time that passes between two consecutive returns of the sun to the vernal equinox. In 1900, the solar year lasted 365 days, 5 hours, 48 minutes, and 46 seconds (shorter than the sidereal year); the length of the solar year is decreasing 0.53 second per century ad the Earth slows down in its orbit.

Solar-terrestrial relations The effects of solar activity on the Earth and its magnetic field. The largest effects originate from magnetic disturbances which travel out from the Sun and produce geomagnetic storms as they interact with the Earth's magnetosphere. A sharp increase in the strength of the Earth's magnetic field at the start of a storm results from a compression of the field on the sunward side. Disturbances occur worldwide and may continue for a day or so. Magnetic substorms are more localized, occurring near the geomagnetic poles, and are associated with auroral displays. Links between solar activity and the Earth's weather have been proposed, notably at the time of the Maunder minimum.

Solstices The time when the Sun is at its maximum declination either north or south of the celestial equator. There are two solstices; summer solstice (around 22nd June each year) when the Sun is 23 and a half degrees north of the celestial equator. At this point the Earth's northern hemisphere will have its longest day and the Sun's noon altitude will be at its highest; and winter solstice (around 22nd December each year) when the Sun is 23 and a half degrees south of the celestial equator. At this point the Earth's southern hemisphere will have its longest day and the Sun's noon altitude will be at its highest.

Solubility The ability or tendency of one substance to dissolve into another at a given temperature and pressure.

Solunar Solunar means pertaining to the rising and setting times of the sun and the moon, the phases of the moon, and of solar and lunar eclipses. Solunar tables list when these various events occur.

Sombrero galaxy The galaxy M104 in Virgo. It is an 8th-magnitude spiral of type Sa or Sb with a large nucleus and is seen edge-on, giving the appearance of a Mexican hat. A dark lane of dust runs across its centre. It is about 30 million L.Y away, closer than the Virgo Cluster.

Sothic cycle A cycle of 1460 years in the calendar of ancient Egypt. The Egyptian calendar had a year of fixed length, 365 days, with no leap years. Consequently over one Sothic cycle the calendar regressed by one year with respect to the seasons, since the calendar year was approximately one quarter of a day shorter than the tropical year. The cycle takes its name from Sothis, the ancient Egyptian name for Sirius, whose heliacal rising marked the start of their year.

South atlantic anomaly A region within the inner Van Allen Belt, which reaches its minimum altitude (250 km) over the Atlantic Ocean off the coast of Brazil because of the offset between Earth's magnetic and geographical axes. Artificial satellites in low-inclination and low-altitude orbits pass frequently through the South Atlantic Anomaly, with a consequent risk to their electronic components (including degradation of solar cells) from energetic trapped particles.

Southern reference stars A list of 20 488 stars in the southern hemisphere designed to supplement the AGK3R in the north, to give a whole-sky coverage of reference stars for calibrating photographic surveys. Observations of stars mostly in the magnitude range 7.5 to 9.5 were made at a dozen observatories between 1961 and 1973, and coordinated at the US Naval Observatory and Pulkovo Observatory. The results were issued in computerreadable form in 1988.

Southern sky survey A photographic survey of the sky south of declination −17°, produced jointly by the European Southern Observatory's 1-m Schmidt camera and the 1.2-m United Kingdom Schmidt Telescope at Siding Spring. Both telescopes have the same focal length as the 1.2-m Oschin Schmidt, which produced the northern-hemisphere Palomar Observatory Sky Survey, so the scale of the plates in each survey is identical. The Southern Sky Survey consists of 606 pairs of red-sensitive and blue-sensitive plates, the red plates being taken with the ESO Schmidt and the blue plates with the UK Schmidt.

Soyuz A Russian spacecraft capable of seating up to three people. On its maiden flight on 1967 April 23 it crashed on landing, killing the sole occupant, Vladimir Mikhailovich Komarov who thereby became the first person to die during a spaceflight. Soyuz is used for ferrying crews to and from space stations, but it has also made separate flights and was used for the Soviet half of the Apollo-Soyuz Test Project. An unmanned version, called Progress, is used for resupplying space stations.

Space motion The linear velocity of a star expressed in kilometres per second as components along three axes. Any three axes can be chosen, but most commonly they are oriented towards the galactic centre, in the direction of the Galaxy's rotation, and towards the north galactic pole. Alternatively, axes based on the directions of the equinox and celestial pole can be used.

Space probe A spacecraft that is sent away from the Earth to examine another celestial body or the conditions in space. The first space probe was Luna 1, launched by the Soviet Union on 1959 January 2.

Space shuttle A reusable winged space transporter, which is launched like a rocket but which glides back to land on a runway like an aircraft. The Space Shuttle orbiter has its own main engines, fuelled at launch by a large external tank which is discarded shortly before reaching orbit. Two external strap-on boosters, which are discarded during ascent, provide additional thrust for launch. The Space Shuttle orbiter can carry up to eight astronauts in its crew compartment and can remain in orbit for a maximum of 18 days. Among its various uses are launching satellites and space probes, carrying the Spacelab space station, and ferry missions to the International Space Station. It can also be used to retrieve objects from orbit and return them to Earth. There are four Space Shuttle orbiters: Columbia, first launched on 1981 April 12; Discovery, first launched on 1984 August 30; Atlantis, first launched on 1985 October 3; and Endeavour, first launched on 1992 May 7.

Space station A space station is an orbiting structure where people can live in space. The first space station, Salyut 1, was launched on April 19, 1971. It led to Mir, which was launched in 1986. Construction (in orbit) on the new International Space station began in 1998. The first crew was launched on a Russian Soyuz (in Russian, soyuz means "union") spacecraft for a three-month stay beginning January 2000.

Space velocity The velocity of a star relative to the Sun. It can be determined from measurements of the star's radial velocity and tangential velocity.

Spacelab A space station built by the European Space Agency which is carried in the cargo bay of the Space Shuttle. Spacelab consists of a pressurized module and structures called pallets, open to the vacuum of space, on which instruments can be mounted. The pressurized module and pallets can fly together or separately. Spacelab made its maiden flight in 1983 November aboard the STS 9 mission.

Space-time A four-dimensional description of the Universe in which the position of an object is specified by three coordinates in space and one in time. According to the theory of special relativity, there is no absolute time which can be measured independently of the observer, so events that are simultaneous as seen by one observer occur at different times when seen from a different place. Time must therefore be measured in a relative manner, as are positions in three-dimensional space, and this is achieved through the concept of space-time. The trajectory of an object in space-time is called its world line. General relativity relates the curvature of space-time to the positions and motions of particles of matter.

Space-time curvature The curvature of space-time is a distortion of space-time that is caused by the gravitational field of matter. The degree of curvature depends on the strength of the gravitational field (which depends on the massiveness of the objects in that part of space). An object traveling in space moves along the curves in space-time.

Special relativity Special relativity supplanted Newtonian mechanics, yielding different results for very fast-moving objects. The Theory of Special Relativity is based on the idea that speed has an upper bound; nothing can pass the speed of light. The theory also states that time and distance measurements are not absolute but are instead relative to the observer's frame of reference. Space and time are viewed as aspects of a single phenomenon, called space-time. Energy and momentum are similarly linked. As a result, mass can be converted into huge amounts of energy, and vice versa, according to the formula $E=mc^2$. Albert Einstein devised the Theory of Special Relativity.

Special theory of relativity A theory proposed by A. Einstein in 1905, based on the proposition that the speed of light in a vacuum is constant throughout the Universe, and is independent of the motion of the observer and the emitting body. A consequence of this proposition is that three things happen as an object's velocity approaches the speed of light: its mass goes up, its length shortens in the direction of motion, and time slows down. Hence, according to special relativity, no object can ever reach the speed of light because its mass would then become infinite, its length would become zero, and time would stand still. In addition, Einstein concluded that the mass of a body is a measure of its energy content, according to the famous equation $E = mc^2$, where c is the speed of light. This equation describes the conversion of mass into energy in nuclear reactions within stars.

Spectral class Spectral classes are groups of stars that have similar characteristic emission lines in their spectra (indicating that they have similar compositions).

Spectral line A spectral line is a bright or dark line found in the spectrum of some radiant source. Bright lines indicate emission, dark lines indicate absorption. A bright spectral line represents light emitted at a specific frequency by an atom or molecule. Each different element and molecule gives off light at a unique set of frequencies. Astronomers can determine the composition of gases in stars by looking for characteristic frequencies (this is called spectroscopy). For example, carbon monoxide (CO) has a spectral line at 115 Gigahertz (equal to a wavelength of 2.7 mm).

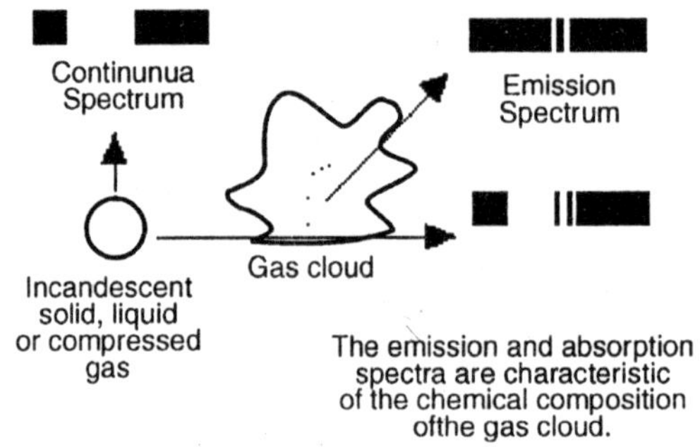

Fig. Spectral line

Spectral type The spectral type of stars is a system of classification of stars based on the stars' spectra, emission lines that correlate with each star's surface temperature (and color). There are seven major spectral types. Stars range from blue and hot to red and cool. The spectral types are: O, B, A, F, G, K, and M (from hottest to coolest). Each of these letters is divided into 10 numerical classes, from hotter to cooler: 0, 1, 2, 3, 4, 5, 6, 7, 8, and 9. For example, our Sun has the spectral type G2.

Spectroheliograph An instrument for photographing the Sun at the wavelength of a strong Fraunhofer line. An image of the Sun produced by a telescope is focused on to a primary slit, and light from this slit passes to a grating or prism. The dispersed light so formed is then intercepted by a secondary slit positioned at a desired wavelength (e.g. a part of the Ha line) and a photographic plate is placed behind this slit. The primary slit is then moved across the image of the Sun and the secondary slit moved to maintain its spectral location, producing a spectroheliogram on the photographic plate.

Spectroscopic binary A binary star that cannot be resolved visually or photographically, but whose duplexity is established by periodic Doppler shifts of lines in its spectrum. All close binaries (including cataclysmic and interacting binaries) fall into this category, as do most eclipsing binaries. The information obtainable about a spectroscopic system depends on whether a radial velocity curve can be obtained for one or both components. With a single-line binary, it is possible to obtain only an _expression known as the mass function, which is a crude measure of the properties of the system. With a composite-spectrum binary or double-lined binary, the relative masses may be derived from the mass ratio. If the orbital inclination can be found (normally only in eclipsing systems), the dimensions, masses, and orbit of the stars can be calculated. With interacting binaries, there are limitations to the latter procedure because of their complex structure.

Spectroscopic parallax A method of estimating the distances of stars by comparing their apparent magnitude with their absolute magnitude. The absolute magnitude of a star is deduced from its spectral type and luminosity class, which are found by examining its spectrum. Spectroscopic parallax is the most common method of determining the distances of stars that are too distant to show a reliable trigonometric parallax.

Spectroscopy The technique of obtaining and studying the spectra of celestial bodies, from which their composition and motions can be determined. The instrument used is a spectrograph, a form of spectroscope that uses a detector (nowadays a CCD) to record the spectrum. There are three main types of spectrum: a continuum, or continuous spectrum, consisting of an unbroken band of colours; an emission spectrum, which consists of bright lines; and an absorption spectrum, consisting of dark lines crossing a continuum. Study of a star's spectrum allows its temperature, luminosity, and size to be determined.The displacement of spectral lines reveals the object's movement along the line of sight (its radial velocity).

Spectrum 1. A range of electromagnetic energies arranged in order of wavelength or frequency. The emission spectrum of a body or substance is the range of radiations it emits when it is heated, is bombarded by electrons or ions, or absorbs photons. The absorption spectrum of a substance consists of dark lines or bands in a continuous spectrum, each line being a wavelength or group of wavelengths at which light is removed from the continuous spectrum by the absorbing medium. These lines and bands are at the same wavelengths as some of the lines and bands in the substance's emission spectrum. Emission and absorption spectra may show a continuous spectrum (also called a continuum), a line spectrum, or a band spectrum. A continuous spectrum contains an unbroken sequence of frequencies over a wide range; continuous spectra are produced by incandescent solids, liquids, and compressed gases.

Speed of light (c) The speed at which electromagnetic radiation propagates in a vacuum; it is defined as c = 299,792,458 m/s (186,000 miles/second). Einstein's Theory of Relativity implies that nothing can travel faster than the speed of light.

Spherical astronomy The description of positions and angular displacements of celestial objects by means of coordinates on the celestial sphere. It is one of the oldest branches of astronomy, and is now also concerned with the tracking of artificial satellites and space probes.

Spicule A spike-like jet in the upper chromosphere of the Sun, visible especially in Ha light. Spicules rise upwards into the corona at velocities of about 30 km/s, reaching altitudes of 9000 km in about 90 s. They last about 15 min, ending their lives by fading rather than descending. They tend to be clustered, giving the appearance of porcupine quills on the solar limb. Seen against the disk, in Ha light, they seem to be identifiable as fine mottles, light or dark depending on their height (although they always appear dark when observed in the red line wing of the Ha line). Fine mottles form clusters with their bases in the coarse mottles which outline the chromospheric network. At the centre of the Sun's disk, clumps of spicules form radiating patterns called rosettes.

Spinar A hypothetical type of supermassive star once proposed as the power source of quasars. They were envisaged as having a mass of about 100 million solar masses, to spin rapidly, and to radiate like a scaled-up version of a pulsar. However, even if such an object could form, it would be unstable and would rapidly collapse into a supermassive black hole.

Spindle galaxy An elongated 9th-magnitude galaxy in Sextans, also known as NGC 3115. It is classified as either a highly flattened elliptical galaxy, or a spiral with a large bulge and hardly any surrounding disk. It lies about 30 million L.Y away.

Spin-down A phenomenon in which the rotation period of a pulsar, P, steadily lengthens as the pulsar loses energy with age. Observed spin-down rates, range from 10^{-12} s per second for the youngest pulsars down to 10^{-19} s per second for recycled pulsars. The quantity P/2 is known as the characteristic age or spin-down age of the pulsar, and is likely to represent an upper limit on the true age of the pulsar.

Spin-orbit coupling A relationship between the orbital period of one body around another and its rotation period on its axis. The relationship has been brought about by tidal forces. For example, the period or rotation of the planet Mercury is two-thirds of its period of revolution. Other examples of the same effect are co-rotation and synchronous rotation.

Spiral arm A curved structure in the disk of spiral (and some irregular) galaxies where young stars, clusters, nebulae (H II regions), and dust are concentrated. Some galaxies have a well-defined, two-arm spiral pattern, while in others the arms may be three or four in number and sometimes fragmentary. The arms are visible because of the recent formation of bright, massive, short-lived stars within them. This star-forming activity is periodic, corresponding to the movement through the disk of a density wave.

Spiral galaxy A type of galaxy with bright arms of stars, gas, and dust that extend in a spiral pattern from a central hub; Hubble type S. There are usually two arms, which can wrap right around the galaxy, but four-armed and even threearmed examples are known. The arms can also be broken into many short sections. At the centre of a spiral galaxy is a spheroidal bulge of old stars of Population II. The bulge is large in type Sa spirals, which have tightly coiled arms, but much smaller and less conspicuous in types Sc and Sd, whose arms are coiled more loosely. Types Sa and Sb are, on average, intrinsically brighter and more massive than types Sc and Sd.

The spiral arms are sites of active star formation, and their appearance is dominated by bright, blue, massive young stars of Population I and gaseous H II regions. In mass they range from about 10^9 to 5×10^{11} solar masses, and in diameter from about 10000 to over 300000 L.Y .

Sporadic meteor A randomly occurring meteor, not associated with any meteor shower. Sporadic meteors can be seen on any clear night, with rates varying from perhaps 3-4 meteors/h in the spring to 8-10 meteors/h in the autumn. In addition to this seasonal variation there is also a nightly variation, which results in higher sporadic rates before dawn than in the evening hours. The sporadic background represents an end-point in meteor stream evolution, reached when the meteoroids in a stream become so dispersed that a meteor shower is no longer detectable. Because they are random, sporadic meteors are often considered a useful control sample in shower studies.

Spörer minimum A period between about AD 1450 and 1550 when activity on the Sun seems to have been unusually low, as judged from historical records of naked-eye sightings of sunspots and aurorae, and from measurements of carbon-14 in tree rings. The Spörer minimum and the similar Maunder minimum coincide with a period of lower temperatures on Earth known as the Little Ice Age. It was discovered by the American solar physicist John Allen Eddy who named it after G. F. W. Spörer.

Spörer's law The drift of the average latitude of sunspots towards the Sun's equator during the 11-year sunspot cycle. Spots of a new cycle are generally located at latitudes 30-40° north or south, but over the course of the cycle the average latitude decreases to 5-10°, as demonstrated by the butterfly diagram. The 'law' is named after G. F. W. Spörer, but was actually first noticed by R. C. Carrington.

Spray An explosive solar prominence, associated with a flare, with very high velocity—up to 2000 km/s, faster than the escape velocity of the Sun. The material of a spray appears fragmented, sometimes in a loop structure. spring equinox Another name for the vernal equinox.

Spring tide The tide raised in the Earth's oceans at full or new Moon, when the Moon and Sun are pulling on the Earth in line. The amplitude of spring tides (i.e. the difference in water level between high and low tide) is the greatest in the monthly tidal cycle.

Sputnik The Soviet Sputnik space missions were the first to orbit the Earth. The word Sputnik means "Traveling Companion" in Russian. The first Sputnik mission was launched on on October 4, 1957; this was Earth's first artificial satellite. Sputnik 1 was about the size of a basketball, weighing roughly 183 pounds. It was sent into an elliptical orbit around the Earth, revolving in about 98 minutes. Sputnik 2 was launched on November 3, 1957 with the dog Laika aboard. Sputnik 3 failed on February 3, 1958. The last Sputnik mission, Sputnik 3 was successfully re-launched on May 15, 1958 and remained in orbit for 2 years.

Square of pegasus A large asterism whose four corners are marked by the stars Alpha, Beta, and Gamma Pegasi, and Alpha Andromedae.

SS cygni star A type of U Geminorum star that exhibits distinct, well-defined outbursts, with an amplitude of 2-6 mag.; abbr. UGSS. The rise time is 1-2 days, with a rather longer decline. The mean interval between outbursts ranges between ten and thousands of days. There are variations in the amplitude, duration, and shape of individual maxima. Although most stars have distinct short and long outbursts, the latter bear little resemblance to the supermaxima of an SU Ursae Majoris star.

Standard atmosphere A model of the vertical structure of the Earth's atmosphere, approximating to a global, year-long average of the changing conditions of temperature and pressure with altitude. The model takes account of the subdivision of the atmosphere into distinct levels, including the troposphere, stratosphere, mesosphere, and thermosphere, and the boundaries between them.

Standard candle In astronomy, a standard candle is an object with a known amount of energy output. By knowing the actual energy output, astronomers can use the observed output to calculate the distance to that object.

Standard epoch A date for which star coordinates and other data are calculated in star catalogues; also known as a fundamental epoch. The standard place of a star is its right ascension and declination referred to the mean equator and equinox of the chosen epoch. By referring all star positions to a standard date, the effects of proper motion, precession, and nutation are removed. The standard epoch currently used is the Julian epoch J2000.0, which corresponds to the date 2000 January 1.5 TDB.

Standard solar model A mathematical description of the Sun's structure that gives the variation of pressure, temperature, and other quantities with radius. It is obtained by considering how a mass of gas simulating the Sun at its birth evolves to its present state, with radius and luminosity matching measured values. The most recent models give a core temperature of 15.6 million K and a density of 148000 kg/m^3.

Standard star A star used to calibrate observations of previously unstudied stars, particularly for photometry. In spectrophotometry, standard stars are compared with a black-body source located close to the telescope so that the amount of radiation emitted at each wavelength is known. In conventional photometry, standard stars have accurately known magnitudes and colours, against which stars under study are compared. Each system of photometry (e.g. Johnson, Kron-Cousins RI, or Strömgren photometry) has its own set of standards which have been carefully intercompared. Standard stars must be bright enough to be easily observable with small telescopes, but not so bright as to saturate photometers on large telescopes.

Star A luminous ball of gas that, at some stage of its life, produces energy by the nuclear fusion of hydrogen to form helium. The term thus not only includes stars such as the Sun, which is currently burning hydrogen, but also protostars, not yet hot enough for such burning to have begun, and various evolved objects such as giant and supergiant stars, which are burning other nuclear fuels, or white dwarfs and neutron stars, which consist of spent nuclear fuel.

The maximum mass of a star is about 120 solar masses, above which it would be blown apart by its own radiation. The minimum mass is 0.08 solar masses; below this, objects never become hot enough at their cores for hydrogen burning to begin, and instead become brown dwarfs. Star luminosities range from about half a million times the Sun's luminosity for the hottest to under one-thousandth of the Sun's for the faintest dwarfs.

Star atlas An collection of maps that marks the positions of stars, nebulae, galaxies, and other astronomical objects on a coordinate system.

Star classification Stars are classified by their spectra (the elements that they absorb) and their temperature. For example, the Sun is a G1 star. There are seven main types of stars. In order of decreasing temperature, they are: O - He II absorption; B - He I absorption; A - H absorption; F - Ca II absorption; G - strong metallic lines; K -bands developing; M - very red. O and B stars are uncommon but very bright; M stars are common but dim. An easy mnemonic for remembering these is: "Oh be a fine girl, kiss me."

Star cluster A collection of stars formed together from the same gas cloud and bound together by mutual gravitational attraction. There are two main types of star cluster. Globular clusters contain from tens of thousands to millions of old stars packed into a near-spherical ball, and are found in the haloes of galaxies. Open clusters are much more sparse and irregular in shape, consisting of no more than a few hundred relatively young stars in the spiral arms of galaxies. A group looser than an open cluster is known as a stellar association.

Star count The number of stars in a specific area of sky and within a given range of magnitudes. Star counts are used extensively in statistical studies of the distribution of stars in the Galaxy.

Star diagonal A star diagonal is a simple mirrored device on some refracting telescope that bends the light 90 degrees. This lets the observer look down through the eyepiece instead of kneeling to look directly through the telescope. The resulting image has left and right reversed.

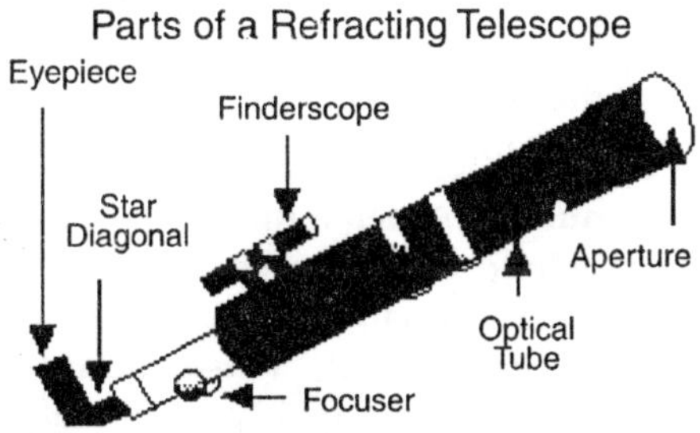

Fig. Star diagonal

Starburst galaxy A starburst galaxy is a bright, blue-glowing galaxy that has a high rate of star formation (sometimes, the star-formation area is only at the core of the galaxy, and these galaxies are called nuclear starbursts). This type of galaxy is filled with glowing hydrogen gas clouds, dense interstellar dust, and many, hot, newly-formed stars. These galaxies have strong radio emissions that are the remnants of recent supernovas. Starburst galaxies were formed by violent gravitational encounters, like near-collisions of nearby galaxies. Examples of starburst galaxies include NGC 253 and the barred spiral galaxy NGC 1808. Starburst galaxies may be related to active galactic nuclei.

Starquake Starquakes are fractures in the crust of neutron stars and pulsars caused by the star's strong magnetic field. As the crust cracks in the starquake, energy is released and x-rays are emitted. Another type of starquake occurs on ordinary stars (like our Sun). In these quakes, violent seismic events on stars are caused by solar flares (jets of hot gases ejected thousands of miles from the surface of a star). When a starquake occurs, energy is released in seismic waves on the relatively fluid surface of the star. These waves radiate in concentric circles from the epicenter of the starquake.

Static limit The region around the event horizon of a rotating black hole within which it is impossible for any object to remain stationary; also known as the stationary limit. Within the static limit, an object would be forced to spin with the black hole, as though in a whirlpool. The static limit touches the event horizon at the black hole's poles, where there are no rotational forces, but becomes larger than the event horizon towards the black hole's equator. The static limit is the outer boundary of the ergosphere. In a non-rotating black hole, the static limit coincides with the event horizon.

Static universe A universe that is neither expanding nor contracting. A. Einstein constructed such a model by adding the cosmological constant to his equations of general relativity. Such a universe is incompatible with the observed Hubble law.

Stationary point 1. A position at which a planet temporarily does not move in right ascension on the celestial sphere. Planets exterior to the orbit of the Earth reach a stationary point when changing from direct motion to retrograde motion and back again.

2. A point in space at which a body can remain in equilibrium with other bodies, maintaining a constant distance from them, for example a Lagrangian point.

Steady-state theory A cosmological model of a universe which is expanding but has the same density at all times due to the continuous creation of matter. The steady-state theory is based on the perfect cosmological principle, which requires the universe to be the same at all times, as well as in all places. The mathematical solution of the equations of general relativity that results from this principle is the de Sitter universe. For many years the steady-state theory was a rival to the Big Bang theory, but it has fallen out of favour because it is inconsistent with the observed properties of the cosmic background radiation and cannot explain the evolution in the appearance of our Universe with time. Unlike the Big Bang theory, the steady-state theory has no initial singularity and does not require the existence of a primordial fireball phase. It was put forward in 1948 by H. Bondi, T. Gold, and F. Hoyle.

Stefan-Boltzmann law A law relating the energy emitted by a black body (such as a star) to its temperature; also known as Stefan's law. According to the law, the total energy radiated in watts per square metre is proportional to the fourth power of its thermodynamic temperature in kelvin; hence a doubling of temperature leads to a sixteen-fold increase in energy output. Expressed mathematically, E = sT4, where a is the Stefan-Boltzmann constant. The total power per square metre can vary from 3 μW for the microwave background radiation, to 75 MW for the Sun, and thousands of gigawatts for hot stars such as white dwarfs. The law was discovered by Joseph Stefan and derived theoretically by Ludwig Edward Boltzmann.

Stellar atmosphere The low-density outer region of a star, transparent to light except at certain wavelengths which it absorbs. Absorption in stellar atmospheres produces the dark Fraunhofer lines in a star's spectrum, which provide information about the chemical composition, pressure, temperature, rotation, and magnetic field strength in stellar atmospheres.

Stellar evolution The series of changes that stars undergo during their lifetimes, the time-scale of which depends strongly on the star's mass and also, to some extent, on its initial composition. The progress of a star during its evolution can be followed on a graph called the Hertzsprung-Russell diagram.

A star is born when a dense region of a cloud of gas collapses under its own gravity. A star first shines because the gravitational potential energy lost in this collapse is released as'heat and light. Eventually the temperature at the centre of the protostar becomes high enough to ignite nuclear reactions involving deuterium (an isotope of hydrogen), and for some time the energy from this is sufficient to prevent further collapse. Once the deuterium has been exhausted, the collapse continues, and the star is classified as a pre-main-sequence object, following a characteristic path on the HR diagram. For a star the mass of the Sun, this phase lasts several million years.

Stellar nursery A stellar nursery is a nebula (a large cloud of hydrogen gas in space) in which star formation is occurring (stars are formed from gas). These nebulae are frequently illuminated by ultraviolet light which is emitted from the newborn stars.

Stellar wind The outflow of gas from the surface of a star. The Sun itself loses about 10^{-14} of its mass through the solar wind each year, but winds are of much greater importance in pre-main-sequence stars (e.g. T Tauri stars) and in giants and supergiants. Extreme stellar winds are found in X-ray binaries, in which stars of spectral type O or B can lose around 10^{-6} of a solar mass per year to a compact companion (a neutron star or black hole).

Stishovite A rare mineral that is produced when a particularly high-pressure shock wave meteorite impact passes through rock containing quartz. It is a highdensity form of silica produced at a pressure of 130000 bar. Stishovite-bearing fragments in the vicinity of large craters are taken as evidence that the crater is of meteoritic origin.

Stony meteorite A meteorite composed primarily of silicate minerals, usually with some nickel-iron; also known as an aerolite. Around 95% of all meteorites observed to fall are stones. However, they are difficult to distinguish from terrestrial rocks, and many go unnoticed. Their fusion crust is thicker than that of irons and is often black in colour, either dull or shiny. The interior of stones is usually grey or dark grey, and granular in texture. Stony meteorites are divided into two main subtypes: chondrites, which comprise 86% of all observed falls, and achondrites.

Stratopause The upper boundary of the stratosphere, immediately underlying the mesosphere at an altitude of 50 km above the Earth's surface. Stratospheric temperatures reach a maximum of about 0 °C at the stratopause.

Stratosphere The layer of the Earth's atmosphere immediately above the troposphere, extending to the mesosphere (i.e. between altitudes of 10-15 km and 50 km). In contrast with the troposphere, temperatures in the stratosphere rise with increasing altitude. The concentration of water vapour in the stratosphere is low, and clouds are rare. During the winter months, nacreous or polar stratospheric clouds may occasionally form at high latitudes, around altitudes of 15-30 km. The ozonosphere, including the ozone layer, lies within the stratosphere, and there is also some overlap with the lower part of the chemosphere.

Strewn field An area on Earth where tektites are found. The largest strewn field covers the whole of southern Australia and Tasmania (the australites). Other major strewn fields are found in the Czech Republic (the moldavites), Africa (the Libyan Desert glass and Ivory Coast tektites), the USA (the bediasites, georgiaites, and Martha's Vineyard tektites), across Southeast Asia (the billitonites, indochinites, javanites, malaysianites, philippinites, and rizalites), and in central Russia (the irghizites).

Strömgren sphere The region of ionized gas surrounding a hot star. B. G. D. Strömgren first derived the relationship between the density of the interstellar gas, the temperature of the star, and the radius of the region which it ionizes. The ionized gas is usually called an H II region; its size is termed as the Strömgren radius.

Strong force The strongest of the four fundamental forces, with the shortest range of all; it holds the quarks together within protons and neutrons, and holds the protons and neutrons together to form atoms.

Subdwarf A star that lies appreciably below the main sequence on the Hertzsprung-Russell diagram, and is bluer than a main-sequence star of the same luminosity. Subdwarfs belong to Population II, being stars from the galactic halo which are passing through the solar neighbourhood. They differ from normal dwarfs in having a low abundance of heavy elements.

Subgiant A star that has exhausted the hydrogen at its centre and is evolving into a giant. They are of luminosity class IV. The subgiants we see are usually less massive than the Sun, because more massive stars move very quickly through this stage into giants. Low-mass stars take many billions of years to evolve this far, so low-mass subgiants are very old. A subgiant branch on the Hertzsprung-Russell diagram, linking the main sequence to the giant branch, is therefore found only for old clusters such as globular clusters.

Subluminous star A star that is less luminous than a main-sequence star of the same mass. Subluminous stars include subdwarfs, white dwarfs, and brown dwarfs.

Submillimetre astronomy The study of the sky at radio wavelengths shorter than 1 mm, specifically 0.3-1.0 mm. As with millimetre-wave astronomy, this part of the spectrum is rich in lines emitted by interstellar molecules. This region was previously considered part of the far infrared, but is now regarded as an extension of the radio spectrum since instrumental techniques developed to investigate it have more in common with radio astronomy than infrared astronomy.

Major submillimetre telescopes include the 15-m James Clerk Maxwell Telescope and the 15-m Swedish-ESO Submillimetre Telescope on La Silla.

Sub-pulse One of several distinct bursts of radiation making up a single pulse from a pulsar. Sub-pulses are themselves composed of many brief flashes (micropulses), which are believed to be the basic components of the pulsed emission. Occasionally a sub-pulse is seen to change its position gradually within the pulse, giving the appearance of drifting, but the cause is poorly understood.

Summer solstice The solstices are days when the Sun reaches its farthest northern and southern declinations. The winter solstice occurs on December 21 and marks the beginning of winter (this is the shortest day of the year). The summer solstice occurs on June 21 and marks the beginning of summer (this is the longest day of the year).

Summer triangle The Summer Triangle is a group of three bright stars (Deneb, Vega, and Altair) which are visible during the summer in the Northern Hemisphere. The Summer Triangle is not a constellation, but an asterism.

Sun The central body of the Solar System, and by far the nearest star – the only one that can be studied in great detail. It is classified as a G2V star: a yellowish star with an effective temperature of 5770 K (spectral type G2) and a main-sequence dwarf (luminosity class V). Its apparent visual magnitude is -26.7, but its absolute magnitude is only +4.83. The Sun is largely hydrogen (71% by mass), with some helium (27%) and heavier elements (total 2%). Its age is estimated to be about 4.6 billion years. The energy produced by nuclear reactions at its core is transferred to the surface layer, or photosphere, by radiation through the inner two-thirds of its radius and thence by convection for the outer onethird. The transfer of energy from the core to the surface takes 10 million years. At its centre the temperature is calculated to be 15.6 million K and the density 148 000 kg/m3. The vast majority of its energy escapes to space from the photosphere.

Sun dog A sun dog is one of two very bright spots in the sky that are sometimes seen in a solar halo (a luminous ring that is sometimes seen surrounding the Sun). The two sun dogs are located on opposite sides of the sun. They are also called "mock suns." The halo and the two sun dogs are produced as sunlight is reflected and refracted through tiny, flat ice crystals in the atmosphere. Sun dogs and halos are always at an angle of 22° away from the sun, due to the hexagonal structure of the ice crystals. Sun dogs are also called mock suns, false suns, or parhelia (meaning "beside the sun" in Greek).

Sun spot Relatively dark spots on the sun. They are the centers of intense, localized magnetic fields.

Sunquake Sunquakes are violent seismic events on the Sun that are caused by solar flares (jets of hot gases ejected thousands of miles from the surface of the Sun). When a sunquake occurs, energy is released in seismic waves on the relatively fluid surface of the sun. These waves radiate in concentric circles from the epicenter of the sunquake. These seismic waves seem to be compression waves (perhaps like "P" waves generated by earthquakes). Sunquakes would rate about 11.3 on the Richter scale. These huge quakes release about 40,000 times more energy than the 1906 San Francisco earthquake. Sunquakes were first observed by Alexander G. Kosovichev (Stanford University) and Valentina V. Zharkova (Glasgow University, UK).

Sunspot A dark area on the Sun's photosphere that is cooler than its surroundings, associated with very strong (0.4 tesla) magnetic fields. Spots generally appear in pairs or groups, the leading and following spots having opposite magnetic polarities. Spot sizes vary from small pores about 300 km across to groups spanning more than 100 000 km. The largest spots usually last longest, up to 6 months; some small spots may last for less than an hour. Sunspots are mostly confined to belts either side of the equator between about 40° and 5° latitude north and south, appearing at higher latitudes at the start of the sunspot cycle and moving towards the Sun's equator as the cycle develops. Well-developed spots have a darker interior, the umbra (2), about 1600 K cooler than the photosphere, and a lighter outer penumbra (2), which accounts for up to 70% of the spot's area and is about 500 K cooler than the photosphere.

Sunspot cycle The variation in the number of sunspots and other forms of solar activity with an average period of about 11 years. In each successive cycle the north and south magnetic polarities of the Sun are reversed, so that there is a 22-year magnetic cycle. The monthly averages of relative sunspot numbers are around 6 in years of sunspot minimum and 116 at sunspot maximum. Not only sunspot number varies, but also the numbers of active regions and flares, and the level of ultraviolet, radio, and other emission associated with such regions. The 11-year periodicity is thought to arise through the action of the solar dynamo. The sunspot cycle may not always have operated in its present observed form, as suggested by the existence of the Maunder minimum.

Supergiant elliptical A very luminous elliptical galaxy in a cluster; also known as a cD galaxy. Supergiant ellipticals may be over 10 times brighter than the next-brightest cluster member, and are often radio sources.

Supergiant The largest and brightest type of star, with a luminosity of about 10000-100000 times the Sun's, and a diameter from 20 to several hundred times the Sun's. Stars of at least 10 solar masses become supergiants when they swell up and leave the main sequence towards the end of their lives. They occupy a band at the top of the Hertzsprung-Russell diagram and are of luminosity class I. Rigel and Betelgeuse are well-known supergiants.

Supergranulation A convection pattern in the Sun's photosphere, consisting of large (30 000 km) convection cells within which there is a mainly horizontal flow outwards to the cell's edge, with a slight upward flow at the cell's centre and downward flow at the cell's edges. The velocity of the horizontal flow is only 0.4 km/s. A typical lifetime of a supergranule cell is a day. Small areas of strong magnetic field are transported outwards from cell centres to the edges, where the field is mostly concentrated. This appears to give rise to the chromospheric network, consisting of clumps of spicules.

Superior planets Mars, Jupiter, Saturn, Uranus, Neptune, and Pluto are called superior planets because they are farther from the Sun than Earth. [Planets that are closer to the Sun than Earth are called inferior planets.]

Supermassive black hole A black hole formed from the collapse of an object, such as a massive gas cloud, significantly more massive than typical stars. By 'supermassive' astronomers usually mean greater than 10^5 solar masses. Such objects can increase their mass by accreting material from their surroundings, and the energy released in this process may be responsible for the activity seen in galaxies and quasars.

Supermassive star A hypothetical star of over about 120 solar masses, so luminous that it would be expected to disintegrate under the outward pressure of its own radiation. Supermassive stars were proposed as an explanation for very bright objects in the Large Magellanic Cloud, but these are now known to be clusters of ordinary O stars.

Supernova An extremely violent explosion of a star many times more massive than our Sun. During this explosion, the star may become as bright as all the other stars in a galaxy combined, and in which a great deal of matter is thrown off into space at high velocity and high energy. The remnant of these massive stars collapse into either a neutron star or a black hole.

Supernovae are classified into Types I and II, the latter type showing hydrogen in its spectrum, whereas the spectrum of a Type I supernova shows no hydrogen. There is further subclassification into Types Ia, Ib, and Ic according to other details of the spectrum. Type Ia supernovae reach a maximum magnitude of about -19, while Types Ib and Ic are about 1.5 magnitudes fainter. Type II have a wide range of peak magnitudes, but on average are similar to Types Ib and Ic. Types II, Ib, and Ic occur in young stars of Population I, and thus are concentrated in the disks of spiral galaxies. Type Ia supernovae occur among old stars of Population II, as are found in elliptical galaxies and the halos of spirals.

Supersynchronous orbit A supersynchronous orbit is a type of transfer orbit (a temporary orbit that a satellite is put into before being injected into its final, desired orbit). A satellite is in a supersynchronous orbit when its orbital apogee (farthest point of the orbit) is much greater than the eventual geosynchronous altitude. Later, the satellite will be maneuvered into its final geosynchronous or geostationary orbit.

Surface temperature The temperature of the surface of a celestial body, as opposed to its underlying layers (for example, the central temperatures of stars are vastly greater than their surface temperatures). The surface temperature a star would have if it radiated exactly like a black body of the same size and energy output is called its effective temperature. The colour temperature of a star is found by comparing its energy distribution at different wavelengths with that of a black body; this usually differs from the effective temperature because stars are not perfect black bodies. In practice, a star's surface temperature is an average of the temperatures of a series of layers at different heights in its photosphere. A planet's surface temperature depends on the relative amounts of solar radiation absorbed by the planet and the radiation emitted by its surface.

Surveyor A series of NASA probes which soft-landed on the Moon, acting as pathfinders for the Apollo manned landings. All the probes carried TV cameras; in addition, Surveyors 3 and 7 had mechanical scoops to test the strength of the soil, and Surveyors 5, 6, and 7 carried equipment to analyse the surface composition.

Symbiotic star A star (in many cases a cataclysmic variable) that exhibits spectral lines at highly contrasting temperatures, such as those typical of a latetype red giant or supergiant (3000 K), and a dwarf B star (20 000 K). Such characteristics indicate that the star is an interacting binary. The group is very heterogeneous, but includes the relatively well-defined Z Andromedae stars, the symbiotic novae or RR Telescopii stars, and objects such as R Aquarii that appear to have an accretion disk and jets emerging approximately perpendicular to the plane of the disk. Mass transfer occurs either through mass streaming or as a stellar wind from the giant secondary.

Synchronous orbit An orbit in which a satellite moves around a planet in the same time as the planet takes to rotate on its axis. In the Earth's case, a body in synchronous orbit has a period of one sidereal day and an average distance of 42 162 km from the Earth's centre. If the orbit is circular and in the Earth's equatorial plane, the satellite remains above a particular point on the equator. Such an orbit is called a geostationary orbit and is used for communication satellites. If the orbit is synchronous but in a plane inclined to the equator, the satellite appears to trace out a figure-of-eight each day, as seen from Earth. The centre of the figure-of-eight lies on the equator, and the loops of the figure-of-eight lie between latitudes equal to the orbit's inclination to the equator.

Syndyname The curve connecting regions in a comet's dust tail which contain particles of identical size; also known as the syndyne. Particles with diameters of around 1 U+0085m are subject to the influence of radiation pressure, which leads to sorting on the basis of size; smaller particles are carried further from the nucleus by radiation pressure in a given time and therefore have a syndyname farther from the Sun than do large particles.

Synodic month The mean period between successive occurrences of identical lunar phases, e.g. from new Moon to new Moon; also known as a lunation. The synodic month is over two days longer than the Moon's orbital period (the sidereal month), because of the Earth's orbital motion around the Sun. The number of synodic months in the year is exactly one fewer than the annual number of revolutions of the Moon around the Earth. The mean length of the synodic month is 29.530 59 days.

Syrtis major The most prominent dark marking (albedo feature) on Mars, roughly triangular, about 1000 km wide and 1200 km long, centred at +10° lat., 290° long. It is the Greek name for the Gulf of Sirte on the north African coast, whose shape it resembles, and means 'the great sand-bank'. Fittingly, it appears to consist of a large area of dark wind-blown dust. It was first sighted by C. Huygens in 1659. The official name of this surface region is now Syrtis Major Planum.

T

T association A loose grouping of very young, low-mass stars of the T Tauri type, still associated with the nebula from which they formed, and often partly obscured by it. They have relative velocities of several kilometres per second and are not bound together by gravity, so the association is likely to break up after only about 10 million years.

T tauri star A very young star, less than 10 million years old, with a mass similar to or somewhat less than that of the Sun. T Tauri stars have diameters several times the Sun's, and are still contracting. On the Hertzsprung-Russell diagram they lie on Hayashi tracks above the main sequence. T Tauri stars are classified according to their spectra as either classical, weak-line, or naked. Classical T Tauri stars have strong emission lines, and are much brighter in the infrared than other stars of similar temperature. This is thought to be due to surrounding disks of warm dust, similar to the early solar nebula, which may be sites for planet formation. Weak-line objects show less evidence of surrounding material, while naked T Tauri stars show none at all. T Tauri stars exhibit irregular variability ranging from ultraviolet flares on a time-scale of minutes to optical variations on time-scales of days, months, or years.

Tail, cometary The part of a comet containing dust and gas released from the comet's head. Many comets fail to develop a tail but, when present, tails are always directed away from the Sun, so that comets move tail-first after perihelion. Tails do not normally develop until a comet is within about 2 AU of the Sun, and are usually most impressive shortly after perihelion. Comet tails have two components: the Type I or gas tail (also known as the ion tail or plasma tail), and the Type II or dust tail. Gas tails consist of ionized gas carried away from the coma by the solar wind, and is more or less straight. They can reach 108 km or more in length. Gas tails appear bluish or greenish and are dominated by emission from singly ionized carbon monoxide (CO^+) at 420 nm wavelength, resulting from excitation by solar ultraviolet radiation. They are subject to disconnection events. The dust tail, by contrast, appears yellowish because it shines by reflected sunlight, and often appears markedly curved.

Tarantula nebula The largest and brightest nebula in the Large Magellanic Cloud, also known as 30 Doradus or NGC 2070. It has a diameter of over 800 L.Y, with faint extensions to 6000. L.Y, and contains half a million solar masses of ionized gas. The ionization is produced by several clusters of O and B stars, including the very powerful and compact cluster R136 near its centre. The nebula's name comes from its spider-like shape.

Taurus A constellation of the zodiac, representing a bull. The Sun passes through Taurus from mid-May to the third week of June. Its brightest star is Aldebaran (Alpha Tauri). Beta Tauri is Elnath. The constellation contains two large, bright open clusters, the Hyades and the Pleiades. Theta Tauri is a wide double in the Hyades, magnitudes 3.4 and 3.8. Within Hind's Variable Nebula lies T Tauri, the prototype of a class of young variable stars. Ml is the famous Crab Nebula, a supernova remnant. The Taurid meteors radiate from the constellation from late October to the end of November every year.

Tebbutt, comet A long-period comet discovered on 1861 May 13 by the Australian amateur astronomer John Tebbutt also called the Great Comet, formerly designated 1861 II. It reached perihelion (0.82 AU) on June 12 and was 0.13 AU from Earth on June 30, when the Earth passed through its tail. Comet Tebbutt reached a peak magnitude of 0, and had a tail of 100° or more. The comet's orbit has a period of 409 years, eccentricity 0.985, and inclination 85°.4.

Technetium star An M star or carbon star containing isotopes of technetium. Since the longest-lived isotope of technetium that can be created by stellar nucleosynthesis has a half-life of only 210 000 years, this material must have been created recently inside the star and then brought to its surface.

Telescope An optical instrument that collects light from faint and distant objects and magnifies their images. Telescopes have an objective lens or mirror which gathers light and focuses it at a focal plane. The resulting image may be observed with an eyepiece or recorded by a photographic emulsion, a charge-coupled device, or some other detector. In a refracting telescope the objective is a lens, and in a reflecting telescope it is a mirror. A telescope that uses both lenses and mirrors is called a catadioptric system. The larger the aperture of a telescope the more light it collects, making it possible to observe fainter stars and to see finer details. The highest magnification that can be used on a telescope of given aperture is restricted by the effects of diffraction. In practice, the finest detail visible with a telescope is set by the Dawes limit. Telescope apertures range from about 25 mm for the smallest refractors up to many metres for the largest reflectors.

Tenth planet No tenth planet beyond Pluto has been directly observed. A few astronomers think that there might be a tenth planet (or companion star) orbiting the Sun far beyond the orbit of Pluto. This distant planet/ companion star may or may not exist. The hypothesized origin of this hypothetical object is that a celestial object, perhaps a hard-to-detect cool, brown dwarf star (called Nemesis), was captured by the Sun's gravitational field. This tenth planet is hypothesized to exist because of the unexplained clumping of some long-period comet's orbits. The orbits of these far-reaching comets seem to be affected by the gravitational pull of a distant, Sun-orbiting object.

Terrestrial planet A small, high-density planet with a solid, rocky surface. In the Solar System, the terrestrial planets are Mercury, Venus, Earth, and Mars.

Tethys The fifth-largest satellite of Saturn, diameter 1060 km, sharing the same orbit as Calypso and Telesto; also known as Saturn III. Its orbital distance is 294660 km, and its orbital period 1.888 days, the same as its period of axial rotation. Tethys was discovered in 1684 by G. D. Cassini. It is densely covered with impact craters; one of these, Odysseus, is 440 km across, nearly half the satellite's diameter. Another prominent feature is Ithaca Chasma, a valley over 1000 km long and 80 km wide. Tethys has a density close to that of water, suggesting that it is composed largely of ice.

Thalassa Thalassa is a tiny moon of Neptune. It was discovered using NASA's Voyager 2 mission in 1989. It orbits 50,070 km from the center of Neptune and is about 80 km in diameter.

Thales of Miletus (625BC-547 BC) Greek philosopher, born in modern Turkey. He was one of the first Greeks to adopt a scientific approach, seeking natural rather than divine causes for natural events. However, his reputed astronomical achievements are known only via the writings of others. He was said to have used the Babylonian saros to predict a solar eclipse that took place on May 28, 585 BC, but there is little evidence for the prediction or even for the Babylonian saros. According to Aristotle, Thales believed that everything was created from a primordial mass of water.

Tharsis montes A large upland region on Mars about 2100 km across, centred at +3° lat., 113° long.; also known as the Tharsis ridge. It includes the three giant volcanoes Ascraeus Mons, Pavonis Mons, and Arsia Mons, each of which rises 27 km high. Tharsis is over 9 km high at the bases of these volcanoes and reaches over 11 km high at Noctis Labyrinthus, a complex network of fractures at the western end of the Valles Marineris.

Thebe Thebe is the fourth-closest moon of Jupiter's 16 moons. Thebe is 68 x 56 miles (110 x 90 km) in diameter and orbits 138,000 miles (222,000 km) from Jupiter. Amalthea and Thebe likely provide the dust for the Gossamer ring. Thebe has a mass of 8 x 10^{17}kg. It orbits Jupiter in 0.6745 day (Earth) days and is in synchronous rotation (always keeping the same side facing Jupiter). Thebe was discovered by Stephen Synnott (Voyager 1) in 1980.

Themis family A Hirayama family of asteroids in the outer part of the main asteroid belt at a mean distance of 3.13 AU from the Sun. It is one of the most populous and best-defined families, consisting of a core of large objects surrounded by a cloud of predominantly smaller ones. The core includes (24) Themis, (62) Erato, (90) Antiope, (468) Lina, (526) Jena, and (846) Lipperta. Most family members have low albedos and are of the carbonaceous C class. Themis itself is a C-class asteroid, diameter 228 km, discovered in 1853 by the Italian astronomer Annibale de Gasparis (1819-92). The orbit of Themis has a semimajor axis of 3.130 AU, period 5.54 years, perihelion 2.71 AU, aphelion 3.55 AU, and inclination 0°.8.

Theory of relativity Albert Einstein formulated the Theories of Special and General Relativity. Special relativity supplanted Newtonian mechanics, yielding different results for very fast-moving objects. The Theory of Special Relativity is based on the idea that speed has an upper bound; nothing can pass the speed of light. The theory also states that time and distance measurements are not absolute but are instead relative to the observer's frame of reference. Space and time are viewed as aspects of a single phenomenon, called space-time. Energy and momentum are similarly linked. As a result, mass can be converted into huge amounts of energy, and vice versa, according to the formula E=mc2. General Relativity expands the theory of special relativity to include acceleration and gravity, both of which are explained via the curvature of space-time.

Third contact In a total solar eclipse, the moment when the trailing limb of the Moon coincides with the Sun's western limb, at which time totality ends; or, in a total lunar eclipse, the point when the Moon's leading limb reaches the eastern edge of the Earth's umbra, ending totality. Immediately after third contact in a solar eclipse, the diamond ring or Baily's beads may become visible. At an annular eclipse, third contact refers to the instant when the Moon's leading limb leaves the Sun's eastern limb, and the annular phase ends.

Tholus A small domed hill on a planetary surface; pl. tholi. The name, which means 'dome', is not a geological term, but is used in the nomenclature of individual hills, for example Ceraunius Tholus on Mars, and Inachus Tholus on Io.

Thuban Thuban (alpha Draconis) is the brightest star in the constellation Draco; it is located in the dragon's tail. Thuban means "dragon or serpent" in Arabic. Thuban was the Earth's pole star about 5,000 years ago. Thuban is white giant star, spectral type A0III. It is about 297 light-years from Earth. Thuban has an apparent magnitude of 3.7 (not very bright) and an absolute magnitude of about 1.1. Draco is near the northern celestial pole; the tail of Draco is between the Big Dipper and the Little Dipper.

Tidal coupling When one celestial body orbits another, tidal forces tend to equalize the rotational periods, so that, after a very long time, they may become identical. This end result is called tidal coupling. Pluto is tidally coupled to its moon Charon; if you were standing on Pluto, its moon Charon would either always be visible or never be visible. As a consequence of the tidal interactions with the Moon, the Earth's rotational period is slowly decreasing. Billions of years from now, the Earth and the Moon will have the same period of rotation, and these will also exactly equal the orbital period of the Moon around the Earth.

Tidal friction A force between the oceans of Earth and the ocean floors caused by the gravitational attraction of the Moon. The Earth tries to carry the ocean waters round with it, while the Moon tries to keep them heaped up under it and on the far side of the Earth Over long periods of time tidal friction decreases the Earth's rate of spin, so lengthening the day. In turn, the Moon has angular momentum added to it in its orbit and gradually spirals away from Earth. Ultimately, when the day equals the Moon's orbital period (each being about 40 times the length of the present day) the process will cease. A new process will then begin in which the Sun's tide-raising power takes angular momentum from the Earth-Moon system. The Moon will then spiral in closer to the Earth until it is torn to pieces when it enters the Earth's Roche limit.

Tides The rise and fall of the Earth's oceans, caused by the gravitational attractions of the Moon and Sun. At the Earth's centre, centrifugal force due to the motion of the Earth about the barycentre of the Earth-Moon system is balanced by gravitational force. On the side of the Earth nearest the Moon, the Moon's gravitational pull is greater than centrifugal force, while on the opposite side the centrifugal force is greater. In both cases a tidal bulge is produced in the oceans. The Sun's tide-raising power is about one-third that of the Moon, as a result of its greater distance from the Earth. The Sun and Moon act together at new or full Moon to produce extra-high and extra-low tides; these are the so-called spring tides. When the Moon is at first or third quarter the Sun and Moon are acting in different directions. The high tides are then at their lowest and the low tides are at their highest; these are the so-called neap tides.

Time dilation The slowing down of time that occurs at speeds approaching that of light, as predicted by the special theory of relativity. A clock moving relative to an observer will appear to slow down by a factor of $v[1 - (v^2/c^2)]$, where v is the relative velocity and c the speed of light. At everyday speeds, as on Earth, the effect is not noticeable, but the slowing increases rapidly as v approaches c. Not just clocks but all processes slow down, so that an astronaut would appear to have aged far less during a high-speed flight than a person who remained on Earth.

Time The dimension that allows events occurring at the same place to be distinguished. In the classical physics of Galileo and Newton, time had an absolute significance and a time-scale could in principle be adopted so that all observers would agree on the time at which any event occurred. Different observers would see the event occurring at different times, but these differences were explained by the travel time of light from the event to the observer. Moreover, in classical physics this common time-scale was in step with each observer's own local measure of time, the proper time. All observers would agree that the time of any event is simply the time recorded by the local clock.

Newton's theory of gravity gives a very accurate description of orbital motion within the Solar System, allowing the calculation of each body's position at any given instant of time.

Time zone One of the 24 longitudinal bands into which the Earth is divided; or, any region of the Earth that keeps the same civil time. The 24 standard time zones are each 15° wide in longitude, and each is assigned a zone time that differs by 1 hour from the adjacent time zone, being 1 hour later to the east and 1 hour earlier to the west, until the International Date Line is reached when a whole day changes. The Earth's prime time zone is centred on the meridian of Greenwich and its standard time is Universal Time. Large countries such as the USA, Russia, and Australia are divided into several time zones according to local geographical or political considerations rather than strict parallels of longitude.

Tio bands Absorption bands due to the molecule titanium oxide that are prominent in the spectra of K and M stars. Only in such cool stars can molecules such as TiO survive.

Titan The largest satellite of Saturn, and the second-largest satellite in the Solar System, diameter 5150 km; also known as Saturn VI. Titan orbits Saturn in 15.945 days at a distance of 1221830 km. Its axial rotation period is the same as its orbital period. At opposition it is magnitude 8.3. It was discovered in 1655 by C. Huygens. Titan is probably composed of rock and ice in about equal parts. It is the only satellite in the Solar System to have a substantial atmosphere. The atmosphere is mainly nitrogen, with 2-10% methane, 0.2% hydrogen and traces of ethane, propane, ethyne, hydrogen cyanide, and carbon monoxide. The atmospheric pressure at Titan's surface is 1.5 bar and the temperature about -180 °C. There may be methane drizzle on the surface, and perhaps methane snow or methane rain. Orange smoggy clouds of hydrocarbons at a height of 200 km veil its surface, while layers of atmospheric haze exist up to 500 km.

Titania The largest satellite of Uranus, 1580 km in diameter; also known as Uranus III. It orbits in 8.706 days at a distance of 435 910 km. Its axial rotation period is the same as its orbital period. It was discovered in 1787 by F. W. Herschel, soon after the discovery of Uranus itself. Titania has an icy surface covered with impact craters, and a number of steep-sided troughs bounded by recently active faulting. The satellite may have expanded as a result of internal freezing.

Torrino scale A Torrino Scale is a risk-assessment scale for showing the potential dangers associated with asteroids and comets that may hit the Earth. The Torrino Scale notes various objects' kinetic energy (incorporating both mass and velocity) vs. the collision probability. The scale was created by Dr. Richard P. Binzel of MIT.

Torus An object shaped like the inner tube of a car tyre or a ring doughnut: a cylinder bent into a circle and the ends joined together leaving a hole in the middle; pl. tori. The cross-section through the cylinder is often circular, but may take other shapes. The Van Allen Belts around the Earth are toroidal in shape. Other examples include accretion disks and the ionized material ejected from Jupiter's volcanic satellite Io, which spreads along Io's orbit to form the Io plasma torus.

Total eclipse A total eclipse happens when the moon entirely blocks the Sun or the Earth's shadow totally blocks our view of the moon. A total eclipse can only happen during a new Moon. During a total eclipse, the Moon's shadow moves across the Earth at speeds over 1,000 mph (1,600 kph). Because of the relative size and distance of the Sun, Earth and the moon, the Earth is the only planet in our Solar System that experiences total solar eclipses.

Toutatis Asteroid 4179, a member of the Apollo group, discovered in 1989 by the French astronomer Christian Pollas. At present, Toutatis can approach to within 0.006 AU (900 000 km) of the Earth's orbit. Of the largest nearEarth asteroids, Toutatis is perhaps the most potentially dangerous to the Earth in the foreseeable future, because of its frequent approaches plus a 3 : 1 resonance with Jupiter which makes the orbit chaotic. Radar observations show that it consists of two irregular, cratered bodies in close contact, with maximum widths of about 4 km and 2.5 km, respectively. Its orbit has a semimajor axis of 2.516 AU, period 4 years, perihelion 0.92 AU, aphelion 4.11 AU, and inclination 0°.5.

Train Anything remaining along the trajectory of a meteor or fireball after the head of the object has passed. The train may consist of light, dust, vapour, or ionized gas. The faintly glowing train of ionization left along the path of a bright meteor is generally termed a wake if it glows for a second or less, or a persistent train if the duration is longer. A train may be bent and broken up by winds in the upper atmosphere. Meteorite-dropping fireballs often leave a train of fine dust.

Transfer orbit The path followed by a spacecraft when moving from one orbit (the departure orbit) to another (the destination orbit). For example, the transfer orbit of a spacecraft from Earth to Mars intersects the Earth's orbit and the orbit of Mars, and involves two firings of the spacecraft's engines, one to put it into the transfer orbit and the other to put it into the destination orbit when it arrives. Such a procedure is also used for transferring communications satellites from low Earth orbit to geostationary orbit. The special case of a transfer orbit that involves minimum expenditure of rocket fuel is known as a Hohmann orbit.

Transparency, atmospheric The extent to which the Earth's atmosphere transmits light from celestial objects. The atmosphere's clarity varies widely, depending on the amount of absorbing material such as water vapour, dust, aerosols (suspended fine particles), and polluting gases. In general, transparency improves with altitude, particularly above inversion layers, which occur where cold surface conditions produce a layer of cold air close to the ground which traps much of the absorbing material. In temperate latitudes, transparent air often follows a cold front.

Triangulum australe The sixthsmallest constellation, lying in the southern sky, and popularly known as the Southern Triangle. Its brightest star, Alpha Trianguli Australis, is Atria. The triangle is completed by Beta and Gamma Trianguli Australis, both of magnitude 2.9.

Triangulum galaxy The galaxy M33 in Triangulum, a member of the Local Group. It is a type Sc spiral seen nearly face-on, up to 1° wide and of 6th magnitude. It lies 2.9 million L.Y away and has a true diameter of 52 000 L.Y

Trifid Nebula The nebula M20 in Sagittarius, also known as NGC 6514. It is a combined emission and reflection nebula, extending for nearly ½°. The Trifid takes its name from three dark lanes of dust that trisect it. It is centred on the multiple star HN 40, the brightest members of which evidently illuminate it. The Trifid lies 5200 L.Y away.

Trigonometric parallax The angle subtended at a star by a standard distance of one astronomical unit, the radius of the Earth's orbit. It is usúally expressed in seconds of arc. The reciprocal of trigonometric parallax is the star's distance in parsecs.

Triton The largest satellite of Neptune, 2706 km in diameter; also known as Neptune I. It lies 354760 km from Neptune, and orbits in 5.877 days in a retrograde direction. Its period of axial rotation is the same as its orbital period. Triton was discovered by W. Lassell in 1846, soon after the discovery of Neptune itself. Triton has a tenuous nitrogen atmosphere with some methane. The surface pressure, however, is only about 16 μbar, little more than 10^{-5} of the Earth's atmospheric pressure, and its surface temperature is -235°C, the coldest surface so far measured in the Solar System. Triton has a young icy surface with few impact craters, and different types of terrain including smooth plains, hummocky plains, cantaloupe terrain, long linear features, and also a large, probably seasonal, polar cap of nitrogen ice. The Voyager 2 space probe photographed dark plumes rising from dark spots on the surface to an altitude of 8 km before drifting downwind.

Trojan asteroid A member of one of two groups of asteroids, at a mean distance of 5.2 AU from the Sun, which share the orbit of Jupiter. They lie clustered around the leading (L 4) and following (L 5) Lagrangian points, 60° ahead of and behind Jupiter. Perturbations by other planets cause the Trojans to oscillate along the orbit of Jupiter about 45-80° from the planet, taking 150-200 years per cycle. The first Trojan to be discovered was Achilles in 1906. Well over 200 Trojan asteroids are now known, most in the leading (L 4) group. Over 60% of the Trojans are of D class (dark and reddish), with P class making up the majority of the remainder, plus a few C-class objects. Their mass distribution is similar to that of main-belt asteroids, indicating that they have been at their present location since the formation of the Solar System. In 1990 the first Martian Trojan, (5261) Eureka, was discovered in the L5 region of Mars's orbit.

Tropic Either of the two latitudes on Earth at which the Sun appears overhead at the summer and winter solstices each year. The tropics lie at latitudes of about 23½°, north and south, an angle defined by the Earth's axial tilt.

Tropical month The mean orbital period of the Moon about the Earth with respect to the vernal equinox. Its length is 27.32158 days. This definition differs from that of the sidereal month only through the inclusion of the precession of the equinoxes. The difference is only 7 s of time, but this very small difference will build up to one complete month over the 25 800-year cycle of precession.

Tropopause The uppermost limit of the troposphere, at its boundary with the stratosphere. The precise altitude of the tropopause depends on latitude, being around 10 km in temperate regions, but closer to 15 km at the equator. Seasonal variations also occur, the height of the tropopause being greater in summer. Temperatures in the troposphere reach their minimum at the tropopause, typically around -40°C.

Troposphere The lowest distinct layer within the atmosphere, extending from sea-level to

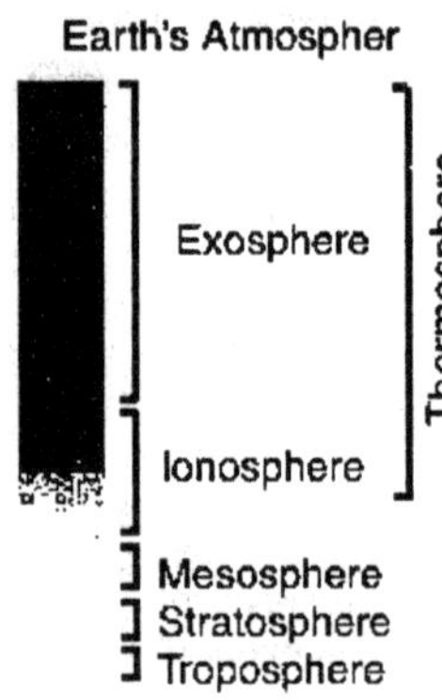

Fig. Troposphere

the tropopause at an altitude of 10-15 km. Weather systems are confined principally to the troposphere, which accounts for about 75% of the Earth's total atmospheric mass. Temperatures fall with increasing height in the troposphere, reaching a minimum at the tropopause.

True equinox The intersection of the ecliptic with the true equator. Its path on the sky consists of oscillations due to nutation superimposed on the progressive motion due to precession.

True field of view The angle of sky seen through an eyepiece when it is attached to a telescope; the true field equals the apparent field divided by the magnification.

True pole The direction towards which the Earth's axis points at a given time. Its path on the sky consists of oscillations due to nutation superimposed on the precessional motion about the pole of the ecliptic.

Trumpler classification A classification of open clusters according to three criteria: degree of central concentration (from I, the most concentrated, to IV, little different from the surrounding stars); range of brightness of individual stars (from 1, small range, to 3, large range); and the total number of stars in the cluster (p meaning poor, less than 50 stars; m, moderately rich, 50-100 stars; r, rich, more than 100 stars). The suffix 'n' indicates that nebulosity is involved with the cluster, as in the Pleiades which is classified as I 3 r n. The system was introduced by the American astronomer Robert Julius Trumpler.

Trumpler star The name given to certain highly luminous stars which were once erroneously believed to be hundreds of times the mass of the Sun.

Truth This is a word best avoided entirely in physics except when placed in quotes, or with careful qualification. Its colloquial use has so many shades of meaning from 'it seems to be correct' to the absolute truths claimed by religion, that it's use causes nothing but misunderstanding. Someone once said 'Science seeks proximate (approximate) truths.' Others speak of *provisional* or *tentative* truths. Certainly science claims no final or absolute truths.

Tsiolkovsky Konstantin Tsio lkovsky (1857-1935) was a Russian schoolteacher who was the first person on record to propose the idea of explorating space using rockets. Tsiolkovsky proposed the idea in 1898 and published it in 1903. He suggested that the rockets should be powered by liquid propellants. Many people call Tsiolkovsky the father of modern astronautics.

Tsunami A tsunami is a huge wave, caused by undersea earthquakes, volcanic eruptions, or, more rarely, by asteroid or meteoroid impact (as in the case of the K-T extinction).

Tucana A constellation of the southern sky, representing a toucan. Its brightest star is Alpha Tucanae, magnitude 2.9. Beta Tucanae is an optical triple, magnitudes 4.4, 4.5, and 5.1. 47 Tucanae is a major globular cluster. The constellation's other notable feature is the Small Magellanic Cloud.

Tully-Fisher relation The Tully-Fisher Relation is a correlation between the luminosity of spiral and irregular galaxies and their rate of rotation (rotational velocity, which is measured from the broadening of its spectral lines since the light from it is both red- and blue-shifted). This means that larger spiral galaxies (more luminous ones) rotate faster than smaller ones. If you know the rate of rotation of a spiral galaxy, you can calculate its intrinsic brightness (intrinsic brightness is how bright it really is - not how bright it is from Earth, which is its apparent magnitude). Then, by comparing the galaxy's intrinsic brightness with its apparent magnitude, you can determine how far away the galaxy is. The physical reason for this empirical relationship is unknown. The relationship was first realized by R. B. Tully and J. R. Fisher in 1977.

Tunguska event An explosion above the Podkamennaya Tunguska (Stony Tunguska)

River region of central Siberia at about 7.30 a.m. local time on 1908 June 30, caused by a large stony meteorite or cometary fragment. A fireball as bright as the Sun detonated in mid-air, producing dust which caused abnormally bright nights across Europe for days afterwards. Expeditions to the site found a vast (2200 km^2) area of devastation, with trees knocked over and scorched up to 40 km away, but no craters were found. The incoming body is thought to have been 50-100 m in diameter, and to have exploded 8-9 km above the ground with the energy of 15-20 megatonnes of TNT.

Twilight The period before sunrise and after sunset during which the sky is not completely dark. Three different periods of twilight are defined according to how far the Sun is below the horizon. During civil twilight, the centre of the Sun is less than 6° below the horizon; during nautical twilight, the centre of the Sun is between 6° and 12° below the horizon; and during astronomical twilight the centre of the Sun is between 12° and 18° below the horizon.

Tycho's star A supernova in Cassiopeia that appeared in 1572 November, and was extensively studied by Tycho Brahe. The maximum magnitude was -4, and the star remained visible until 1574 April-May. The light-curve shows that it was a Type I supernova. The radio source 3C 10, faint optical nebulosity, and a weak X-ray source comprise the supernova remnant, about 8000 L.Y away.

U

U burst A fast radio burst spectrum of a FLARE. It has a U-shaped appearance in an intensity-vs.-frequency plot.

U geminorum star A type of cataclysmic binary showing abrupt outbursts with typical amplitudes of 2-6 mag. that occur at unpredictable intervals; abbr. UG. They are also known as dwarf novae. A U Geminorum star is an interacting binary, with mass transfer from the main-sequence or slightly evolved secondary to an accretion disk around the white-dwarf primary. Eclipses are observed in some systems, together with a hump caused by a hot spot (1), located either where the gas stream hits the disk or at the inner edge of the disk. Orbital periods are 3-15 hours, and both components have masses similar to that of the Sun. Specific subtypes are SS Cygni stars, SU Ursae Majoris stars, and Z Camelopardalis stars.

U-class asteroid A class of asteroids included in older classification schemes, where the 'U' stood for unclassifiable, for asteroids which could not be allocated to one of the five compositional types C, S, M, E, and R. However, more recent schemes group the asteroids into 9, 11, or 14 different classes, depending on the technique used for analysis. In the 14-class scheme the original U class has been subdivided into R, A, M, P, Q, E, V, and T classes, in addition to the C, G, B, F, D, and S classes used. In the new scheme, the letter U is appended to the classification of objects that possess an unusual spectrum for their class (e.g. CU, MU).

Ultraviolet astronomy The study of the Universe in the ultraviolet region of the electromagnetic spectrum, approximately 91.2-350 nm. These wavelengths are largely blocked by the Earth's atmosphere, so observations became possible only with the use of rockets after World War II. Balloons were also used, but the altitude they could achieve allowed observations only in the near ultraviolet, longer than 200 nm.

The Orbiting Astronomical Observatory series of ultraviolet missions commenced in 1968, the first success being OAO-2. This was followed in 1972 by OAO-3, also known as the Copernicus satellite, which revealed some of the detailed structure of interstellar matter, in particular its patchiness. TD-1A, a European satellite, made an ultraviolet survey from 1972 to 1974 at 135-290 nm. ANS, the Astronomical Netherlands Satellite, made photometric observations of a large number of stars at 155-330 nm, also in the 1970s.

Ultraviolet excess star A star that shows an excess of ultraviolet radiation compared to normal stars. An ultraviolet excess can be used to identify hot O and B stars, white dwarfs, and objects surrounded by an accretion disk, such as neutron stars and black holes.

Ulugh beg Title assumed by Muhammad Taragi,Mongol ruler and astronomer, born in modern Iran. In 1420 he established an observatory at Samarkand, in modern Uzbekistan. He equipped it with a huge (radius 40 m) sextant of stone that allowed observations accurate to a few seconds of arc to be made. The result was the first original star catalogue since those of Ptolemy, which it surpassed in precision.

Ulysses A joint ESA-NASA spacecraft launched in 1990 October to study the solar wind, in particular from the unexplored regions around the Sun's poles. The spacecraft passed Jupiter in 1992 February, which swung it on to a trajectory that took it over the Sun's poles. Ulysses passed through the solar wind dominated by the south polar coronal hole, measuring wind speeds in the region of 700 km/s, in 1994 June-November, and the corresponding region over the Sun's north pole in 1995 June-September.

Umbra 1. The conical zone of darkness directly opposite the Sun from a planet or satellite, within which the Sun's disk is completely obscured. Passage of a body through this shadow results in its eclipse as, for example, when the Moon traverses Earth's umbra. The dark umbra is surrounded by a broader penumbra (1) in which the Sun's disk is only partially obscured.

2. The central, darkest part of a sunspot. It is also the coolest part, with a temperature of about 4200 K Umbrae are not uniformly dark but contain light umbral dots. These are smaller versions of photospheric granules, only about 300 km across, and last somewhat longer than granules, up to 25 min. They seem to be penumbra! grains that have just entered the umbra.

Umbriel Umbriel is one of the larger of the 18 moons of Uranus, and the darkest. It has a heavily-cratered surface, indicating an old surface. It has a bright ring at its top (probably a crater floor) called a fluorescent cheerio. Umbriel was discovered by Wm. Lassell in 1851. Umbriel has a diameter of 1,170 km and orbits, on average, 265,970 km from Uranus.

Undina family A small family of asteroids in the outer part of the main asteroid belt, at a mean distance of about 3.2 AU from the Sun. The largest member is (94) Aurora, diameter 190 km. The family is named after the M-class (92) Undina, diameter 184 km, discovered in 1867 by the Danish-American astronomer Christian Henry Frederick Peters (1813-90). Its orbit has a semimajor axis of 3.196 AU, period 5.71 years, perihelion 2.92 AU, aphelion 3.47 AU, and inclination 9°.9.

Unit distance A distance taken as a convenient unit of length. Within the Solar System the unit distance is the astronomical unit,the average distance of the Earth from the Sun. For stellar distances the light year (L.Y) or the parsec (pc) is used. Larger units such as the kiloparsec (1000 pc) or the megaparsec (1000000 pc) are used for distances within our Galaxy or to other galaxies.

Universal gravitational constant The universal gravitational constant (abbreviated G) is the constant of proportionality in Newton's equation (formulated in 1666) that describes the gravitational attraction between objects; their gravitational attraction (F) depends only on their masses and the distance between them, according to the formula $F = Gm_1m_2 / r^2$. Henry Cavendish, in 1798, determined the numerical value of G to be 6.668×10^{-8} dynes cm^2/g^2.

Universal time A world-wide standard time-scale, the same as Greenwich Mean Time. Universal Time is the mean solar time on the meridian of Greenwich. It is defined as the Greenwich hour angle of the mean sun plus

12 hours, so that the day begins at midnight rather than noon. It is closely linked to Greenwich Mean Sidereal Time,since the mean sidereal day is a precisely known fraction of the mean solar day. In practice, UT is determined by a formula from GMST, which in turn is derived from the observations of the meridian transits of stars. The version of UT derived directly from such observations is designated UT0, which is slightly dependent on the observing site. When UT0 is corrected for the variation in longitude due to the Chandler wobble, a version of Universal Time, UT1, is derived which has genuine world-wide application. When UT1 is compared with International Atomic Time,it is found to be losing approximately a second a year against TAI.

Universe Everything that exists, including space, time, and matter. The study of the Universe is known as cosmology. Cosmologists distinguish between the Universe with a capital 'U', meaning the cosmos and all its contents, and universe with a small 'u', which is usually a mathematical model derived from some physical theory. The real Universe consists mostly of apparently empty space, with matter concentrated into galaxies consisting of stars and gas. The Universe is expanding, so the space between galaxies is gradually stretching, causing a cosmological redshift in the light from distant objects. There is growing evidence that space may be filled with unseen dark matter that may have many times the total mass of the visible galaxies. The most favoured concept of the origin of the Universe is the Big Bang theory, according to which the Universe came into being in a hot, dense fireball about 10-20 billion years ago.

Upsilon andromedae Upsilon Andromedae is a star in the constellation Andromeda. Astronomers Geoffrey W. Marcy and R. Paul Butler discovered a massive planet orbiting this star (with each orbit taking only 4.6 days.) in 1996. Recently, 2 even more massive planets have been discovered orbiting this star. The three planets orbit within 2.5 Astronomical Units of the star.

Uranometria The first star atlas to cover the entire sky, published in 1603 by J. Bayer. Each constellation was depicted on a page of its own, with star positions from the catalogues of Ptolemy and Tycho Brahe, with the exception of the 12 newly described southern constellations invisible from Europe, which were combined on a single page. This atlas introduced the system of labelling the brightest stars by Greek letters, known as Bayer letters.

Uranus The seventh planet from the Sun. It is blue-green in colour due to the absorption of red light by methane in the upper atmosphere. Its mean opposition magnitude is +5.5, making it just visible to the naked eye under favourable conditions, but it was unknown until 1781 when discovered telescopically by F. W. Herschel. Uranus is distinctly ellipsoidal in shape (equatorial diameter 51 118 km, polar diameter 49 946 km). Its density, 1.3 g/cm^3, is the lowest of the planets except Saturn. Its rotation axis is tilted at over 90° to the upright, so that its rotation is retrograde, and it presents its poles and its equator alternately towards the Sun as it orbits. The rotation period of the visible surface ranges from about 16 hours at 70° south to about 18 hours near the equator, but radio observations indicate that the core rotates in 17h 14m. Uranus has a thick atmosphere composed of 83% hydrogen, 15% helium, 2% methane (molecular percentages), and traces of ethyne.

Urca process A cycle of nuclear reactions in which an electron is absorbed by a nucleus and is subsequently re-emitted as a beta particle (a fast electron) with the generation of a neutrino-antineutrino pair. The process makes no change to the composition of the nucleus, but removes energy from it in the form of the neutrino and antineutrino. While insignificant in ordinary stars, it is important at very high densities and temperatures such as those which give rise to supernovae. The process is named after the Urca Casino in Rio de Janeiro, where it was said that money disappeared as rapidly as energy from the stars.

Ureilite A class of calcium-poor achondrite meteorite; also known as olivine-pigeonite achondrites. They consist mainly of millimetre-sized grains of olivine and pigeonite minerals enclosed in a matrix of carbonaceous veins, with some metal also present. They are named after the meteorite that fell at Novo Urei, Russia, on 1886 September 10. The ureilites are noted for their carbon content (1% in Novo Urei), much of which is in the form of diamond, possibly produced by shocks from impacts while in space. The ureilites probably originated within the solar nebula, which would make them the most primitive of the achondrites.

Ursa major The third-largest constellation, lying in the northern sky and better known as the Great Bear. Its seven main stars make up the familiar shape known as the Plough or Big Dipper. Its two brightest stars are Dubhe (Alpha Ursae Majoris) and Alioth (Epsilon Ursae Majoris), both magnitude 1.8. Mizar (Zeta Ursae Majoris) is a famous naked-eye double. Xi Ursae Majoris is a binary of magnitudes 4.3 and 4.8 and a period of 60 years; it was the first binary to have its orbit computed. Lalande 21 185 is the fourth-closest star to the Sun. M81 and M101 are spiral galaxies of 7th and 8th magnitudes. M82 is an 8th-magnitude peculiar galaxy, now thought to be a spiral seen edge-on that is encountering a dust cloud. M97 is the Owl Nebula.

Ursa minor The constellation that contains the north celestial pole, popularly known as the Little Bear. Its brightest star is Polaris (Alpha Ursae Minoris), the north pole star. The stars Kochab (Beta Ursae Minoris) and Pherkad (Gamma Ursae Minoris) are popularly known as the Guardians of the Pole. The constellation itself is sometimes referred to as the Little Dipper.

UU herculis star A yellow giant or supergiant semiregular variable (type SRD) in which two different periods may alternate at irregular intervals. Other periods are sometimes encountered.

UV ceti star A type of eruptive variable (often known as a flare star) exhibiting sudden flares with a rise time of seconds to tens of seconds, and decay times of minutes to tens of minutes; abbr. UV. Amplitudes of up to 6 magnitudes have been recorded. X-ray and radio flares also occur, sometimes coincident with the optical flares. The stars are K-M emission-line dwarfs with very strong magnetic fields, and those in associations are rapidly rotating, young stars. Tidal effects in a binary system may create the strong magnetic fields of other UV Ceti stars. The flare activity is similar to that found on the Sun, but appears more prominent because of the intrinsic faintness of these cool red dwarfs.

UX ursae majoris star A nova-like variable that appears to be a Z Camelopardalis star permanently at standstill. It may represent a form of postnova. A subgroup of these stars occasionally drop to minimum for a short time, and may therefore be regarded as Z Camelopardalis stars with exceptionally long standstills. Because their light-curves look like an inverted version of dwarf-nova light-curves, this second group is sometimes known as anti-dwarf novae.

V

V magnitude The magnitude of a star in yellow-green light, the centre of the range of wavelengths to which the eye is sensitive, on the system of Johnson photometry. The V filter has a central wavelength of 545 nm and a bandwidth of 88 nm. It is the photoelectric equivalent of the older photovisual and visual magnitudes. There are also V filters in Geneva photometry, Vilnius photometry, Walraven photometry, and the six-colour system, but the context should always make it clear when these are intended.

Valles marineris A large system of canyons on Mars, centred at about -12° lat., 71° long. It is the largest canyon in the Solar System, 4100 km long, up to 500 km wide, and over 4 km deep in places. It is really a complex of several parallel troughs, the central ones being the Ius, Melas, Coprates, and Eos Chasmata, with Tithonium, Ophir, Candor, and Gangis Chasmata to the north. Hebes and Juventae Chasmata are separate from the main complex, lying further north. They are all thought to have been caused by tensional faulting within Mars, and all have been modified and enlarged by the action of ancient water channels running down their walls. Valles Marineris, named after the Mariner 9 spacecraft that discovered it in 1971, is visible from Earth as the dark streak Coprates. Its darkness may be due to dust covering part of the canyon floor.

Van allen belts The Van Allen radiation belts are two doughnut-shaped belts of ionized gas (plasma) that circle the Earth. The belts are are caused by the interaction between the Earth's magnetic field and the solar wind (stream of ions, electrically charged particles, that are given emitted by the sun). The particles (plasma) is trapped by the Earth's magnetosphere. When stray particles from these belts collide with air, they emit light, causing an aurora.

Van Allen, James Alfred American space scientist. In 1945 he began high-altitude rocket research, initially using captured German V2 missiles. He was responsible for instruments aboard the first American satellites, the Explorer series, with which the radiation zones now called Van Allen Belts were discovered. In all Van Allen worked on 24 missions, including Pioneers 10 and 11, studying planetary magnetospheres (Saturn's in particular), solar X-ray emission, and the solar wind.

Variable star Any star that varies in brightness. Two broad categories are recognized: extrinsic variables, which vary for a mechanical reason (e.g. rotation); and intrinsic variables, which undergo a real change in luminosity of either an individual star or some element in a binary system. Certain stars may combine both forms of variation. The standard reference for the classification of types of variable stars and their nomenclature is the General Catalogue of Variable Stars.The current total of designated variables is over 30 000.

The classification of variable stars was originally based upon the form of their light-curve, its amplitude (1), and periodicity (or lack of it). Increasingly, however, the physical mechanisms that underlie the different forms of variation, or the physical structure of the stars (and binaries) are used to define six groups of variables, each with specific features.

Variation 1. A periodic disturbance in the Moon's celestial longitude caused by changes in the Sun's gravitational attraction as the Moon orbits the Earth. It has an amplitude of 40'and a period of half a synodic month.

2. The yearly change in the coordinates of a star due to the effects of precession and proper motion; also known as annual variation.

Varuna Varuna is a large Kuiper object - a body in our Solar System that orbits far from the Sun. Varuna (2000 WR106) is a rocky body about 900 km in diameter. It is roughly 43 AU from the Sun (assuming a circular orbit). One orbit around the Sun takes about 285 years. It has an apparent magnitude of 20. This Transneptunian object (TNO) was discovered from the Kitt Peak Observatory by Robert S. McMillan on November 28, 2000. Varuna was named for the oldest of the vedic (Hindu) deities, the maker and upholder of heaven and the earth.

Vastitas An extensive lowland plain on a planetary surface; pl. vastitates. The name, which means 'desolation' or 'vastness', is not a geological term, but is used in the nomenclature of individual features, for example Vastitatis Borealis on Mars.

V-class asteroid A rare class of asteroid distinguished by a moderately high albedo and a reflectance spectrum displaying a strong absorption feature at wavelengths shorter than 0.7 μm. There is also a strong absorption feature at wavelengths near 0.95 μm, in the near-infrared. This dip is characteristic of the silicate mineral pyroxene. The 'V' is for Vesta, the first asteroid to be assigned to the class.

Vega The star Alpha Lyrae, magnitude 0.03, the fifth-brightest star in the sky. It is an A0 dwarf 25. L.Y away. The Infrared Astronomical Satellite discovered that it is surrounded by a disk of dust and gas that may be the site of a forming planetary system. Vega probes Two identical space probes launched by the former Soviet Union in 1984 December, each consisting of a Venus lander and a comet Halley fly-by probe. The name 'Vega' was a contraction of the Russian words Venera ('Venus') and Gallei ('Halley'). In 1985 June the spacecraft released balloons into the atmosphere of Venus and landers on to the planet's surface before flying on to comet Halley, which they reached in 1986 March. They passed 8900 km and 8000 km respectively from Halley's nucleus, photographing it and analysing the comet's dust and gas.

Veil nebula A beautiful filamentary nebula which is part of the Cygnus Loop, a large supernova remnant. The brightest part of the Veil is designated NGC 6992-5; NGC 6960 and NGC 6979, to the west, are fainter parts of the same structure. The filaments are gas which has cooled from a supernova explosion some 30 000 years ago.

Vela A constellation of the southern sky, representing the sails of the ship Argo Navis. Gamma Velorum is a wide double star of magnitudes 1.8 and 4.3, the brighter of which is the brightest known Wolf-Rayet star. The constellation contains the Gum Nebula and the Vela Pulsar.

Vela pulsar A pulsar in the Vela Supernova Remnant; also known as PSR 0833-45. It has a period of 89.3 ms and is slowing down at a rate of 10.7 ns per day. The Vela pulsar is one of the youngest pulsars known, with a maximum age (spin-down age) of about 11 000 years, and one of the few to have been positively identified with a supernova remnant. In 1977 it was discovered to be flashing at visible wavelengths, making it the second known optical pulsar.

Venera A series of space probes to the planet Venus, launched by the former Soviet Union. Veneras 4-6 ejected capsules which sent back information on the atmosphere of Venus before being crushed by the intense pressure. Venera 7 made the first successful landing on Venus; the Venera 9 lander took the first pictures of the planet's surface, while its orbiter section was the first craft to orbit Venus. Venera 13 made the first analysis of the planet's soil.

Venus The second planet from the Sun. It has the most circular orbit of all the planets. Its mean geometric albedo, 0.65, is the highest of all the planets, a result of its unbroken cover of white clouds. At its brightest it reaches magnitude -4.7, far brighter than any other planet. Its rotational axis is tilted at nearly 180° to the upright, so that its rotation is retrograde. It rotates on its axis every 243 days, and so presents the same face to the Earth every time the two planets are at their closest. Venus has a dense atmosphere composed (by volume) of about 96.5% carbon dioxide and 3.5% nitrogen, with traces of sulphur dioxide, water vapour, argon, hydrogen, and carbon monoxide. The pressure of the atmosphere at the surface is about 92 bar (i.e. 92 times the Earth's sea-level pressure). Surface temperatures average about 460 °C due to the greenhouse effect in the planet's atmosphere. Lightning is very frequent.

Vernal equinox The point on the celestial sphere at which the Sun passes from south to north of the celestial equator each year. It is also known as the spring equinox or the first point of Aries. It occurs on or around March 21. At the vernal equinox, as with the autumnal equinox six months later, night and day are equal in length the world over. Right ascension is measured from the vernal equinox, which by definition has zero right ascension and declination.

Vesta asteroid 4, The fourth asteroid to be discovered, by H. W. M. Olbers in 1807. Its mean diameter is 576 km, making it the third-largest main-belt asteroid. Vesta's mass is 2.76×10^{20} kg and its mean density is 3.3 g/cm^3. Its orbit has a semimajor axis of 2.362 AU, period 3.63 years, perihelion 2.15 AU, aphelion 2.57 AU, and inclination of 7°.1. Vesta's colour changes slightly as it rotates every 5.34 hours, indicating that its surface is not uniform in composition. The Hubble Space Telescope has shown bright and dark features as small as 80 km across on Vesta, suggesting a geologically diverse terrain with an exposed mantle, ancient lava flows, and impact basins. The average albedo is 0.38. Vesta's spectrum is characteristic of the mineral pyroxene, common in lava flows. Vesta may be a source of the eucrite meteorites and some other small asteroids. It is sufficiently different from other asteroids to be assigned a special class, V.

Viking Two US space probes to Mars, each of which consisted of an orbiter and a lander. The orbiters studied the planet and its satellites, while the landers photographed the surface around them, took weather readings, and analysed the composition of the soil. Of particular interest was the possible presence of living organisms on Mars, but none were found.

Violent relaxation The rapid evolution of a star cluster or galaxy that has formed in a configuration far from equilibrium. During violent relaxation, the orbits of individual stars change dramatically because of changes in the gravitational potential of the system. It is thought to play an important part in shaping elliptical galaxies during the first billion years of their evolution.

Virginid meteors A meteor shower, producing low rates (ZHR around 5) from various radiants around the ecliptic in Virgo throughout March and April. Several maxima occur, with the main one perhaps around April 12, from radiants at RA 14h 04m, dec. -9° (east of the Virgo bowl) and RA 13h 36m, dec. -11° (close to Spica). Virginid meteors are slow, often long, and sometimes bright. The Virginids are sometimes regarded as part of a continuous flux of activity from around the ecliptic during the spring, extending to the Ophiuchids and Alpha Scorpiids in May and June.

Virgo The largest constellation of the zodiac, and the second-largest constellation of all, representing the goddess of justice. It lies on the celestial equator. The Sun passes through Virgo from the third week of September to the end of October, and hence lies within the constellation at the autumnal equinox. Its brightest star is Spica (Alpha Virginis). Gamma Virginis is a double star known as Porrima. The constellation contains numerous members of the Virgo Cluster of galaxies. M104, the Sombrero Galaxy, is not part of the cluster. Virgo also contains the brightest quasar, 3C 273, magnitude 12.9, about 3000 L.Y away.

Virial theorem A way of estimating the total mass of an object such as a galaxy or a cluster of galaxies from the movement of its individual members. The theorem states that the average gravitational potential energy of the constituent objects is twice their average kinetic energy. Calculations with the virial theorem show that galaxies and clusters contain up to ten times as much mass as can be seen telescopically, providing strong evidence for the existence of large quantities of dark matter. A modified version of this theorem, called the cosmic virial theorem, applies on cosmological scales. It relates the statistics of galaxy motions and the correlation function (which describes the way galaxies cluster in space) to the average density of the Universe. Since the first two quantities are measurable, the density parameter can thus be estimated. The usual result obtained is around 0.2, indicating that there is dark matter on cosmological scales, but not enough to reach the critical density.

Visual binary A binary star in which the two components are sufficiently well separated to be individually detectable visually or photographically. The brighter star is known as the primary, and the fainter as the secondary, companion, or comes. Even when orbital motion is extremely slow, visual binaries may sometimes be identified from the common proper motion of their components.

Vogt-Russell theorem The theorem, found valid except in rare circumstances, that there is only one internal structure possible for a star of given mass and chemical composition. The calculation of that structure depends on knowing how quantities such as pressure, rate of energy production, and opacity depend on local gas properties such as temperature and chemical composition. The mass-radius and mass-luminosity relations in main sequence stars are among the theorem's consequences. It is named after the German astronomer Heinrich Vogt and H. N. Russell.

Void A region of space containing far fewer galaxies than average, or even no galaxies at all; also known as a cosmic void. Voids with less than one-tenth the average density of the Universe on scales of up to 200 million L.Y have been detected in large-scale surveys. These regions are often (but not always) approximately spherical. The first large void was detected in Boötes in 1981; it has a radius of about 180 million L.Y and its centre is approximately 500 million L.Y from the Milky Way. The existence of large voids is not surprising, given the existence of clusters of galaxies and superclusters on very large scales.

Voigt profile The profile of a strong absorption line in a stellar atmosphere which combines the broadening effects of thermal (Doppler) motions of the atoms in the atmosphere with those of atomic collisions (pressure broadening) which dominate the line wings. It is named after the German physicist Woldemar Voigt.

Volans An unremarkable constellation of the southern sky, representing a flying fish. Its brightest star is Beta Volantis, magnitude 3.8. Gamma and Epsilon Volantis are attractive double stars.

Voskhod A multi-seater spacecraft launched by the former Soviet Union, a modification of the single-seater Vostok. Only two Voskhod flights were made. Voskhod 1, launched on 1964 October 12, carried three cosmonauts on a day-long flight. Voskhod 2, launched on 1965 March 18, was a two-man flight during which Alexei Arkhipovich Leonov made the first spacewalk.

Vostok A single-seater spacecraft, spherical in shape, launched by the former Soviet Union. The first human spaceflight was made in Vostok 1 by Yuri Alekseyevich Gagarin,who orbited the Earth once on 1961 April 12. Later missions lasted up to five days. The sixth and last Vostok flight was made by Valentina Vladimirovna Tereshkova,the first woman in space, on 1963 June 16-19.

Voyager Two US probes to the outer planets. Voyager 1, which was launched after Voyager 2 but on a faster trajectory, visited Jupiter and then Saturn, passing close to Saturn's largest satellite, Titan. Its trajectory was swung out of the ecliptic by the Saturn encounter, so that it could not reach other planets. Voyager 2's e members of a virtual particle pair move too far apart to rejoin they may become real particles, as happens in Hawking radiation from a black hole; the energy to make the particles real is extracted from the black hole. The lifetime of a virtual particle pair increases as the mass or energy involved decreases. Thus an electron and positron can exist for only up to about 4×10^{-21} s, but a pair of radio photons with wavelengths of 300 000 km could last for up to one second.

Vulcan A hypothetical planet supposedly lying within the orbit of Mercury, invoked in 1845 by U. J. J. Le Verrier to explain the advance of Mercury's perihelion (later accounted for by S. Newcomb's improved calculation of perturbations by the other planets, and A. Einstein's general theory of relativity). It is now known not to exist.

Vulpecula A faint constellation of the northern sky, representing a fox. Its brightest star, Alpha Vulpeculae, is of magnitude 4.4. It contains the striking star cluster known popularly as the Coathanger. M27 is the Dumbbell Nebula, a large planetary nebula. The first pulsar was discovered in Vulpecula.

W

W ursae majoris star A type of eclipsing binary with a period of less than 1 day; abbr. EW. The system is a contact binary, with ellipsoidal components, generally of spectral type F-G (or later). Light variation is continuous throughout the orbit, such that the beginning and end of eclipses cannot be defined. Primary and secondary minima are usually of similar amplitude, which is normally below 0.8 mag. A W Ursae Majoris star is thought to evolve into a common envelope star.

W virginis star A type of Population II Cepheid variable; abbr. CW. They are giants that undergo radial pulsation in the fundamental mode, with periods from 0.8 to 35 days and amplitudes of 0.3-1.2 mag. Their light-curves generally resemble those of the classical Cepheids, particularly if they have a period of 3-10 days, but outside this range the curves have different amplitudes or shapes. In addition, W Virginis stars have lower masses (0.4-0.6 solar masses) than classical Cepheids and obey a different period-luminosity relation, with absolute magnitudes that are fainter by 0.7-2 mag. There are two subtypes: CWA, with periods of more than 8 days; and CWB, with periods of less than 8 days. The latter are sometimes known as BL Herculis stars.

Weight The 'heaviness' of an object, the amount of attraction between two or more masses. Weight is often measured in Kilogrammes but the proper (SI) unit of weight is the Newton. Weight is not the same as mass which is a measure of how much matter or inertia an object has. Weight on the other hand is dependant on two or more masses and is a measure of the force of gravity acting on those two objects. For example, an astronaut will experience weightlessness in space because he is distant (though not entirely free) from the Earth's gravitational influence. His mass however will still be the same as it would be on Earth. He is just as massive as he was before - but he weighs little or nothing.

Weightlessness Weightlessness (or free fall) is the state in which an object appears to have no weight (but the object's mass remains the same). During weightlessness, an object's gravitational pull is negligible (close to zero).

Weird terrain An informal name for hilly and lineated terrain found on Mercury and the Moon. On Mercury, weird terrain is found on the opposite side of the planet to the large Caloris Basin, and is thought to have formed when the seismic waves from the Caloris impact converged, a process known as seismic focusing. Weird terrain on the Moon, not so clearly defined as on Mercury, occurs in two places, one directly opposite the centre of the Imbrium Basin, the other opposite the centre of the Orientale Basin.

Weizsäcker, Carl Friedrich Von German theoretical physicist and astrophysicist. In 1938, independently of H. A. Bethe, he proposed that stars generate their energy via the carbon-nitrogen cycle, converting hydrogen into helium by nuclear fusion. In 1944 he set out a modern version of the nebular hypothesis proposed in the 18th century, first by I. Kant and later by P. S. de Laplace, to account for the origin of the Solar System.

Werner lines A sequence of absorption or emission lines due to molecular hydrogen. They lie at ultraviolet wavelengths in the range 100-123 nm, the same region of the spectrum as the Lyman series of atomic hydrogen. They are named after their discoverer, the Danish physicist Sven Theodor Werner.

West, comet A long-period comet discovered in 1975 November by the Danish astronomer Richard Martin West;; formerly designated 1976 VI. Comet West reached perihelion (0.20 AU) on 1976 February 25. In early March, when closest to Earth at 0.8 AU, the comet reached a maximum brightness of magnitude -1. The broad fan-shaped dust tail was 30-35° long. A few days after perihelion, the nucleus split into four fragments. The comet's orbit has a period of about 500 000 years, eccentricity 0.99997, and inclination 43°.1.

Whirlpool galaxy The 8th-magnitude galaxy M51 in Canes Venatici, also known as NGC 5194. It is an Sc-type spiral seen nearly face-on, and was the first galaxy in which spiral structure was clearly seen, by Lord Rosse. It is interacting with a small irregular galaxy, NGC 5195, which appears to lie at the end of one of its spiral arms but is actually slightly behind it. The Whirlpool Galaxy lies about 25 million L.Y away.

White dwarf A small, dense star that is the end-result of the evolution of all but the most massive stars. White dwarfs are thought to form from the collapse of stellar cores once nuclear burning has ceased there, which are exposed to view when the outer parts of the star are driven off in a planetary nebula. Such a core contracts under its own gravity until, having reached a size similar to that of the Earth, it has become so dense (5×10^8 kg/m^3) that it is supported against further collapse by the pressure of electron degeneracy. White dwarfs are formed with high surface temperatures (above 10 000 K) because of the heat trapped within them, released both by previous nuclear burning and through gravitational contraction. They gradually cool, becoming fainter and redder. White dwarfs may constitute 30% of the stars in the solar neighbourhood, but because of their low luminosity (typically 10^{-3} to 10^{-4} of the Sun's) they are very inconspicuous.

White giant A white giant is a rare, huge white star of spectral type A.

White hole The time-reverse of the collapse of an object into a black hole. The equations of general relativity which describe such a collapse are time-symmetric, so there is no theoretical reason why they should not run in reverse. A white hole would therefore be a place where matter spontaneously appeared into our Universe. However, no such object has been observed.

White-light corona The Sun's corona as seen at visible wavelengths during total solar eclipses and with coronagraphs. The white-light emission arises from light from the Sun's photosphere that is scattered by free electrons (the K corona) and dust (the F corona). A small amount of visible light is from emission lines (the E corona).

Wien's displacement law The relation between the wavelength of peak emission from a black body and its temperature. At low temperatures black-body radiation is confined mainly to the infrared region of the spectrum, but at progressively higher temperatures the peak of the emission is displaced to progressively shorter wavelengths. Although celestial bodies are not perfect black bodies, the displacement law is still useful for predicting the wavelengths near which most of their radiation is emitted.

Wild 2 Comet Wild 2 (aka Comet 81P) is the first comet that has been visited by a spacecraft. NASA's Stardust Mission rendezvoused with comet Wild 2 in January, 2004. Comet Wild 2 is a short-period comet that was discovered by the Swiss astronomer Paul Wild on January 6, 1978. The comet's nucleus is about 3 miles (5 km) across. Wild 2 orbits the Sun every 6.39 years; its elliptical orbit ranges from about Mars' orbit to Jupiter's orbit.

Wild duck cluster The 6th-magnitude open cluster M11 in Scutum, also known as NGC 6705. Visually it appears somewhat fan-shaped, like a flight of ducks, covering nearly ¼°. It lies 5600 L.Y away.

Wild's triplet A group of three disturbed barred spiral galaxies with luminous connecting bridges, about 200 million L.Y away in the constellation of Virgo. The bridges are probably the result of gravitational tidal interactions between the galaxies. The trio is named after the British-born Australian astronomer (John) Paul Wild,who investigated it in the early 1950s.

Wilson effect The apparent displacement of a sunspot's umbra towards the Sun's centre as the spot approaches the limb, named after the Scottish astronomer Alexander Wilson,who first observed it in 1769. The effect gave rise to the belief that sunspots were saucer-shaped depressions. However, good examples of the effect are few. Sometimes, displacement of the umbra away from the Sun's centre is seen (the reverse Wilson effect). The effect could be explained by the fact that spots rarely have a circular outline.

Winter solstice The solstices are the days when the Sun reaches its farthest northern and southern declinations. The winter solstice occurs on December 21 and marks the beginning of winter (this is the shortest day of the year) in the Northern Hemisphere. The summer solstice occurs on June 21 and marks the beginning of summer (this is the longest day of the year) in the Northern Hemisphere.

Wolf diagram A graph of star counts at various apparent magnitudes, from which the distance to a dark nebula can be determined. Max Wolf first plotted star counts for a dark nebula and a comparison field on the same graph, and used the magnitude at which the two counts started to diverge to estimate the distance at which the obscuration began.

Wolf sunspot number An obsolete name for relative sunspot number.

Wolf, (Johann) Rudolf Swiss astronomer. Following the discovery of the solar cycle by S. H. Schwabe, Wolf set about collecting all available data on sunspot numbers. He established the dates of all sunspot maxima and minima going back to the year 1610, and calculated the length of the solar cycle to be 11.1 years. He devised the system of measuring solar activity by counting sunspots.Wolf was one of several astronomers to note that variations in terrestrial magnetism and auroral activity mirrored the solar cycle.

Wolf-Lundmark-Melotte system An 11th-magnitude dwarf Magellanic-type irregular galaxy, a member of the Local Group; also known as DDO 221. It was discovered in 1909 by Max Wolf and later rediscovered by K. E. Lundmark and the English astronomer Philibert Jacques Melotte. It lies about 3 million L.Y away in Cetus.

Wollaston prism A device used for analysing and producing polarized light, used in some designs of polarimeter. It consists of two prisms of either quartz or calcite cemented together. The Wollaston prism divides incoming light into two linearly polarized beams with perpendicular planes. It is named after the British scientist William Hyde Wollaston.

Wolter telescope Any of a series of designs for grazing-incidence telescopes by the German physicist Hans Karl Herman Wolter. To obtain a true image over an extended field of view requires photons to undergo two successive reflections from combinations of paraboloid-hyperboloid or paraboloid-ellipsoid surfaces. For imaging the commonly used design is the Wolter I telescope; for spectroscopy, where a smaller field of view but very high spatial resolution is required, the Wolter II design is appropriate.

World line The trajectory followed by an object in space-time. The fact that space-time is four-dimensional makes world lines difficult to visualize but, if the Universe had only one dimension in space and one in time, the world line could be drawn on a graph with time plotted vertically and distance plotted horizontally. A particle at rest with respect to the coordinate system would have a world line which runs along the vertical axis, while moving particles would have world lines which are curves or straight lines sloping upwards. In the real Universe, the path of a moving particle is a curved line in space-time.

Wormhole A hypothetical hole or tunnel in the fabric of space-time. Standard cosmological theories are based on the assumption that space-time is smooth and simply connected. To give a three-dimensional analogy, space-time is assumed to be like a sphere rather than a torus; a sphere is said to be simply connected, whereas a torus is not. In quantum cosmology it is thought that, on scales of the order of 10^{-35} m, space-time has a very complicated, multiply connected structure in which 'tunnels' and 'handles' supply short cuts between apparently distant points. In principle, sufficiently large wormholes might allow one to travel to a distant part of the Universe much more quickly than light and, in some situations, to travel in time. The existence of wormholes is, however, highly speculative.

WZ sagittae star A form of dwarf nova with an extremely long period, comparable with that of a recurrent nova (i.e. decades). WZ Sagittae itself erupted in 1913, 1946, and 1978. Small-amplitude variations (less than 0.3 mag.) occur between the major eruptions, and Algol-like eclipses are sometimes visible. The outbursts resemble those of an SU Ursae Majoris star with superhumps. Observations suggest that an excretion disk of matter is being ejected from the system.

X

X-rays X-rays are a type of electromagnetic radiation (between ultraviolet light and gamma rays in wavelength, frequency, and energy) - basically, it's light that is way past the blue-violet end of the visible spectrum - we cannot see it. They have short wavelength (and high frequency) as compared to visible light. Each photon of X-ray radiation has a lot of energy. X-rays can go through most solid objects. X-ray images of celestial objects are one way of learning about their high-energy properties. For example, the sun's corona emits X-rays, especially over sunspots. The Einstein X-ray satellite was launched in 1978 to survey celestial X-ray sources.

X-ray astronomy The observation of X-ray emission from celestial objects, which allows us to study violent and energetic processes in the Universe. The subject was born in 1949, when X-rays from the Sun were discovered by a rocketborne experiment. The first celestial X-ray source, Scorpius X-1, was detected in 1962, also by a sounding rocket. Rockets and high-altitude balloons during the 1960s revealed other individual sources, the existence of a diffuse X-ray background, and the variability of several objects.

The first dedicated X-ray astronomy satellite was Uhuru, launched in 1970, which carried out the first X-ray survey of the sky. Other satellites with X-ray instruments included the Copernicus satellite and the Astronomical Netherlands Satellite. Balloon and rocket experiments also continued throughout this period. The UK's Ariel 5 and the US SAS-3 satellites launched in 1974 and 1975 extended the catalogues of known X-ray sources and monitored their variability.

X-ray background A diffuse X-ray emission not associated with known individual sources. At least part of this emission arises from hot gas within the Galaxy, particularly at soft X-ray energies. However, the origin of its higher-energy component is still uncertain. It may be an as yet unresolved population of point sources, or it could be a truly diffuse thermal background.

X-ray binary star X-ray binary stars are a special type of binary star in which one of the stars is a collapsed object such as a white dwarf, neutron star, or black hole. As matter is stripped from the normal star, it falls into the collapsed star, producing X-rays.

X-ray bright point A small, bright area visible in X-ray images of the Sun's corona. About 40 may be visible at any one time, distributed all over the Sun but more numerous near low latitudes. They are small coronal loops, with footpoints in the photosphere, filled with very hot (about 3 million K) gas. A single bright point lasts about a day or less, and is generally only a few thousand kilometres across. They are not associated with sunspot groups.

X-ray multi-mirror mission A planned ESA satellite for X-ray astronomy, carrying three grazing-incidence telescopes with CCD cameras and grating spectrometers for imaging and spectroscopic observations. XMM will be placed into a 24-hour orbit to allow continuous observation of sources. A 0.3-m optical telescope will check for visual variability.

X-ray nova A nova-like outburst at X-ray wavelengths, which may also have an optical counterpart. The main difference between X-ray and optical novae may be that the compact object is a neutron star or black hole rather than a white dwarf.

X-ray pulsar A regularly variable X-ray binary, in which the pulsation is associated with the spin period of the compact companion, a magnetized neutron star. Periods range from a few seconds to a few minutes. These pulsations are thought to be caused by the magnetic field channelling the accreting gas on to the poles of the star, producing localized 'hot spots' which move in and out of view as the star spins. An example of such a system is Hercules X-1.

X-ray transient A burst of X-ray emission that, after rising to a maximum brightness, fades until no longer detectable. Examples are novae, supernovae, cataclysmic variables, and stellar flares. Many sources originally detected as transients have subsequently been found to be weak quiescent X-ray sources.

Y

Yarkovsky effect A change in orbital momentum experienced by small rotating particles as a result of absorption and re-emission of solar radiation, similar to the Poynting-Robertson effect. Whereas the Poynting-Robertson effect is equal in all directions, the re-emission which causes the Yarkovsky effect is anisotropic. The object may be accelerated or decelerated by the momentum carried away by the radiated photons, depending on the spin direction. The effect is named after the Russian scientist Ivan Osipovich Yarkovsky.

Year The Earth's period of orbital revolution around the Sun, and hence by extension the orbital period of any planet. Astronomically the Earth's year can be defined in several ways. The actual orbital period with reference to the fixed stars is the sidereal year, 365.256 36 days. However, the positions of the stars are gradually changing because of precession. The interval between successive passages through the mean equinox is termed the tropical year, 365.242 19 days, which takes precession into account. This is the most commonly adopted definition of the year as it is the one that relates directly to seasonal changes. A further form of year is the anomalistic year, which is the average interval between successive passages of the Earth through the perihelion point of its elliptical orbit, 365.259 64 days. This differs slightly from the sidereal year due to small perturbations of the Earth's orbit by the gravitational influence of the other planets.

Year, anomalistic The period for successive perihelion passages of the Earth, a little less than 5 minutes longer than the sidereal year.

Year, calendar The mean length of the year according to the Gregorian calendar, 365 days, 5 hours, 49 minutes and 12 seconds.

Year, sidereal The period taken by the Earth to complete one orbit of the Sun, 365.26 days.

Year, tropical The period taken for successive passages of the Sun across the Vernal Equinox, 365.24 days.

Yellow dwarf Yellow dwarfs are small, main sequence stars. The Sun is a yellow dwarf.

YY orionis star A subtype of the irregular Orion variables, related to the T Tauri stars; abbr. IN(YY). Their spectrum shows absorption on the long-wave side of the emission lines, indicating an infall of material.

Z

Z andromedae star A type of cataclysmic variable with irregular brightness fluctuations, exhibiting the spectra typical of a hot star (20 000 K), a cool star (3000 K), and extended nebulosity excited by radiation from the hot component; abbr. ZAND. Interaction occurs either through mass transfer or a stellar wind from the cool giant secondary. Currently, a heterogeneous collection of objects is classified by this type (often known as symbiotic stars), including the symbiotic novae (RR Telescopii stars), and stars with an accretion disk and jets, such as R Aquarii.

Z camelopardalis star A type of U Geminorum star that occasionally exhibits a standstill at intermediate light, but which otherwise behaves like an SS Cygni star; abbr. UGZ. A standstill normally begins on the decline from an outburst and may last for a few days or persist for years. During this state there is increased mass transfer between the components. The amplitude of normal outbursts is similar to those of the UGSS type (2-5 mag.), but the mean interval tends to be shorter (10-40 days).

Zanstra's theory A theory from which the temperature of the central star in a planetary nebula can be estimated. The basic assumption is that the nebula absorbs all the ultraviolet light from the star which can cause ionization. For each ultraviolet photon absorbed a Ha photon is emitted when the ionized hydrogen subsequently recombines with an electron. Thus the strength of the Ha line is. related to the ultraviolet magnitude of the star. The strength of the red continuum (broad-band) emission underlying the Ha line gives the red magnitude and hence a colour temperature for the star. This method is not affected by interstellar absorption, since both colours are effectively measured at the same wavelength. It is named after the Dutch astrophysicist Herman Zanstra.

Zeeman spitting The splitting of a spectral line due to a magnetic field. It is named after the Dutch physicist Pieter Zeeman.The Zeeman effect is widely used for determining magnetic fields in astronomical objects, notably the Sun. However, field strengths are often insufficient to cause actual line splitting (sunspot fields, of about 0.4 tesla, are an exception); instead, line broadening occurs. The strengths of weak fields can be deduced by measuring the polarization of each line wing, which is the principle of the solar magnetograph, an instrument for measuring magnetic fields.

Zenith The point on the celestial sphere directly above an observer. A line to the zenith is at right angles to the plane of the horizon. The zenith is in the opposite direction from the nadir. Strictly speaking, this definition refers only to the astronomical zenith; two other, slightly different forms of zenith may also be defined. The geocentric zenith is the direction indicated by a line from the centre of the Earth through the observer. The geodetic zenith is at right angles to the geoid at the observer's location. All three forms of zenith differ slightly because of the nonspherical shape of the Earth. If used without qualification, 'zenith' means the astronomical zenith.

Zenith telescope An instrument for measuring the zenith distances, and/or times of meridian transit, of stars close to the zenith. The telescope is mounted so that it points vertically, and so does not distort under its own weight, as it would if it were not vertical. In addition, the effects of atmospheric refraction are smallest at the zenith. Hence star positions can be measured at the zenith with maximum accuracy.

Zenithal attraction An effect in which the observed radiants for meteors appear closer to the zenith than their true position. It occurs because meteoroids impacting on the upper atmosphere are subject to the Earth's gravitational pull, which increases their geocentric velocities. The change in apparent orbital velocity is most marked for slow meteoroids. The effect must be corrected for when analysing positional observations to determine meteor shower radiants and heliocentric orbits.

Zenithal hourly rate An index of the activity of a meteor shower. The ZHR gives the activity in meteors per hour which would be expected by an experienced visual observer under a perfectly clear sky (limiting magnitude +6.5) with the radiant overhead. In practice, observed rates are almost always lower than the derived ZHR due both to sky conditions and the fact that the radiant is seldom directly overhead. Meteor showers with ZHRs lower than about 5 are usually difficult to distinguish from the background level of sporadic meteors.

Zero-age horizontal branch That part of the Hertzsprung-Russell diagram occupied by low-mass stars that have just undergone the helium flash. The scatter in properties of stars on the zero-age horizontal branch is due to the range of stellar masses resulting from mass loss on the giant branch.

Zero-age main sequence The diagonal strip on the HertzsprungRussell diagram occupied by stars that have just begun to convert hydrogen to helium in their cores. As this conversion proceeds, stars become slightly redder and more luminous, so they move upwards and to the right on the HR diagram. As a result, the main sequence is a broad band that is displaced slightly from this zero-age strip.

Zeta aurigae star An eclipsing binary with a composite spectrum, in which the components are a late-type bright giant or supergiant and an early-type star. Immediately before and after primary eclipse the hot star is visible through the outermost layers of the secondary. If the spectrum shows features which indicate the presence of an M-type supergiant, emission, or mass transfer (but not necessarily eclipses), the star is sometimes known as a VV Cephei star. In Zeta Aurigae and VV Cephei themselves the orbital periods are 972 and 7430 days, respectively. The secondary in VV Cep is estimated to have a diameter 1600 times that of the Sun, making it one of the largest stars known.

Zodiac The strip of sky up to 8° either side of the ecliptic against which the Sun, Moon, and major planets appear to move. The strip is divided into twelve signs of the zodiac, each 30° long. These signs were named by the ancient Greeks after the constellations of the zodiac that occupied the same positions some 2000 years ago. As a result of precession, the constellations have moved eastwards by over 30° and no longer coincide with the signs.

Zodiacal catalogue Popular name for the Catalog of 3539 Zodiacal Stars published in 1940 by the Nautical Almanac Office of the US Naval Observatory. It lists the positions, magnitudes, spectral types, and proper motions of stars within 8° of the ecliptic, mostly brighter than magnitude 8.5, for epoch 1950.0. It is used for identifying stars that are occulted by the Moon.

Zodiacal dust The population of small (1-300 μm) particles of cometary and asteroidal origin lying roughly in the ecliptic plane out to about 5 AU from the Sun; measurements from the Pioneer and Voyager probes have indicated much lower dust concentrations beyond the orbit of Jupiter. The distribution of the zodiacal dust is controlled mainly by Jupiter's gravitational influence. Reflection of sunlight from the zodiacal dust gives rise to the gegenschein and zodiacal light. The total mass of zodiacal dust is estimated to be similar to that of an average comet nucleus.

Zodiacal light A faint diffuse glow, comparable in intensity to the fainter parts of the Milky Way, produced by the reflection of sunlight from zodiacal dust particles in the ecliptic plane. From temperate latitudes, the zodiacal light is best seen on spring evenings about 90 min after sunset, or in the autumn about 90 min before sunrise; at these times, the ecliptic, along which the light appears to extend for some 60° to 90°, lies at a steep angle relative to the horizon. An observing site free from light pollution is essential. The intensity of the zodiacal light may vary with the solar cycle, being greatest around sunspot minimum when the interplanetary solar wind is dominated by fast-flowing particle streams from coronal holes.

Zond A series of space probes launched by the former Soviet Union for testing spacecraft equipment and flight techniques. Zonds 1 and 2 in 1964 were launched towards Venus and Mars to test long-distance communications, but contact was lost with both. Zond 3 in 1965 July took photographs of the parts of the Moon's farside not covered by Luna 3. Zonds 4-8 in 1968-70 were rehearsals for a planned manned Soyuz mission around the Moon and back that was cancelled after the success of Apollo.

Zone catalogue A catalogue that gives positions of stars within a restricted band of declination but extending over a large range of right ascension. Historically, zone catalogues were compiled from meridian observations; they are now usually compiled from measurements of a series of wide-angle photographs centred on a particular declination.

Zone of avoidance The band around the sky in which very few galaxies are seen. It is a result of obscuration by interstellar dust in the plane of the Milky Way, although the zone is actually inclined at 2° to the galactic equator. The detailed structure of the zone is irregular, varying in width between 38° (towards the galactic centre) and 12°, with some fairly transparent regions known as galactic windows.

Zone time The civil time that is kept within one of the 24 time zones into which the Earth's surface is divided. It differs from Universal Time by a whole number of hours, being ahead of UT in time zones to the east of Greenwich and behind UT in time zones to the west.

Zwicky catalogue Popular name for the Catalogue of Galaxies and of Clusters of Galaxies published in six volumes in 1961-8 by F. Zwicky and his collaborators. It contains 31350 galaxies and 9700 clusters recorded on the Palomar Observatory Sky Survey photographs. Its system of classifying clusters as compact, medium compact, or open is still used.

Zwicky, Fritz Swiss astronomer, born in Bulgaria, who worked in America. In 1933 he found that the mass-luminosity ratio for clusters of galaxies was fifty times that for lone galaxies – an early indication of missing mass, which he attributed to intergalactic matter and dwarf galaxies. In 1934 he and W. Baade coined the term 'supernova' to distinguish such enormous eruptions from the less powerful novae, and proposed that a supernova explosion leaves behind a neutron star, confirmed when the Crab Pulsar was identified in 1968. Zwicky observed supernovae in other galaxies, and began a long study of galaxy clusters.He also drew attention to intergalactic matter, in particular to bridges of stars between neighbouring galaxies.

ZZ ceti star A type of variable white dwarf that undergoes non-radial pulsation, with minimal variation in size but significant temperature changes; abbr. ZZ. Periods range from 30 s to 25 min, and generally several periods exist simultaneously. Amplitudes are 0.001-0.2 mag. There are three subtypes: ZZA, with DA (hydrogen absorption) spectra; ZZB, with DB (helium absorption) spectra; and ZZO, sometimes called GW Virginis stars, with extremely hot, DO spectra. Some ZZ Ceti stars exhibit flares, thought to be caused by a companion UV Ceti star..